FRONTAL NANOTECHNOLOGY RESEARCH

FRONTAL NANOTECHNOLOGY RESEARCH

MARTIN V. BERG
EDITOR

Nova Science Publishers, Inc.
New York

For permission to use material from this book please contact us:
Telephone 631-231-7269; Fax 631-231-8175
Web Site: http://www.novapublishers.com

Library of Congress Cataloging-in-Publication Data
Available upon request

ISBN 13 978-1-60021-212-3
ISBN 10 1-60021-212-3

Published by Nova Science Publishers, Inc. ✢New York

CONTENTS

PREFACE

Nanotechnology is a 'catch-all' description of activities at the level of atoms and molecules that have applications in the real world. A nanometer is a billionth of a meter, about 1/80,000 of the diameter of a human hair, or 10 times the diameter of a hydrogen atom. Nanotechnology is now used in precision engineering, new materials development as well as in electronics; electromechanical systems as well as mainstream biomedical applications in areas such as gene therapy, drug delivery and novel drug discovery techniques. This book presents the latest research in this frontier field.

Since the discovery of carbon nanotubes (CNTs), extensive studies on this new carbon allotrope have formed a new epoch in various fields of science and technology with their excellent physical and electrical properties, and chemical functionality. In particular, understanding of their electrical properties has aroused many research groups in academy and industry to explore the nano-scale electronics based on CNTs such as transistor, memory, interconnect, logic array, and so on. Although the CNT-based electronics is regarded as a strong potential candidate for the next generation electronic devices, many technological barriers are still remained.

First of all, pure nanotubes should be separated from a mixture of single-walled, multi-walled nanotubes, and carbonaceous impurities or should be grown without impurities for reliable device performance. Another critical barrier to integrate nanotubes in devices is uncertainty and uncontrollability in positioning nanotubes on a right region. In addition, transport mechanism in CNTs should be more understood to modulate band gap of them to fabricate various device elements.

In Chapter 1, the authors present recent efforts to overcome the barriers focusing on transistors based on single-walled carbon nanotubes (SWNTs). Experimental and theoretical studies to control the CNT electronic structure are covered for CNT-field effect transistors. The large scale fabrication of CNT electronic elements are also given for the selective growth of CNT that is extendable to device integration. For the purity issue of nanotubes, sono-chemical route to single-walled carbon nanotube and purification method of nanotubes integrated in transistors via sulfidation are demonstrated. For the positioning problems, direct photolithographic method using a mixture of catalytic precursors and conventional resists as a catalytic resist is introduced with other positioning methods of nanotubes.

Nanostructures of metals and their oxides, chalcogenides have been of extensive interest for their widespread use in catalysis, photography, optics, electronics, optoelectronics, information storage, biological and chemical sensing, and surface-enhanced Raman scattering

(SERS). By tailoring the size, shape and/or structure, one can, in principle, tune their intrinsic properties for a range of applications. To this end, coordination structures have been rationally applied as precursors to fabricate diversified nanostructures in solution. In Chapter 2, the authors discuss the synthetic strategy and related reaction mechanism through the illustration of some detailed cases developed by our group in the recent years. The simple process, mild synthetic conditions, clean reactions, excellent reproducibility and high product quality make our strategies significant. It is believed that introduction of coordination structures into the synthesis of metal, oxide and chalcogenide nanocrystals may provide a versatile tool to control their growth and assembly into well-defined nanostructures. We also expect that the related synthetic strategy is applicable to extend to chemical vapor transport or laser-assisted catalytic growth.

With the growing interest for miniaturization in electronic and optoelectronic devices, scientists are more and more interested in fabricating different materials in nano-scale dimensions with different morphologies with a view to use them as the basic building block for nano-scale devices. One-dimensional (1-D) nanostructures such as nanorods, nanowires, nanobelts or nanoribbons are of special importance because of their improved fundamental, physical and chemical properties, which make them potential candidate for nano-scale devices. Zinc sulfide (ZnS) is a unique luminescent material well known for its photoluminescence, electroluminescence, etc., which enable wide applications in the fields of displays, sensors and lasers. Chapter 3 reviews the synthesis, structure and properties of 1-D ZnS nanostructures. ZnS 1-D nanostructures have been fabricated by using different physical (vapor-solid, vapor-liquid-solid etc.) and chemical (solvothermal etc.) routes. Optical properties such as optical absorption, photoluminescence and Raman spectra of the 1-D ZnS nanoforms are also discussed here. Wherever possible, the possible applications of these nanostructures are also included.

As explained in Chapter 4, nanocrystalline ZnO thin films on silica glass substrates were prepared by spin coating-pyrolysis process from a zinc naphthenate precursor. As-deposited films were heat treated at 500°C ~ 1000°C. The ZnO thin films were analyzed by X-ray diffraction, field emission-scanning electron microscope, scanning probe microscope, and ultra violet-visible-near infrared spectrophotometer. Only (002) peak is observed for all samples on silica glass substrates. With the increase of heat-treatment temperature, the peak intensity of (002) plane increases. No evidently aggregated particles are present. From SPM analyses, three-dimensional grain growth, which is thought to be due to inhomogeneous substrate surface and *c*-axis oriented grain growth of the ZnO phase, is independent on heat treatment temperature. The most homogeneous surface of the highly oriented ZnO film was observed at the heat-treatment temperature of 800°C. All the films exhibit a high transmittance (>80%) in the visible region except film heat-treated at 1000°C and show a sharp fundamental absorption edge at about 0.38 ~ 0.40 μm. The estimated energy band gap for all the films are within the range reported for ZnO films and single crystal. ZnO films, consisting of densely packed grains with smooth surface morphology, were obtained by heat treatment at 600 ~ 800°C, expected to be ideal for practical usage.

Transition metal chalcogenides, including sulfides, selenides, and tellurides are important optoelectronic, luminescent, and semiconducting materials. With regard to all of the applications of chalcogenides, highly crystalline particles with almost monodisperse size distribution and regular morphology are required. The synthesis of chalcogenides nanocrystallites under mild conditions is of great interest in both basic research and engineering applications. In Chapter 5, the synthesis of chalcogenides nanocrystallites, such as nanoparticles, nanoplatelets, nanorods, nanowires, and their assembled patterns, via the simple, rapid and convenient refluxing polyol method was reviewed based on our works. Several techniques including powder X-ray diffraction (XRD), scanning electron microscopy (SEM), transmission electron microscope (TEM), X-ray photoelectron (XPS), Raman, photoluminescence (PL), etc. have been used to characterize the produced nanocrystallites. Possible growth mechanism of the refluxing polyol technique to transition metal chalcogenides nanocrystallites is discussed. Combining with other techniques, the refluxing polyol method could be extended to synthesize other important inorganic materials nanocrystallites. Given the generality of this simple refluxing polyol method, coating and modification of the surfaces of transition metal chalcogenide nanocrystallites may also be fulfilled.

Nanostructured materials have attracted tremendous attention in both fundamental studies and applications due to their unique physical and chemical properties. Of the various types of nanostructured materials, TiO_2 nanomaterials are extensively studied for their broad range of applications. A number of methods have been developed for preparing TiO_2 nanoparticles, nanorods, whiskers, nanowires, nanotubes and mesoporous TiO_2 films. In this chapter, two novel methods recently developed in the laboratory to grow TiO_2 nanostructured materials directly on titanium (Ti) substrate are described. Dense and well-aligned TiO_2 nanorod arrays form on a Ti substrate when using organic molecules, such as ethanol, acetaldehyde, acetone and dibutyltin dilaurate, as the oxygen source in the oxidation of Ti at high temperatures. In contrast, polycrystalline grains are formed when pure oxygen and formic acid are used as the oxygen source; elongated micro-crystals are produced when water vapor is used as the oxygen source; whereas the use of argon with a low concentration of oxygen produces random nanofibers growing from the ledges of the TiO_2 grains. Measurements of the water contact angle of these samples show that the wetting properties of the TiO_2 films are dictated by their structure, varying from a hydrophilic surface (polycrystalline grains) to hydrophobic (random nanofibers), and towards a highly-hydrophobic surface (well-aligned nanorod arrays). Secondly, we have synthesized large-scale $Na_2Ti_6O_{13}$ nanoplates, nanowires, and continuous nanowire network films directly on Ti substrate hydrothermally. The morphology of the formed $Na_2Ti_6O_{13}$ nanostructures is easily tuned by varying the experimental parameters of temperature, reaction duration and NaOH concentrations. Our study shows that the synthesized $Na_2Ti_6O_{13}$ nanostructures are easily converted into $H_2Ti_3O_7$ nanostructures, a desirable precursor for the fabrication of various TiO_2-based nanomaterials such as anatase, a mixture of anatase and rutile, and rutile TiO_2 nanomaterials, with shape-preservation through an ion-exchange process. The approaches described in Chapter 6 provide simple and novel methods to synthesize various large-scale TiO_2-based nanostructured materials. This tunable growth of TiO_2 nanostructures is desirable for the promising applications of TiO_2 nanomaterials in the development of photo-catalysts, optical devices, sensors and self-cleaning coatings.

One definition of wear in a tribological sense is that it is the progressive loss of mass from the operating surface of a body as a result of relative motion at the surface. Generally, tribology may be defined as the study of friction wear and lubrication, but in comparison with friction forces or lubrication methods, very little work has been done on the extremely important area of wear and surface damage. For systems consisting of common materials (metals, polymers, ceramics), there are at least four main mechanisms by which wear and surface damage can occur between solids in relative motion: abrasive- and/or adhesive-, chemical- or corrosive-, fatigue-wear mechanisms. The occurrence of such different wear mechanisms involves the fundamental mechanics of molecular and supra-molecular scale. As a consequence, the understanding of surface properties on nanoscale level should be generated to have also a satisfactory knowledge of responses of materials and machine components on macrometer scale. Investigation of the fundamental characteristics of wear at the nanoscale is complicated by some factors and forces that act on such scale level have not yet been addressed in the tribology of macro-systems. Since these forces are sensitive to the environment and surface condition of the specimens, it is quite difficult to determine them accurately. Further, quantification of wear is not straightforward since the amount of wear is often too small to be detected by surface-sensitive instruments. Nanowear measurements have been object of rapid development of precise measuring tools following to the introduction of scanning probe microscopies, such as atomic force microscopy (AFM). The advent of AFM opened new opportunity to study wear mechanisms on nanoscale, making possible to simulate one surface sliding on other one and controlling many physical parameters during the contact dynamics. Further, AFM made possible to a have a direct connection between experimental data and theoretical models.

Chapter 7 is devoted to give a panoramic lecture of some experimental results and models on the onset wear occurrence on nanoscale.

In Chapter 8, the potential of using nanoparticle surface chemistry to tune the dielectric properties of polymer nanocomposites is explored. Composites with 5 wt% of titania (TiO_2) nanoparticles and low-density polyethylene (LDPE) were fabricated by melt mixing. Both polar (N-(2-aminoethyl)3-aminopropyl-trimethoxysilane (AEAPS)) and nonpolar (decyltrichlorosilane (DCS)) silane coupling agents were used to modify the nanoparticle surface to yield either hydrophilic or hydrophobic surfaces. Two coating methods were employed and the resulting surface structures were characterized in detail. It was found that the surface structure was controlled primarily by the experimental conditions of the coating process. In addition, a previously un-described surface reaction that resulted in the formation of $Si\text{-}CH_3$ bonds was proposed, which may be due to the reaction of AEAPS with oxygen vacancies on the TiO_2 nanoparticle surface. A large observed change in the chemical shift (over 40 ppm) in the ^{29}Si NMR and the –2.3 ppm chemical shift in ^{13}C NMR supported this conclusion. The method leading to a "dry" compact monolayer surface structure was chosen for preparing TiO_2 nanoparticles with either polar or non-polar surfaces for composite sample preparation and subsequent electrical studies. In both cases, surface modification had a positive effect on dielectric breakdown strength although the possible underlying mechanisms were different. Compared to the uncoated TiO_2 filled samples, AEAPS-TiO_2 and DCS-TiO_2 filled samples showed a 35% and 43% increase in the short-term Weibull breakdown strength, respectively. The space charge distribution patterns were also clearly different among the samples. The microstructure characteristics of the composite samples, associated with different TiO_2 surfaces, were studied and correlated to the corresponding dielectric properties.

In: Frontal Nanotechnology Research
Editor: Martin V. Berg, pp. 1-39
ISBN: 1-60021-212.3

Chapter 1

PROPERTIES, SYNTHESIS, PURIFICATION, AND INTEGRATION OF CARBON NANOTUBES FOR THE ELECTRONIC DEVICE APPLICATIONS

Wanjun Park[1] and Yo-sep Min
Samsung Advanced Institute of Technology, Yong-In, Kyeongki-Do, 449-712, Korea

ABSTRACT

Since the discovery of carbon nanotubes (CNTs), extensive studies on this new carbon allotrope have formed a new epoch in various fields of science and technology with their excellent physical and electrical properties, and chemical functionality. In particular, understanding of their electrical properties has aroused many research groups in academy and industry to explore the nano-scale electronics based on CNTs such as transistor, memory, interconnect, logic array, and so on. Although the CNT-based electronics is regarded as a strong potential candidate for the next generation electronic devices, many technological barriers are still remained.

First of all, pure nanotubes should be separated from a mixture of single-walled, multi-walled nanotubes, and carbonaceous impurities or should be grown without impurities for reliable device performance. Another critical barrier to integrate nanotubes in devices is uncertainty and uncontrollability in positioning nanotubes on a right region. In addition, transport mechanism in CNTs should be more understood to modulate band gap of them to fabricate various device elements.

In this article, we present recent efforts to overcome the barriers focusing on transistors based on single-walled carbon nanotubes (SWNTs). Experimental and theoretical studies to control the CNT electronic structure are covered for CNT-field effect transistors. The large scale fabrication of CNT electronic elements are also given for the selective growth of CNT that is extendable to device integration. For the purity issue of nanotubes, sono-chemical route to single-walled carbon nanotube and purification method of nanotubes integrated in transistors via sulfidation are demonstrated. For the positioning problems, direct photolithographic method using a mixture of catalytic precursors and conventional resists as a catalytic resist is introduced with other positioning methods of nanotubes.

[1] E-mail address: wanjun@samsung.com

1. INTRODUCTION

Since the discovery of carbon nanotubes (CNT) [1], their various aspects have been explored in different fields of science and technology. In particular, the unique one-dimensional geometry and the excellent transport property have made CNT the most promising building blocks for the next generation electronic devices [2, 3]. After finding of the current modulation by the gate bias with CNT transistor structure [4], many extensive works have been tried to realize the nano-scale electronics with various types of CNT electronic elements –resistor, capacitor, memory cell, logic array, and so on.

Although carbon nanotube can be a strong candidate for the next-era technology based on nanotechnology after the current semiconductor electronics with its nano-sized feature and unique electronic properties, there are many technology barriers such as control of individual CNTs in the array form to build the IC systems as well as control of CNT placement on the right position. The first hurdle for nanotube electronics is in synthesis or growth of nanotubes [5, 6]. To utilize nanotubes as an electronic material, pure nanotubes should be separated from a mixture of single-walled, multi-walled nanotubes, and carbonaceous impurities [7] or should be grown without impurities for reliable device performance. The electrical nature of nanotubes due to their chirality and diameter should be controlled to be semiconducting or metallic for particular applications. Another critical barrier for nanotube electronics is uncertainty and uncontrollability in positioning nanotubes on a right region to integrate as a device element. One more concern for device integration is the growth temperature should be lowered to be compatible to current CMOS process. Fundamental point of view, transport mechanism in CNTs should be more understood to modulate band gap of them to fabricate various device elements.

In this article, we shall discuss the recent progresses in the carbon nanotube electronics toward a view point of integration.

In section 2, experimental and theoretical studies on carbon nanotube transistors are discussed. The general current-voltage characteristics of CNTFETs in intrinsic nanotube are expressed with a ballistic nature in CNT and the Schottky barrier at electrode and nanotube junction. Then molecular adsorption is also discussed with a special emphasis on suppression of ambipolar behavior that is attributed to the Schottky barriers.

In section 3, low temperature of single walled CNT growth is reviewed and discussed. For high purity and uniform distribution of single walled nanotubes, sonochemical route is introduced. On the other hand, a remote plasma enhanced chemical vapor deposition on substrates is discussed as an essential technique for the CNT integration.

In section 4, a direct photolithographic method using a mixture of catalytic precursors and conventional resists is introduced as a fundamental processing technology for large scale device integration with other positioning methods of nanotubes. Purification method of nanotubes integrated in transistors via sulfidation which is useful for removal of undesired electrical leakage path is also demonstrated.

Finally, we summarize this article in section 5.

2. ELECTRIC PROPERTIES OF CNTFET

The atomic structure of a single-wall carbon nanotube (SWNT) is specified by the chiral vector, $C_h = na_1+ma_2$, that can be expressed by the real space unit vectors a_1 and a_2 of the hexagonal lattice. The electronic structure of SWNT is well developed with a simple tight-binding calculation since its one-dimensional nature. Indices (n, m) of the chiral vector indicate either metallic or semiconducting, i.e. if n-m is multiple numbers of 3 including zero, nanotube is metallic, otherwise, nanotube is semiconductor with the energy gap opening in electronic density of states (DOS).

The ballistic nature of carbon nanotubes is attributed to the potential applications for the electronic devices with high driving current capability and less heat dissipation that are essential for the trend of scaling down in either the active or the passive device elements of the integrated circuit systems. Details of fundamental properties for electronic structure and transport of carbon nanotubes are elsewhere [8].

Of special interest is the carbon nanotube field effect transistor (CNTFET) that is expected as the most important building block for the future electronic system. CNTFET has an analogy of MOSFET in device structure and I-V characteristics but operational basics are different. While early devices have shown poor device characteristics, improvements were achieved by using doped CNTs [9] or high-k materials [10]. The contact between metal and CNT can be of ohmic [11] or Schottky type [12, 13]. CNTFETs operate by modulating the transmission coefficient of the Schottky barriers at the contact between the metal and the CNT.

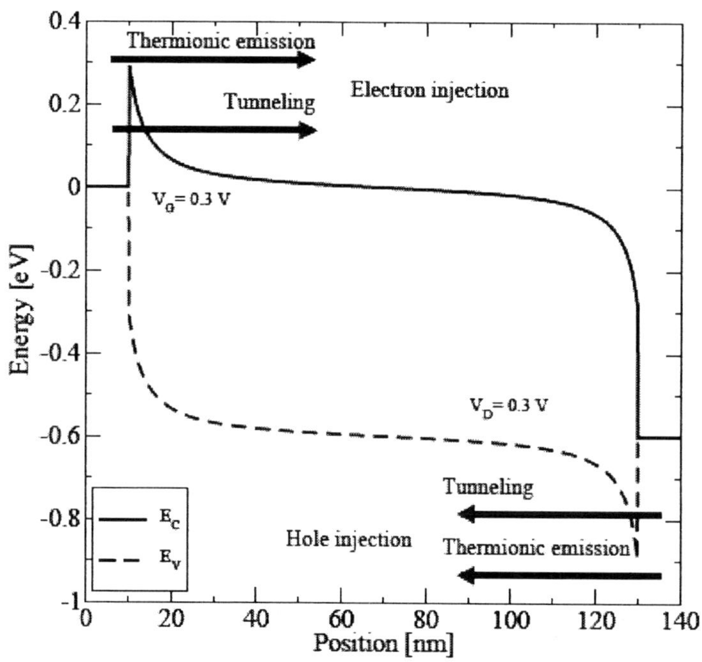

Figure 2.1. Band edge profile with the Schottky barriers at the CNT/metal junctions for source and drain of the CNTFET.

2.1. Intrinsic Characteristics of CNTFET

Without external charges, the band edge profile for CNTFET is shown in Figure 2.1. Positive gate voltages near the source increase the tunneling current of electrons, which is desirable for n-type devices. By decreasing the gate voltage the tunneling current of electrons decreases, but the thermionic emission current of electrons does not vary. If the gate voltage decreases further to negative values the thermionic emission current of electrons also decreases. On the other hand by applying positive voltages higher than the gate voltage to the drain, the Schottky barrier near the drain is suppressed and consequently hole injection at the drain increases, an undesirable phenomenon for an n-type device. Especially in the off regime this would result in an intolerably high off-current.

The resulted current-voltage characteristics are shown in Figure 2.2 where the current is symmetric with respect to the gate voltage and the gate voltage of valley depends on both of metal work function and the drain voltage. This ambipolar behavior of Schottky barrier CNTFETs limits the performance of the CNTFET since the I_{on}/I_{off} ratio is limited [14, 15].

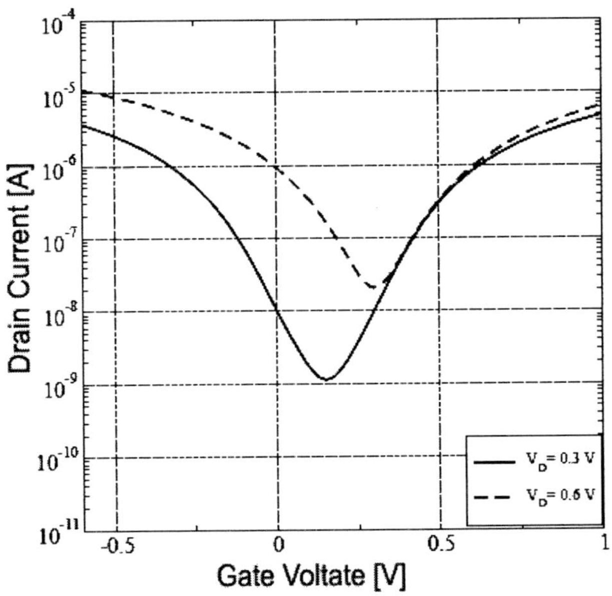

Figure 2.2. Current-voltage characteristics of the CNTFET without external charge.Assuming ballistic transport, the drain current was calculated using theLandauer-Büttiker formula with equilibrium Fermi functions at the source anddrain contacts and the transmission coefficient through the device. The twofold band and twofold spin degeneracy were used. This work was focused on ambipolar devices, where the metal Fermi level is located in the middle of the CNT band gap at each contact. The transmission coefficient was evaluated using the WKB approximation and an idealized band structure. The lattice constant of CNT was taken as 0.246 nm and the band gap energy set here to 0.6 eV corresponding to a CNT of a diameter of 1.4 nm. (See Ref. [16] for more details).

In order to suppress the ambipolar behavior, an asymmetric structure of CNTFET has been proposed, where the gate is not extended to the drain and the control of the gate over carrier injection from the drain is reduced. As a result the ambipolar behavior decreases to some extent.[15] A double-gate (DG) device is also proposed [17], where the first gate controls carrier injection at the source contact and the second one independently controls carrier injection at the drain contact. In this device structure the band-edge profiles near the source and drain are controlled separately by two different gate biases. In order to suppress the tunneling current of holes at the drain it is necessary that the voltage of the drain side gate be equal or higher than the voltage of the drain. In consequence the parasitic tunneling current of holes is suppressed and the parasitic current is limited to thermionic emission of holes over the Schottky barrier at the drain, as shown in Figure 2.3. While electron injection at the source contact can be controlled through the first gate (source side), the second gate (drain side) suppresses parasitic hole current at the drain contact. The same discussion holds for *p*-type devices. By applying negative voltages to the first gate, hole injection at the source contact can be controlled and the second gate suppresses parasitic electron current at the drain contact. the thermionic emission current of holes at the drain contact will decrease exponentially and consequently, a lower parasitic current can be achieved This work showed that by applying a proper voltage range to the second gate, parasitic carrier injection at the drain can be avoided, and improved device characteristics are achieved.

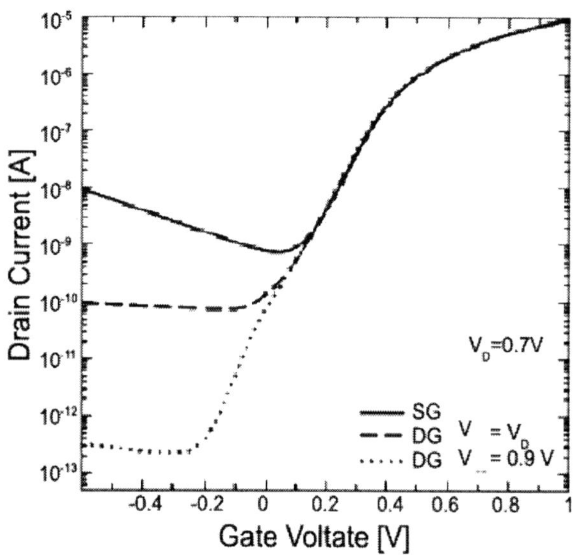

Figure 2.3. Calculated transfer characteristics for the single gate CNTFET (solid line), the DG-CNTFET with $V_{G2}=V_D$ (dashed line), and the DG-CNTFET with V_{G2}=0.9 Volt (dotted line). For both devices V_D=0.7 Volt [17].

2.2. Oxygen Molecule Effects as an External Charge on Electronic Structure of Carbon Nanotube

We studied the carrier suppression on CNT with an asymmetric structure where the modulating potentials on each Schottky barriers of source and drain are independently controlled. This geometric asymmetry is useful for understanding fundamental I-V characteristics of CNT-FET but limited to build the actual nano-scale devices. One dimensional constituent such as nanotube or nano-wire is sensitive to the external charge sources because of large surface area. Specially, CNT transport is extremely affected because of tube geometry.

Common observations of the *p*-type behaviors of the nanotube devices have been attributed to the doping by the adsorbed oxygen from the ambient air [18, 19, 20]. It was shown theoretically [21] and experimentally [22] that electronic properties of CNT-FET are sensitive to the absorption of oxygen molecules attributed to charge transfer and gas induced charge fluctuation. Meanwhile, Derycke et al. have suggested the main effect of the oxygen adsorption lies in the adjustment of the Schottky barrier rather than the bulk doping of the CNT body [9, 23], which is consistent with subsequent studies showing a negligible charge transfer between oxygen adsorbator and the nanotube wall [24, 25, 26]. Despite this apparent controversy, bundle or networks of the nanotube may react differently from the individual nanotube upon the oxygen adsorptions. The thermoelectric power measurement proved that the networks of CNT could be doped as *p*-type upon oxygen adsorption, irrespective of the contact metal [18, 19]. On the other hand, the transfer character of the transistor with a single and short nanotube could sensitively depend on the Schottky barrier rather than doping [9, 23].

Irrespective of whether the binding energy of the O_2 on the wall is strong or not, *ab initio* calculation should predict the existence of the LUMO level of O_2 just above the valence band edge of the semiconducting nanotube [27, 28]. Here we performed *ab initio* density-functional calculations using the Vienna simulation package (VASP) [29, 30, 31]. The spin-polarized gradient-corrected approximation is adopted for the exchange-correlation potential [32]. The ionic potentials are described by the projector augmented wave (PAW) method as provided with VASP [33], and the energy cutoff for the plane-wave basis is set to 400eV [34].

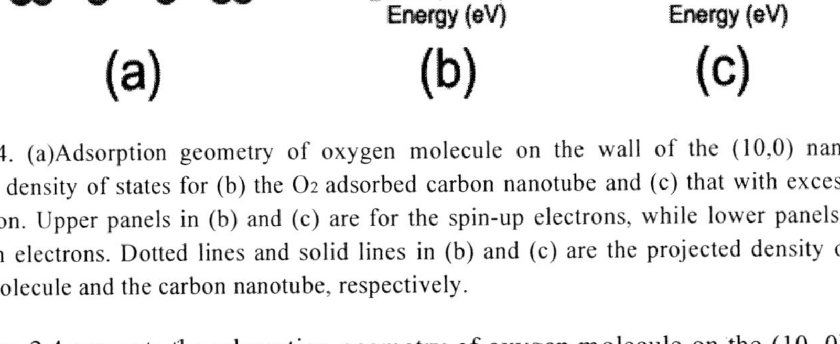

Figure 2.4. (a)Adsorption geometry of oxygen molecule on the wall of the (10,0) nanotube. The electronic density of states for (b) the O_2 adsorbed carbon nanotube and (c) that with excess charge of 0.1 electron. Upper panels in (b) and (c) are for the spin-up electrons, while lower panels are for the spin-down electrons. Dotted lines and solid lines in (b) and (c) are the projected density of states for oxygen molecule and the carbon nanotube, respectively.

Figure 2.4 presents the adsorption geometry of oxygen molecule on the (10, 0) nanotube and the electronic density of states. To simulate an isolated single nanotube, we assumed a large two-dimensional lattice in the perpendicular plane to the tube axis, where the distance between the nanotube and its replica is larger than 10 Å. The unit-cell along the axial direction is twice the minimal unit-cell of the zigzag carbon nanotube (≈8.49 Å). The atomic positions are relaxed with residual forces smaller then 0.05eV/ Å with the given unit-cell parameters. The upper panels in Figure 4(b) and 4(c) are for the majority-spin states and lower panels are for the minority-down states. The dotted peaks in the upper panel at about -2 eV and that in the lower panel at just above 0 eV shows the triplet ground state of the oxygen molecule. The LUMO level of O_2, which is the *ppπ** with the minority-spin state, is just above the valence band maximum of the (10, 0) nanotube. In other word, the Fermi level of the oxygen adsorbed nanotube sits at the valence band edge rather than at the midgap as shown in Figure 2.5.

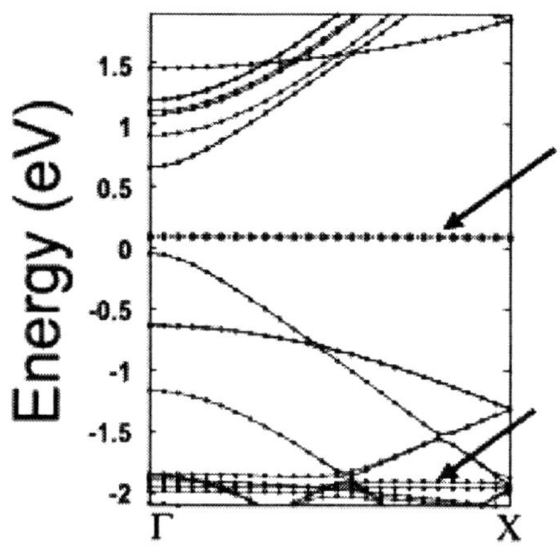

Figure 2.5. Adsorption geometry of oxygen molecule on the wall of the (10,0) nanotube and *ab initio* band structure with the oxygen adsorbed nanotube system. Arrows indicate oxygen molecular states. The LUMO level of O_2 adsorbators could provide the charge trapping centers, irrespective of the possibility of a practical hole-doping by the oxygen adsorption. These charge trapping centers could screen the electric field from the positively biased gate, resulting in the suppression of electron channel formation [22].

We increased the electron occupation to have 0.1 excess electrons in the unit cell (80 carbon atoms and 2 oxygen atoms). We find that the excess electron occupy the LUMO of O_2 rather the conduction band of the carbon nanotube. Thus, the turn-on voltage of the *n*-channel current will be increased by the presence of the oxygen molecule on the nanotube wall. This shift of the Fermi level, and the consequent unease in turning on the *n*-channel current, should be interpreted as p-type doping of the nanotube, regardless of whether the charge transfer is strong or not. Moisture can be considered as another source of charge transfer for CNT under ambient air. However it is found that the LUMO level of H_2O is about 4 eV above the Fermi level of the carbon nanotube. Thus, it is believed that the effect of water vapors on the aforementioned suppression of the n-channel formation is insignificant. Since the HOMO-LUMO gap in the DFT is not so rigorous, TDDFT calculations are necessary for the H_2O effect using the Gaussian package. The HOMO-LUMO gap of H_2O is found to be 6.4 and 7.5 eV in DFT and TDDFT, respectively. Thus, the DFT artifact would not affect our model discussed here.

2.3. Current-Voltage Characteristics of CNTFET

Figure 2.6 shows that a typical p-channel behavior in I_{ds}-V_{gs} curves of the CNT-FET measured with various source and drain voltages (V_{ds}) under the exposure of ambient air without passivation. It could be described as a device operating in p-channel depletion mode with an analogue of MOSFET, whose turn-on voltage (V_{th}) is about V_{gs} = 4V. This turn-on voltage depends on the gate dielectric thickness. This asymmetric change in the I_{ds}-V_{gs} curve, inducing less current for one polarity of gate bias and more current for the opposite polarity is attributed to the charge sources in ambient air. Intrinsic transport was investigated after removing the possible charge sources such as N_2, O_2, H_2O, and so on.

Figure 2.7 shows complete conversion of transport property of CNTFET from unipolar to ambipolar at vacuum of 3×10^{-6} Torr as discussed above for transport of CNTFET without the external charge. These external charge sources are volatile at vacuum even without the extra heat treatment since the overall conversion process can be reversible and can reproduce the patterns.

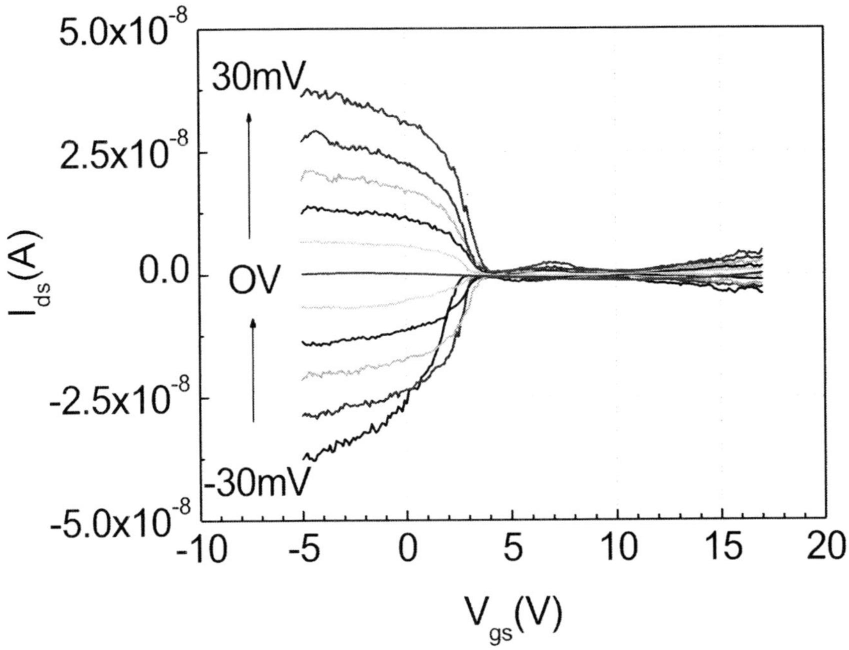

Figure 2.6. I_{ds}-V_{gs} curves of the CNTFET under the influence of ambient air condition with various source-drain voltages of 5mV steps. DWNTFET was chosen to maximize ambipolar behavior since it is mainly ascribed to the smaller band gap (V_g=0.2eV) arising from the large diameter (2nm) of outer shell. Purified DWNTs were suspended in a solvent and spin coated on the SiO_2 grown on heavily doped p-type Si substrate. Scanning Probe Microscopy (SPM) locates individually separated nanotubes on the substrate. After generating patterns for metal contact by electron beam lithography, metal electrodes with 100nm of Pd were defined by the lift-off process. Typical system resistances without gate bias were in the range of several hundreds of KΩ to MΩ. [22].

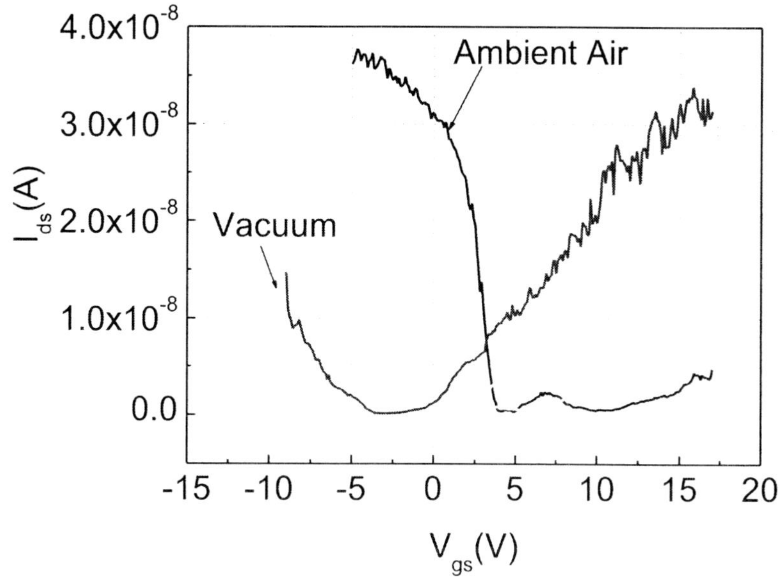

Figure 2.7. Transfer characteristics of CNTFET under ambient air, vacuum of 3×10^{-6} Torr, The n-channel starts to emerge and the FET shows an ambipolar behavior after vacuum environment with complete conversion of transport property. The overall conversion process is reversible and reproducible.

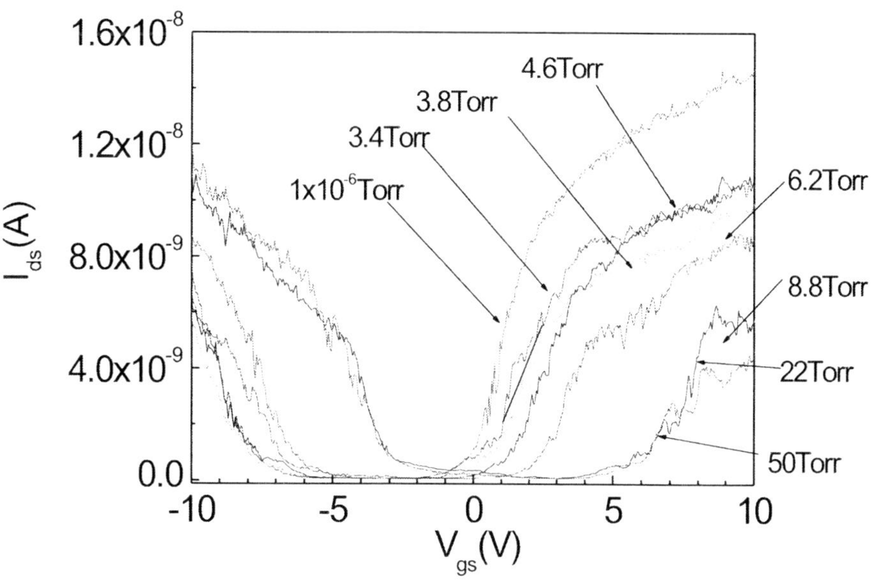

Figure 2.8. The I_{ds}-V_{gs} curves of the CNTFET under oxygen. Single-walled carbon nanotube was grown by the patterned catalyst method on the thermally grown SiO_2 of 500nm. CH_4 gas was introduces in a thermal CVD furnace at 905oC for 10 min with hydrogen gas after loading the wafer. After defining the metal contact patterns with the conventional photolithography technique, the Ti/Au electrodes were deposited on the patterns by using thermal evaporation. The second lift-off process was carried out to form metal contacts with CNT. The distance âåôùåâí source and the drain electrode was fixed at 10μm [35].

The absorption of oxygen molecules in ambient air is dominant to electronic properties of CNT as discussed by ab initio calculation. However it also leads the work function variation in the electrode and may result the ambipolar behavior [23]. To investigate the effect of O_2 adsorption on the transport properties, a long individual single-walled carbon nanotube is studied for the transport measurement. I_{ds}-V_{gs} curves shown in Figure 2.8 indicate that main effect on the transport is O_2 adsorption on the CNT channel. Under the high vacuum of 1×10^{-6} Torr, the n-channel almost turns on at $V_g = 0$. This could be ascribed to the fact that the Fermi level of Ti aligns with the conduction band edge of the nanotube. The n-channel current is more suppressed, while the p-channel current is enhanced as the O_2 pressure is higher. These observations bring important discussions about the origin of the change whether doping-like effect on CNT or metal work function change in metal electrode. The slope of the I_{ds}-V_{gs} curves change upon the Schottky barrier variation, while the curves shift with negligible slope change upon the doping [23]. In this regard the observed I_{ds}-V_{gs} curves from the oxygen exposure are rather close to doping-like effect on CNT than the modulation of Schottky barrier.

The comparison of the curves of 1×10^{-6} Torr and 6.2 Torr shows that the n-channel current is reduced while the p-channel current is almost unchanged. That means there has been no clear demonstration of the O_2-induced *p*-type doping of an individual carbon nanotube and should be ascribed to the n-channel current suppression by the trapped negative charges on the LUMO level of the O_2 molecules nearby nanotube surfaces or the surface of the gate dielectric.

The oxygen exposure of the CNT-FET could induce the work function increase in the metal surface, *p*-channel current suppression, as well as the hole-doping like effect of the CNT body. However, such a doping effect would be dominant as the length of the nanotube increases.

The unipolar behavior of CNT-FET in ambient air needs more comprehensible arguments. It is reasonable to assume that there are significant molecular adsorptions on the wall of CNT as well as on the surface of metal electrode. Provided that the LUMO levels of the adsorbators sit somewhere in the gap of the carbon nanotube, electrons would occupy the molecular levels under a positive gate voltage, as shown in Figure 2.9(a) and 2.9(b). As a result, the electrostatic potential generated by the negatively charged adsorbators screens the gate field and, consequently, suppresses the electron channel formation. The nanotube bands in the channel region would resist to down-shift because of the presence of charged adsorbators. Two upward arrows in Figure 2.9(b) describe the effect of such an electrostatic potential generated by the charged adsorbators. On the other hand, conduction bands of a clean nanotube shift down allowing electron current under such a positive gate voltage, as shown in Figure 2.9(c).

To find a molecule for suppressing holes to build the n-type CNT-FET is still challenging issue for non-volatile adsorption to be acceptable for integration. Actually, for real device application for the CMOS-like logic, carrier suppression by a molecular adsorption may not be sufficient since the molecules on CNT surface is not expected to be stable under the operational environment. Atomic doping in CNT may be a key solution. But there are fundamental issues for degradation of ballistic nature of CNT since dopants act as source of defects as well as for the synthetic method of doped SWNT. Encapsulation is also considered but the technology is very early stage for device application.

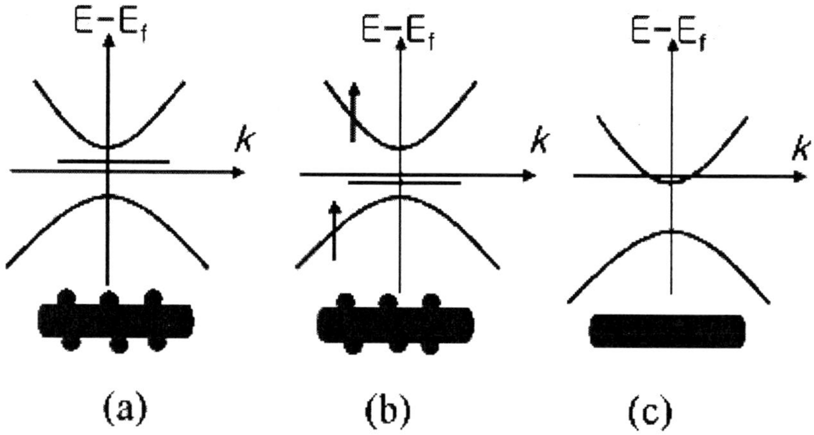

Figure 2.9. Schematic band structure of the carbon nanotube and the LUMO level of the adsorbators, with respect to the Fermi level.

3. Growth of Single Walled Carbon Nanotubes for the Electronic Device Applications

To fabricate nanotube-based FETs, SWNTs should be grown without multi-walled carbon nanotubes (MWNTs) and carbonaceous impurities. However, the growth of SWNTs is generally accompanied by the undesired carbon materials and MWNTs that make electrical short and unexpected leakage. In this section, we will discuss two different ways to obtain high purity SWNTs.

One way to build the CNT array for IC fabrication is to place the individual nanotubes on the desired positions from nanotube powder with an appropriate dispersion method [36]. Syntheses of single-walled carbon nanotube (SWNT) powders have been achieved by several methods such as arc discharge, laser ablation, and HiPCo process [38, 39]. SWNTs synthesized by these techniques should be subsequently purified to fabricate nanotube-based device elements. We introduce a simple way to obtain high purity SWNT powder in this article to overcome the various CNT purifying steps that are inevitable for the previous synthetic methods.

On the other hand, controlled positioning of SWNTs on a defined region has been regarded as a crucial step for device integration [40]. SWNTs grown by chemical vapor deposition (CVD) can be more easily integrated in devices than the synthesized nanotubes, since nanotubes are directly and selectively grown on catalysts such as iron, cobalt, and nickel [41]. Before the SWNT growth, the catalyst is pre-patterned as films or nanoparticles on a defined region by various techniques such as lithographic lift-off, soft lithography, offset printing, template methods, and direct lithography [41-45].

In these two synthetic approaches to SWNTs, there are common efforts to lower the synthesis temperature. For the SWNT powder synthesis, the growth temperature was recently lowered to room temperature by a sonochemical route. For the direct growth of SWNTs on substrates by CVD, remote plasma-enhanced chemical vapor deposition (PECVD) has shown a clue for the low temperature growth of SWNTs.

3.1. Sonochemical Synthesis of SWNTs at Room Temperature

It is well known that cavitation bubbles produced in liquid solution during sonication can instantaneously generate local spots of several thousand degrees Celsius of temperature and several thousand atmospheres of pressure [46]. Inspired by the chemical effects of ultrasound, the sonochemical route to SWNTs was achieved in a liquid solution at atmospheric pressure and room temperature [47]. In addition, this method offers a facile route to high-purity SWNTs without specialized equipment and dose not require multi-step purification process to eliminate the amorphous carbon and multi-walled carbon nanotubes.

The typical process of this method for SWNT growth is depicted in Figure 3.1. The procedure begins with the preparation of a mixture solution of ferrocene and *p*-xylene. Silica powder was then added in this solution. Ferrocene was chosen as a precursor of Fe catalyst for nanotube growth. The carbon source for SWNT growth was provided by *p*-xylene as well as by ferrocene [48], while silica powder acted as a nucleation site for SWNT growth. Ultrasonication was performed under ambient conditions to grow SWNTs. Finally, the resulting powder was collected on filter membrane. Initially, 2 grams of silica powder (diameters of 2.5 mm) were added to 0.01 mol % ferrocene-dissolved *p*-xylene solution (50 mL). The solution was then sonicated by a 1/2 inch titanium tip 200-W probe pulsed 65 % at atmospheric pressure and room temperature for 20 minutes. During this period, the color of silica powder turned into gray, indicating a carbon deposition on its surface. It was also found that silica powder broke into small pieces below several hundreds of μm. This could be attributed to interparticular collisions of silica powder triggered by ultrasound irradiation [49]. The resulting powder, collected on a filter membrane, shows that fiber-like structures exist on the surfaces of silica powder as shown in Figure 3.2.

After filtration to remove the silica powder by using HF solution, fiber-like structures remained on the filter membrane as shown in Figure 3.3(a). This morphology is similar to purified SWNT bundles produced by the laser ablation method [50]. High-resolution transmission electron microscope (HR-TEM) confirmed that these structures mainly consisted of SWNT bundles as shown in Figure 3.3(b). The absence of multi-walled carbon nanotubes was also confirmed by HR-TEM observations. The average diameter of SWNTs based on TEM data was 1.5 ± 0.1 nm as shown in an inset image of Figure 3.3(b).

Control experiments were carried out to determine if the presence of silica powder is essential for producing SWNTs. For a sample sonicated without silica powder, no SWNTs were shown in the resulting HR-TEM images. Concentration effect of ferrocene in *p*-xylene solution was also studied to achieve the best conditions for the SWNT growth. It was found that high-purity SWNTs were obtained at relatively low concentration of ferrocene (0.01 mol %), showing no noticeable evidence for the existence of multi-walled carbon nanotubes and amorphous carbon. Once Fe nanoparticles are produced by decomposition of ferrocene during sonication, there would be two possible competing reactions; SWNT growth by catalytic action of fine Fe catalyst and Fe particle growth by agglomeration of fine Fe catalyst. However, it is expected that the former reaction occur dominantly under our experimental conditions, reducing the chance to form the larger Fe particles, which may result in multi-walled carbon nanotube growth. These are probably caused by very low concentration of ferrocene in our experiment.

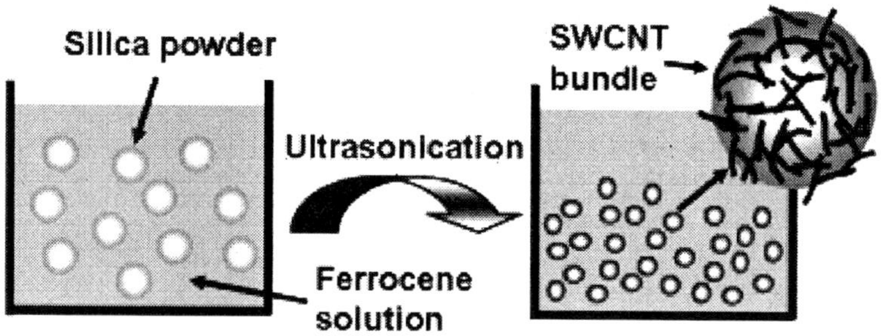

Figure 3.1. Schematic diagram of sonochemical route to SWNTs. Silica powder is immersed in the solution (ferrocene-xylene mixture). Ultrasonication then was done for 20 min at room temperature and atmospheric pressure. Sonication produces high-purity SWNTs on the surface of silica powder [47].

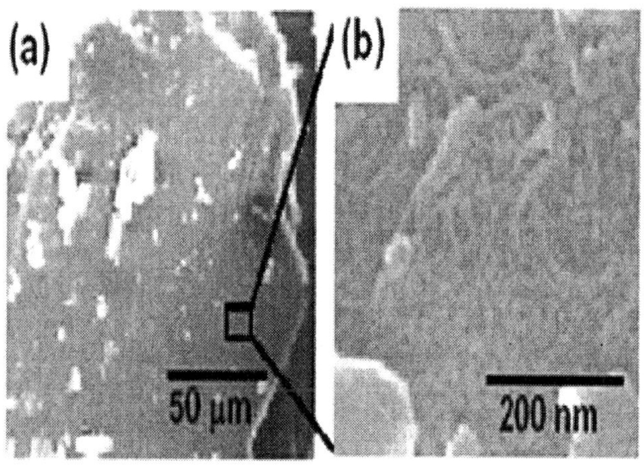

Figure 3.2. SEM images of silica powder after the sonochemical synthesis.

Raman spectroscopy has also confirmed sample homogeneity as shown in Figure 3.4. Spectra were recorded at 10 randomly selected positions and averaged. It shows the typical spectral characteristics of SWNT structure, as characterized by the strong G band (tangential mode) and the presence of the sharp radial breathing mode (RBM) in the lower wave number region (100-300 cm^{-1}). RBM spectra were further analyzed to give the diameter distribution of SWNTs. According to the equation: $\omega_{RBM} = 235/d + 9$ for bundled SWNTs [51], where d (nm) is the diameter and ω (cm^{-1}) is the Raman shift of a SWNT, the radial breathing mode (RBM) region in Figure 3.4(a) yields SWNT diameters of 1.1-2.1 nm with an average diameter of 1.55 nm. The diameters of most SWNTs (>90%) were in the range of 1.3.1.8 nm. These estimates agree well with the measured values from HR-TEM images. A weak peak at 1340 cm^{-1} of D-band (related to disordered graphite or amorphous carbon) and strong peak at 1572 cm^{-1} of G-band (tangential mode) are shown in Figure 3.4(b). The relatively low intensity of D band relative to the G-band indicates that only a small amount of amorphous carbon or few defects exist in these SWNTs [52].

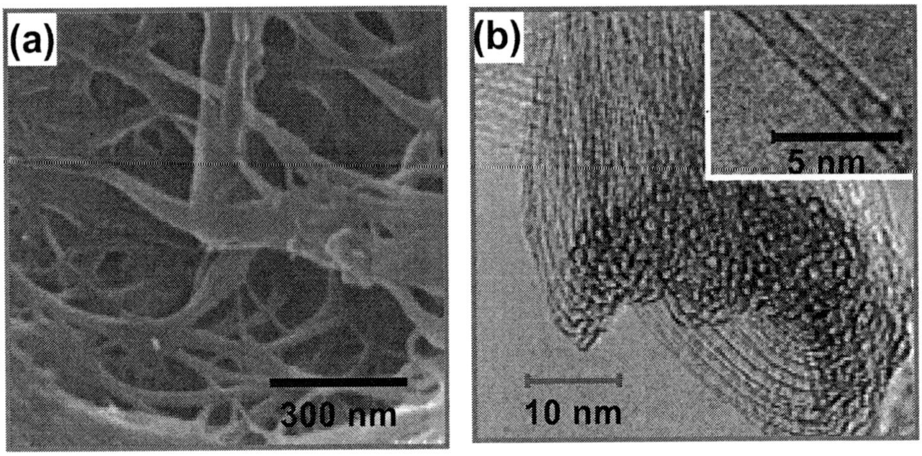

Figure 3.3. (a) SEM image of SWNT bundles on polycarbonate filter membrane. (b) HR-TEM images of SWCNTs. Most of SWNTs exist in bundles. For the measurements of diameters of SWNTs, individual SWCNTs (inset) are also imaged by HR-TEM [47].

Although details of SWNT growth mechanism are still under investigation, we believe that these sonochemical treatments both facilitate decomposition of ferrocene and *p*-xylene, and provide the required energy to synthesize SWNTs. Unlike previous methods for producing SWNTs [37-39, 41], which require severe conditions such as high temperature, high vacuum, or high voltage arc discharge, our method provides a facile route to high-purity SWNTs under ambient conditions.

3.2. Low Temperature Growth of SWNTs by Remote PECVD

Despite the promise of nanotubes grown by CVD, the high growth temperature (> 800 °C) of CNTs is a barrier for device fabrication. Since lower growth temperatures significantly enhance the compatibility of CNT growth with current complementary metal-oxide-silicon (CMOS) technology for CNT-based electronics, plasma enhanced chemical vapor deposition (PECVD) has been recently utilized to grow nanotubes at lower temperatures [53.57]. Unfortunately, the grown nanotubes at lower temperatures by PECVD are generally multi-walled carbon nanotubes (MWNTs) [53, 54]. Using the microwave plasma chemical vapor deposition method, MWNTs were grown at ~ 450 °C by Wilson and his coworkers [54]. Kato and coworkers used radio-frequency magnetron PECVD to grow SWNTs at 550 °C with zeolites as a supporting template for the Fe/Co catalyst [55]. Dai's group demonstrated that SWNTs can be grown at 600 °C without any template [56, 57]. Recently, it was demonstrated that SWNTs can be grown even at 400 ~ 450 °C without any template by remote PECVD [58, 59]. The key process parameter to enable the low temperature growth is low plasma power, which should be held at a minimum.

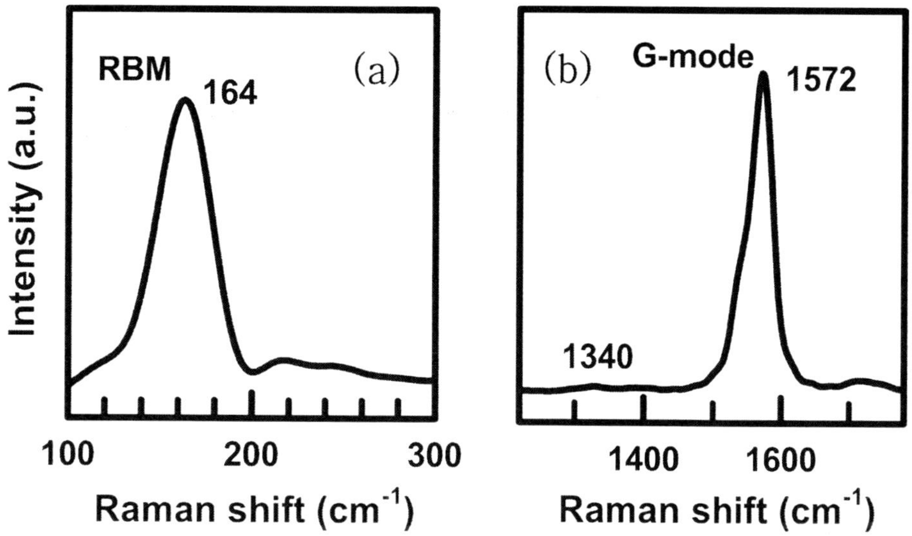

Figure 3.4. Raman spectra (Ar^+ laser: λ=514.5 nm) of synthesized SWNTs. (a) RBM and (b) G-line mode (right graph) are presented for the average of 10 spectra recorded at random points.

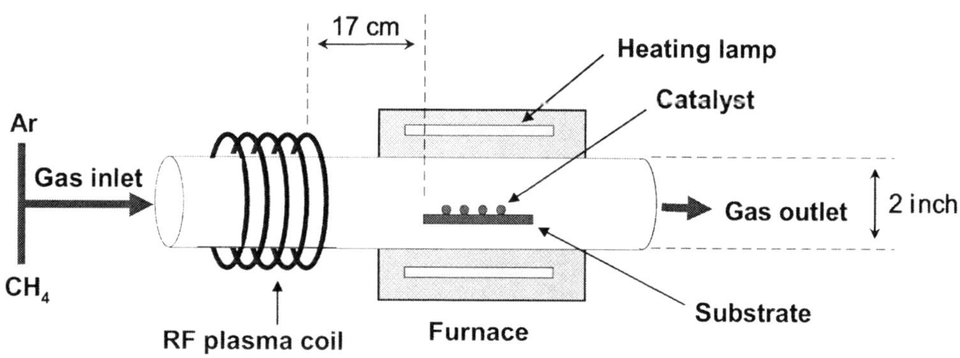

Figure 3.5. A schematic diagram of the remote PECVD system.

Figure 3.5 shows a schematic diagram of the homemade radio frequency (rf, 13.56 MHz) remote PECVD system. A quartz tube with a 2-inch diameter was surrounded by a copper coil in a helical manner from the gas inlet to the sample holder which was located at a distance of 17 cm from the end of the copper coil. Substrates were heated by halogen lamp and the growth temperature was calibrated with a thermocouple-implanted quartz holder. The calibrated temperature was not influenced by the remote plasma ignition.

Catalyst-deposited substrates were placed on a quartz holder and then heated up to the growth temperature for 300 sec in an argon atmosphere (100 sccm). After an interval for the stabilization of substrate temperature, methane (60 sccm) was introduced with argon (15 sccm) in the quartz tube and the subsequent plasma was ignited to grow CNTs. During the CNT growth (10 min), the working pressure was maintained at ~ 0.4 Torr.

To optimize the process condition for SWNT growth, we monitored the presence of radial breathing mode (RBM) peaks due to SWNTs and the D-band to G-band (I_D/I_G) ratio in Raman spectra. Since the G-band and D-band are due to tangential stretch modes of nanotubes and disordered graphite or amorphous carbons, respectively, the ratio generally increases with an increasing amount of disordered or amorphous carbons [60].

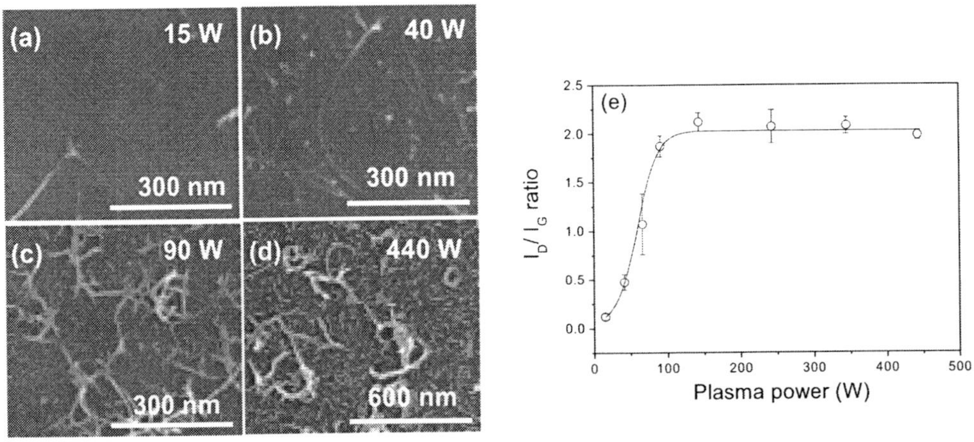

Figure 3.6. SEM images of nanotubes grown at 450 °C with plasma powers of 15 (a), 40 (b), 90 (c), and 440 W (d). (e) I_D/I_G ratio variations with plasma power [58].

The plasma power effect on nanotube growth was also shown with nanotubes grown at 450 °C in Figure 3.6. In the low power region (< 40 W, Figure 3.6(a-b)), CNTs are sparsely grown, but in the high power region, nanotubes with larger diameters are densely grown with short nanotubes or carbonaceous particles as shown in Figure 3.6(c-d). The plasma effects are more clearly revealed in Figure 3.6(e). The I_D/I_G ratio dramatically decreases from ~ 1.9 at 90 W to ~ 0.1 at 15 W as the plasma power decreases. However, the ratio is nearly constant in the high power region. It should be noted that the plasma lowers the growth temperature of nanotubes but the power should be held at a minimum to avoid the increase of I_D/I_G ratio.

Figure 3.7(a) shows I_D/I_G ratio as a function of growth temperature for nanotubes grown with 15 W, 40 W and 65 W plasma powers. Solid circles denote that RBM peaks were obtained from all of scans. Process conditions showing RBM peaks from some scans positions were designated with open circles. Cross symbols represent no RBM peaks. Figure 3.7(b) clearly reveals that the plasma power should be low to obtain SWNTs showing RBM peaks with low I_D/I_G ratio at low temperature. The lowest growth temperature showing RBM peaks from all of scans was 425 °C but their I_D/I_G ratios were higher than those of nanotubes grown at 450 °C. From the nanotubes grown at 400 °C with plasma powers of 15 W and 40W, the RBM peaks respectively appeared from three and two positions of five different scans. SWNT growth may be suppressed by high plasma power due to the bombardment effect [61]. The importance of low plasma power is also supported by a previous report by Delzdit. In their work, MWNTs were grown at low power levels and nanofibers were formed at high power levels, of which transition occurred at 30 ~ 40 W [53].

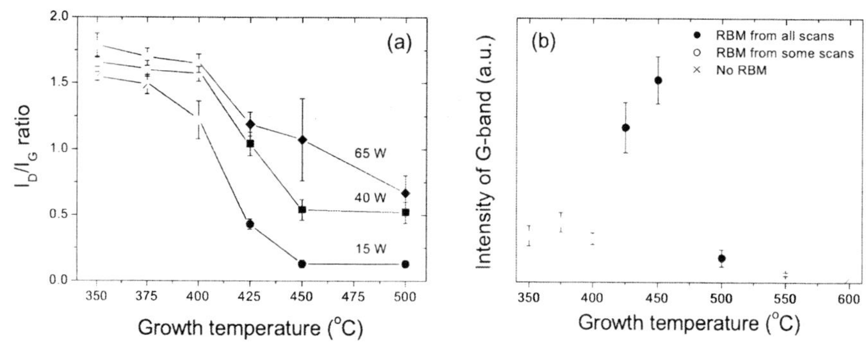

Figure 3.7. (a) ID/IG ratio variation with growth temperature. Solid or open circles designate that RBM peaks appear from all or a few of five different scans, respectively. Cross symbols indicate that RBM peaks were not observed from several scans. (b) Intensity variation of G-band with growth temperature. The nanotubes were grown at 450 °C with a plasma power of 15 W.

Temperature effect on the intensity of G-band was investigated with nanotubes grown by using a low plasma power of 15 W. As the growth temperature increases, the intensity of G-band shows an increasing tendency up to 450 °C and then decreases at higher temperatures than 450 °C as shown in Figure 3.7(b). Eventually, the G-band disappears at 600 °C wherein any peaks due to the carbon materials are not observed in Raman spectra. The disappearance of the Raman peaks may be attributed to an etch effect by plasma. To verify the etch effect by the weak plasma power; we treated nanotubes at 450 °C under argon plasma (15 W) without any flow of methane gas. As the plasma treatment time increased, G-band was gradually weakened and nearly disappeared after 10 min. Therefore, such a temperature dependency of the G-band in Figure 3.7(b) can be explained by a competition between growth and etch reactions of sp^2 carbon structures such as SWNTs, MWNTs, and graphene sheets. As the growth temperature increases, the contribution of the etch reaction may increase. When the etch effect is small, the G-band shows the increasing tendency with the temperature. However, as the etch effect increases, the G-band intensity gradually decreases with the growth temperature (> 450 °C). At 600 °C, the growth of sp^2 carbons is eventually overcame by the etch effect, to result in the disappearance of the G-band. It should be noted that the RBM peaks with a strong G-band are obtained from the specimens grown at 425 ~ 450 °C wherein the etch effect of sp^2 carbons is considerably high. This means that the etch effect by the low plasma power plays a critical role for SWNT growth.

Although the remote PECVD using a low power of argon plasma gives SWNTs at 450 °C as shown in Figure 3.8, the observed nanotubes detached from the substrates for TEM analysis are mainly bundles and defective in their surface. In addition, MWNTs or double-walled carbon nanotubes (DWNTs) were also observed with SWNTs. Figure 3.8(c) shows a DWNT with an outer diameter of 1.33 nm and an inner diameter of 0.61 nm. The interlayer distance of the DWNTs is ~ 0.36 nm, which is similar with reported values [62]. Figure 3.8(d) shows an image of SWNTs with carbonaceous materials.

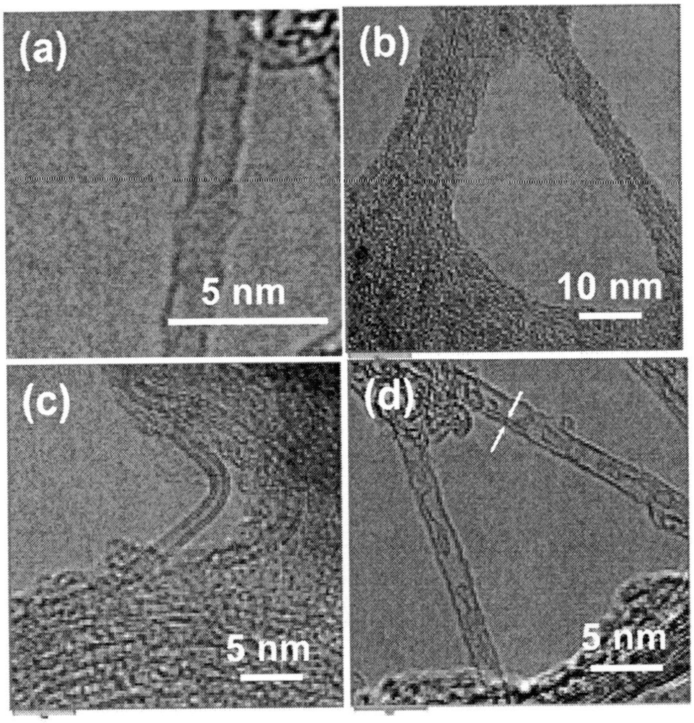

Figure 3.8. HRTEM images of nanotubes grown by remote argon PECVD. (a) Singlewalled nanotuebs. (b) Bundles. (c) Double-walled nanotubes. (d) Single-walled nanotubes with carbonaceous materials inside. Nanotubes were grown at 450 °C with a plasma power of 40. Dark lines correspond to the sidewalls of nanotubes [58].

To fabricate nanotube-based FETs, SWNTs should be grown without multi-walled carbon nanotubes (MWNTs) and carbonaceous impurities. However, the growth of SWNTs is generally accompanied by the undesired carbon materials, especially during chemical vapor deposition (CVD) process, which is the preferred growth method of nanotubes for integrated electronic devices. Although sulfidative purification of nanotubes was recently reported to selectively remove carbonaceous impurities from nanotubes integrated in devices, the undesired MWNTs would be problematic for reliable transistor performance. Therefore, preferential growth of SWNTs without both carbonaceous impurities and MWNTs is essential in CVD for device applications.

The preferential growth of SWNTs over MWNTs and carbonaceous impurities can be achieved by water-plasma CVD. Water plasma lowers the growth temperature down to 450 ^{o}C as argon plasma does, and the grown nanotubes are pure SWNTs without MWNTs.

Several groups have widely investigated an effect of water on purification and synthesis of carbon nanotubes. Tohji, et al. reported that SWNTs could be purified with water by hydrothermal treatment [63]. Yoshimura et al. demonstrated that MWNTs were synthesized in the absence of metal catalyst by hydrothermal treatment of amorphous carbon in pure water [64]. Wang et al. introduced one-step water-assisted synthesis of MWNTs only by dipping hot graphite rod into cool water in the absence of catalyst [65]. Hata, et al. showed a thermal CVD synthesis of SWNTs where the activity and lifetime of the catalysts were enhanced by water. The most common finding is that all of the nanotubes prepared from the mentioned approaches are pure without carbonaceous impurities. Inspired by the water effect, water plasma was chosen as both an oxidant for high purity of nanotubes and an energy source for low temperature growth.

Iron catalyst nanoparticles were deposited on SiO_2 (400 nm)/Si substrates with a mixed solution of ferrocene and conventional photoresist as described in the section 4. A homemade radio frequency (13.56 MHz) remote PECVD system was used for nanotube growth. Water was evaporated from a canister and carried to the quartz tube by the vacuum without any carrier gas. The amount of water vapor was regulated by a metering valve and monitored by pressure increase in the quartz tube.

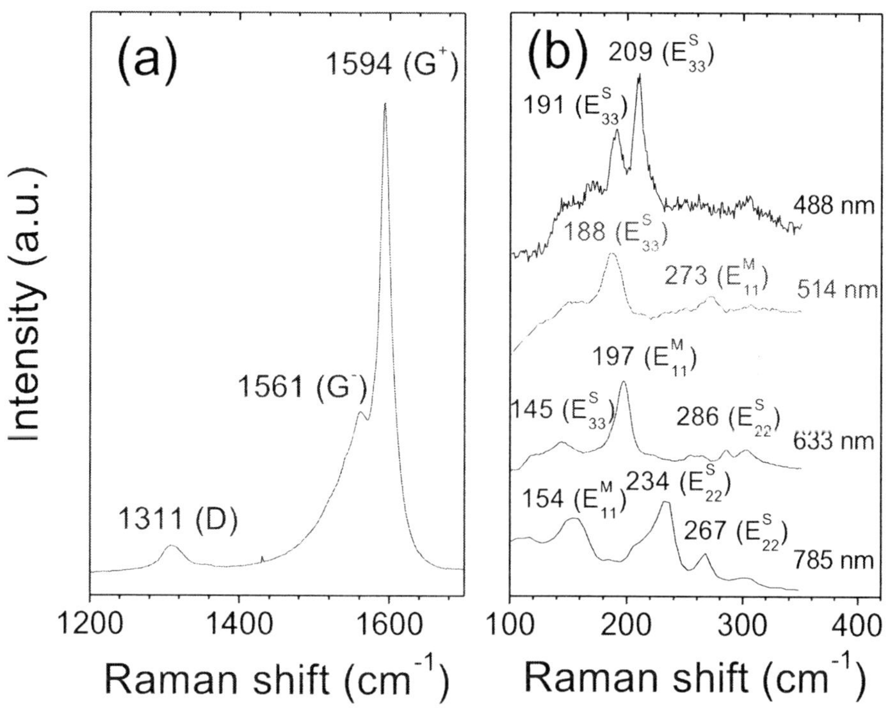

Figure 3.9. Raman spectra of the grown nanotubes. (a) G-band and D-band obtained from 633 nm laser excitation. (b) RBM peaks from 488, 514, 633, and 785 nm laser excitations. All spectra were measured on 5 different positions and averaged. Raman shifts were calibrated with a silicon peak at 521 cm^{-1}.

SWNTs were grown on catalyst-deposited substrates at 450 °C with methane gas (60 sccm) and water plasma (15 W) in the remote PECVD system. Raman spectra from laser excitations of various wavelengths (488, 514, 633 and 785 nm) reveals that the grown nanotubes were single-walled. Figure 3.9(a) and (b) show the typical Raman spectra of SWNTs, as characterized by the strong G band (tangential stretch mode) and the presence of the sharp radial breathing mode (RBM). Most RBM peaks appear in frequency region between 120 and 250 cm^{-1}, which roughly corresponds to the diameter range of 1 ~ 2 nm.

According to the Kataura plot (transition energies as a function of nanotube diameter) [66], the grown nanotubes are a mixture of metallic and semiconducting tubes as designated in Figure 3.9(b). All spectra were measured on 5 different positions and averaged. Raman shifts were calibrated with a silicon peak at 521 cm^{-1}.

To evaluate the quality of nanotubes from the amount of disordered and amorphous carbons, we monitored a Raman intensity ratio (I_D/I_G) of D-band to G-band as growth time increased. The ratio of the nanotubes grown by water plasma is compared to that of the nanotube by argon plasma (15 W, methane 60 sccm) as shown in Figure 3.10. Since water plasma with a weak power (15 W) plays a role of oxidant which removes carbonaceous impurities and disordered carbons, SWNTs grown by water plasma shows lower I_D/I_G ratio than those by argon plasma over all growth time. The I_D/I_G ratio decreases in the initial stage of nanotube growth with the growth time and then keeps nearly constant during a period and subsequently increases possibly due to deformation of nanotubes by active species such as radical or ions generated by plasma. It should be noted that growth and etch reactions of nanotubes compete more severely in plasma than in thermal CVD.

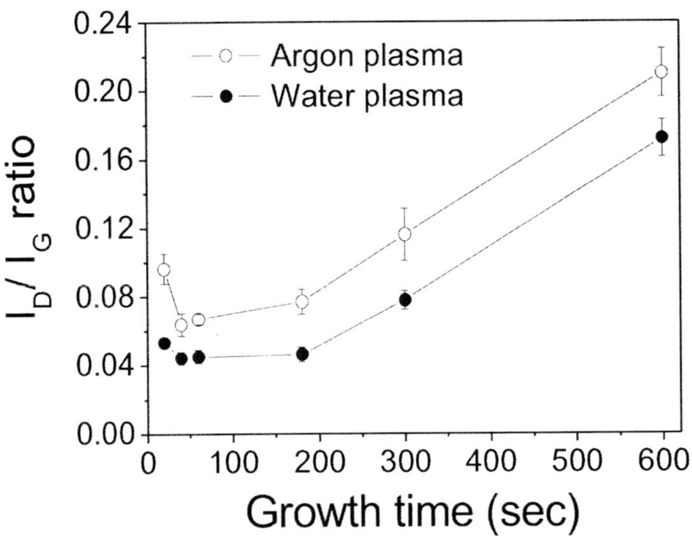

Figure 3.10. Variations of I_D/I_G ratios with growth time. Open and solid circles were plotted with Raman intensity ratio (633 nm) of SWNTs grown by argon and water plasma, respectively.

Figure 3.11(a) shows HRTEM image of the grown SWNTs of which diameters are mainly in a range of 1 ~ 2 nm (Figure 3.11(b)) as expected in Raman spectra. The observed nanotubes were single-walled or bundled. MWNTs were rarely observed as the growth time is short. Among the nanotubes grown for 180 seconds, we could not find MWNTs in HRTEM images. Considering that SWNTs were less frequently observed than MWNTs among the nanotubes grown by argon plasma CVD, the water plasma effect on the growth of pure SWNTs without MWNTs is obvious.

To investigate electrical nature of SWNTs grown by water plasma, we fabricated FETs with random network of SWNTs as a channel. The source-drain channel length was 5 μm and the channel width was 40 μm. The number of SWNTs per unit area was ~ 30/μm^2.

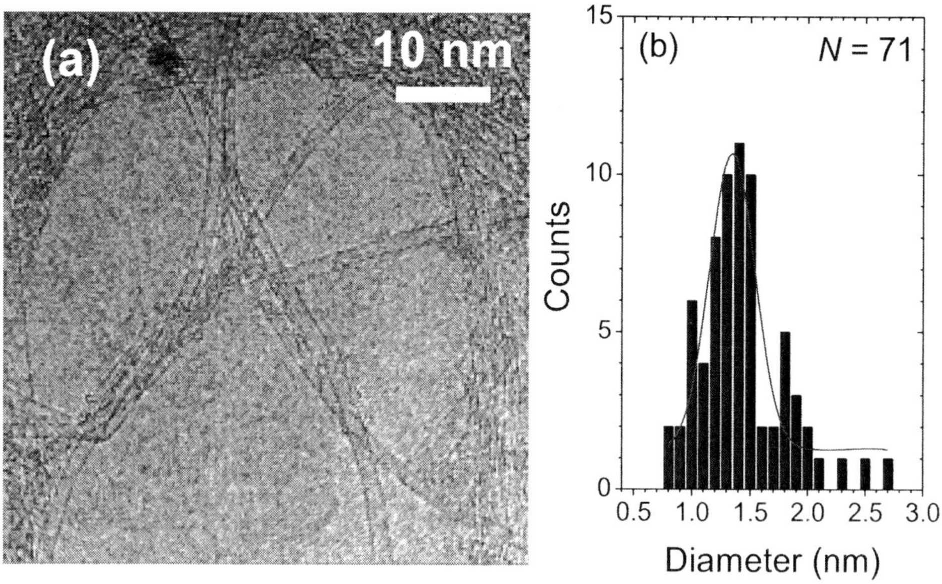

Figure 3.11. (a) HRTEM image of SWNTs grown at 450 oC with a water plasma power of 15 W. (b) The histogram of the diameter of SWNTs determined from HRTEM images [59].

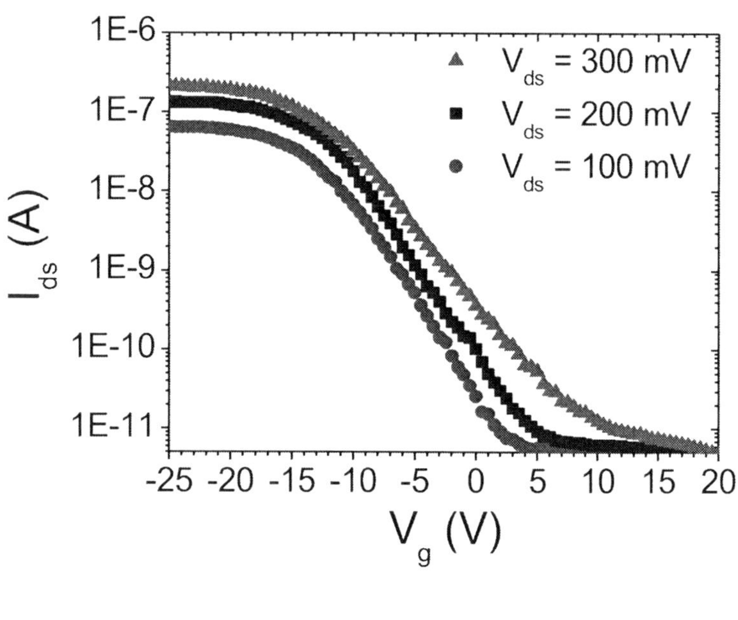

Figure 3.12 Drain-source current (Ids) versus gate voltage (Vg) characteristics of a FET with random network of SWNTs [59].

It is reported that, for SWNTs with densities exceeding a percolation threshold, FETs in the random network are electrically conducting [67]. Above the percolation threshold, which is $0.3/\mu m^2$ for nanotubes with a length of 1~3 μm, the SWNTs will intersect with one another and form continuous conduction path with metallic tubes. However, even though the nanotube density is much higher than the percolation threshold, the FET with network of SWNTs grown by water plasma shows typical p-type semiconducting characteristics with an On/Off ratio of ~ 10^4 (Figure 3.12). This can be explained by some higher portion of semiconducting nanotubes over metallic tubes than the 2:1 ratio expected for random chirality distribution. Dai et al. reported that nearly 90 % of the grown nanotubes by remote PECVD are semiconducting, while the HiPco and laser ablation nanotubes consist of 61 and 30 % semiconductors, respectively [56, 57].

The semiconducting nature of the network FETs also supports the preferential growth of SWNTs over MWNTs. Since MWNTs are generally metallic due to their large diameters, the network transistors should show metallic nature unless the preferential growth was used.

This water plasma CVD method for the low temperature growth of pure SWNTs facilitate to fabricate nanotube-based devices for fundamental characterizations and industrial applications.

4. Carbon Nanotube Placement and Fabrication for the Electronic Devices Integration

Since carbon nanotubes (CNT) are one-dimensional systems which can be metals or semiconductors depending on its chirality, these new carbon allotropes have been considered as the material to replace conventional silicon-based electronics in future. However, the uncertainty and uncontrollability of CNT positioning is one of the most critical barriers for CNT integration. To successfully build upon an integrated circuit (IC) with CNT-based active and passive elements, CNTs should be selectively and individually grown or located only on a defined region.

From a methodological point of view, there are two pathways for positioning CNTs on a defined region: top-down and bottom-up approaches. For an example of the top-down approach, a self-assembly method for CNT positioning has been attempted with an interacting force between CNTs and organic molecular marks pre-patterned on a surface [46]. Currently, this approach is in the early stages of building ICs over a large area and has difficulties in extending the organic molecules to various surfaces of oxides or metals for real applications.

On the other hand, many bottom-up approaches have been tried using pre-patterned catalysts before a subsequent chemical vapor deposition of CNTs, since it is well known that nanoparticles or thin films of some transition metals, such as Fe, Co and Ni, play the role of catalyst for CNT growth [68]. The catalyst pre-patterns have been obtained from various techniques such as lithographic lift-off [41], soft lithography [42], and offset printing [69]. However, these pre-patterning methods are not compatible with current microelectronic processing for IC fabrication. For an example of the bottom-up approach, a pre-patterning method using a catalyst-dispersed solution with nanoparticle supporters is the most popular way to obtain CNTs with high electrical performance [41]. However, since the dispersion solution of catalyst and supporting particles are not uniform in concentration, this method is not reliable for mass production.

Recently, several groups demonstrated direct photolithographic methods for selective growth of CNTs using photoresists or electron beam resists containing catalyst precursors [45]. The modified resist plays the role as both catalyst and resist for growth and alignment of CNTs, respectively. The modified resist patterns are obtained by following general lithographic procedures. After removal of organics from the resist pattern, catalytic particles are deposited on the defined region and then CNTs are selectively grown on the pre-patterned catalyst. Huang, et al used poly (vinyl alcohol) as a polymer resin with metal nitrate precursors [45]. Hinderling et al. demonstrated that iron-containing block copolymer polystyrene-poly (ferrocenylsilane) can be used for CNT growth [70]. Ishida, et al. reported electron beam lithographic method using a calixarene resist modified with iron(III) acetylacetonate ($Fe(acac)_3$) [71]. However, those modified resists were not successful to grow single-walled carbon nanotubes (SWNTs).

Here, we report a combination for the direct photolithographic route to selective growth of SWNTs using a simple mixture (Cat-PR) of ferrocene and a conventional photoresist (AZ 5214E, Clariant Co) [72]. From the Cat-PR pattern, we obtained catalytic nanoparticles of which diameters are in the range of 3 – 10 nm and are mainly smaller than 5 nm. Nanotubes with diameters of 1.0 ~ 1.5 nm are grown on the catalytic nanoparticles as individual SWNTs or bundles of them. It is also demonstrated that nanotubes suspended between three-dimensional structures could be obtained by utilizing the Cat-PR pattern as an etch mask.

4.1. Selective Deposition of Catalytic Nanoparticles

Ferrocene was chosen as the catalyst precursor in creating a homogeneously dissolved and chemically stable solution with commercially available PR (AZ 5214E). Since ferrocene has a sandwich structure consisting of two cyclopentadienyl and an iron ion, it is moderately stable and soluble to various organic solvents [73]. The addition of ferrocene to the PR does not significantly deteriorate the photo-active properties of PR. Furthermore, ferrocene has been recently utilized as a catalyst to grow CNTs in gas phase or liquid phase [47, 74].

Different amounts of ferrocene were dissolved in AZ 5214E to obtain various concentrations of Cat-PR solutions. Standard photolithographic procedures were then followed.[17] The Cat-PR was spin-coated on substrates and baked, followed by the exposure step with a mask aligner in order to define where the CNTs will be positioned. After the photolithographic development step, the Cat-PR pattern with ferrocene was obtained as shown in Figure 4.1.

Before the growth of CNTs, the burning step is added to general photolithographic procedures in order to remove organics from the Cat-PR pattern. The burning temperature was determined to be 550 ^{o}C which is higher than the temperature of a strong exothermic peak in differential scanning calorimetric thermogram obtained in air. The Cat-PR shows a strong exothermic pick at ~ 460 ^{o}C accompanying weight loss due to the removal of organics, and the residue at 550 ^{o}C is around 0.8 %. Supposing that the residue is Fe_2O_3 without any organic substance, the residual weight should be theoretically 0.73 %. The good agreement between the experimental and the theoretical values of the residue supports that nanoparticles were formed without organic residue by the burning process. This complete removal of organics from the Cat-PR pattern is crucial to obtain catalytically active nanoparticles. The exothermic peak and weight loss in TG-DSC should appear in order to remove the organics and to deposit catalyst particles on the defined regions during the burning process. For example, a mixture of polymethylmethacrylate (PMMA) and ferrocene shows no exothermic reactions. Consequently, catalysts are not deposited and CNTs are not grown from the combination.

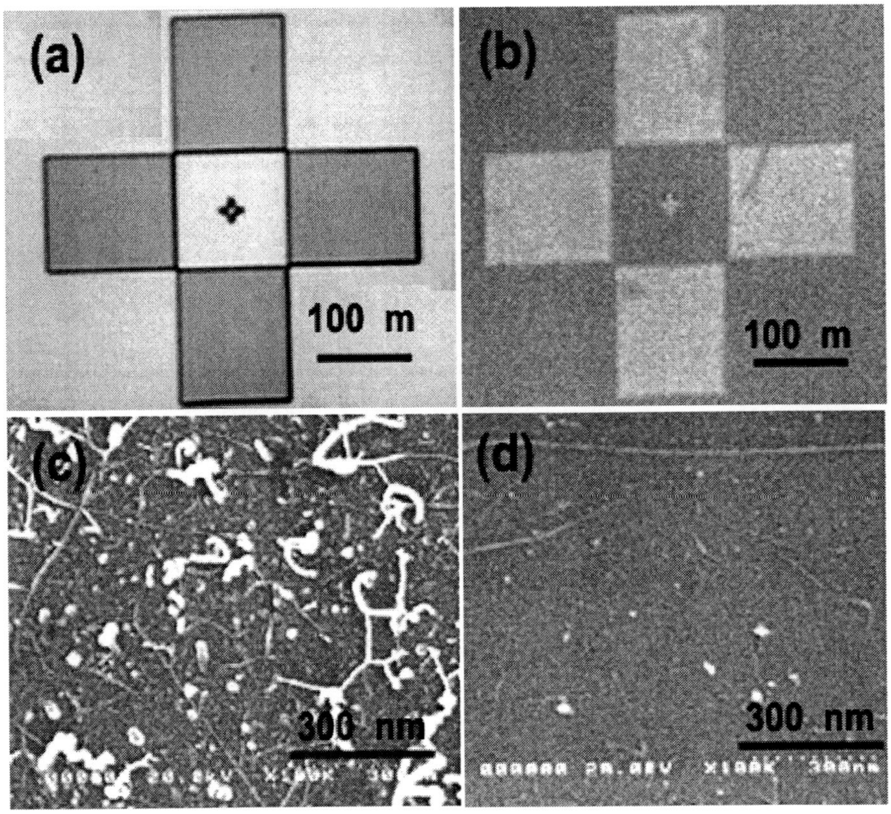

Figure 4.1. CNT selective growth by the direct photolithography using a simple mixture (Cat-PR) of ferrocene and conventional photoresist (AZ 5214E). (a) Optical microscopic image of Cat-PR patterns. (b) SEM image after CNT growth on catalytic particles prepared from 0.3 M Cat-PR solution. (c) SEM image of CNTs grown on catalyst particles prepared from 0.3 M Cat-PR solution. (d) SEM image of CNTs grown on catalyst particles prepared from 0.05 M solution [72].

Figure 4.2 shows low- and high-resolution TEM images of catalytic nanoparticles deposited with 0.1 M Cat-PR. Most particles are randomly and separately distributed on the surface of SiO_2 substrate and well crystallized. However, it was also observed that a few of them were coalesced together (Figure 4.2(a)). Diameters of catalytic particles were in the range of 3 – 10 nm and were mainly smaller than 5 nm (Figure 4.2(b)). Root-mean-square (rms) roughnesses of the catalyst deposited substrate were determined to be 0.8 nm and 1.3 nm from 0.05 M and 0.3 M Cat-PRs, respectively, by atomic force microscope (Seiko,

SPA500). The rms roughness increased with the concentration of ferrocene in the Cat-PR solution.

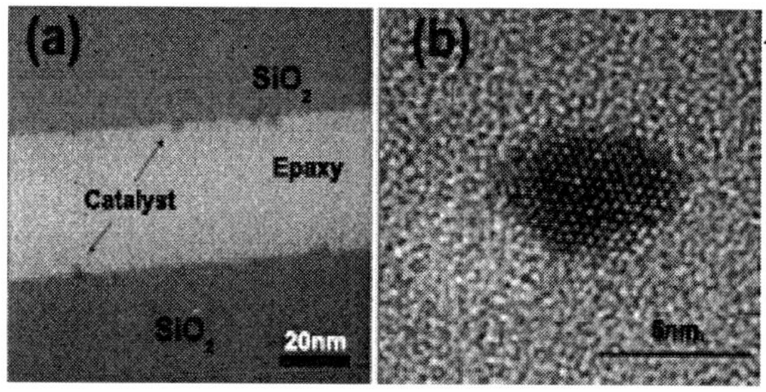

Figure 4.2. TEM micrographs of catalytic nanoparticles prepared from 0.1 M Cat-PR solution after the burning process. (a) Cross-sectional image. (b) High-resolution image of a catalyst particle.

To investigate the chemical state of the deposited catalytic nanoparticles after the burning process, we analyzed the catalytic particles with XPS without removing adventitious carbons of which peak (284.8 eV) was utilized as an internal reference. Fe 2p peaks (Figure 4.3) from catalytic particles are located at 711.5 eV for a $2p_{3/2}$ component and at 724.9 for $2p_{1/2}$. The binding energy of Fe $2p_{3/2}$ peak is similar with the literature values of iron oxyhydroxide (FeOOH) or iron oxide (Fe_2O_3) of which $2p_{3/2}$ peaks are located at 711.6 eV and 710.9 eV, respectively [75]. Although the O 1s region of the oxyhydroxide would be expected to show a doublet with one peak due to oxide and the other for hydroxide [76], since the catalytic particles are located on SiO_2 underlying layer, the O 1s region could not be used to distinguish FeOOH and Fe_2O_3.

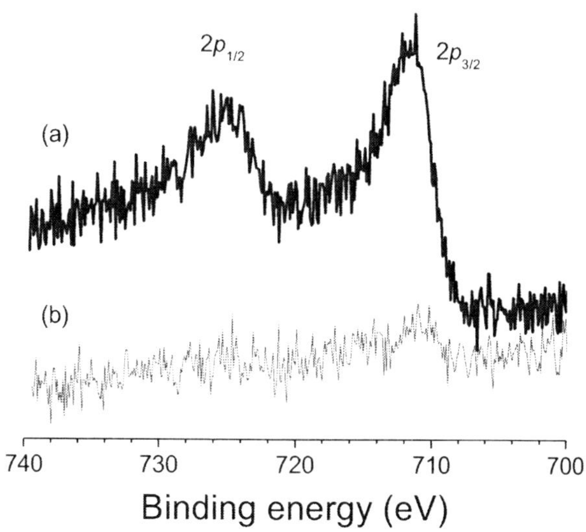

Figure 4.3. Fe *2p* core level photoemission spectra. (a) From the specimen with catalytic nanoparticles. (b) From the specimen of which catalytic particles are removed by exposal and development processes. The binding energies were referenced to C_{1s} at 284.8 eV.

Ishida et al. indicated that their catalyst particles, after removal of residual carbons, formed a mixture of α-Fe, Fe_2O_3, and Fe_3C.[11] However, the $2p_{3/2}$ peak in Figure 4.3 reveals that metallic iron (707.0 eV) or Fe_3C (708.1 eV) are not main components of catalyst particles [75, 77]. It is believed that the catalytic nanoparticles mainly consist of oxyhydroxide or oxide forms of iron without organic residues, which is possibly reduced to iron particles by hydrogen gas during the subsequent CVD process.

For comparison, a specimen coated with the Cat-PR was exposed all over the area without photomask and then developed to remove catalyst. From the catalyst-removed specimen, Fe 2p peaks (Figure 4.3(b)) were not distinguishable from noises, since the iron concentration was very low. This means that catalyst particles can be pre-patterned by the Cat-PR method.

4.2. SWNT Growth and Characterization

CVD for nanotube growth was performed with acetylene gas on the catalyst pre-pattern, which was formed by the burning process. As shown in Figure 4.1(b), CNTs were grown only on the defined region by the Cat-PR pattern (Figure 4.1(a)). Figure 4.1(c) and (d) show SEM images of CNTs grown from 0.3M and 0.05M solutions, respectively. The density of the grown CNTs increases with the ferrocene concentration of the Cat-PR. As shown in Figure 4.1(c) and (d), particles and nanotubes with large diameters more frequently appear as the concentration of Cat-PR increases.

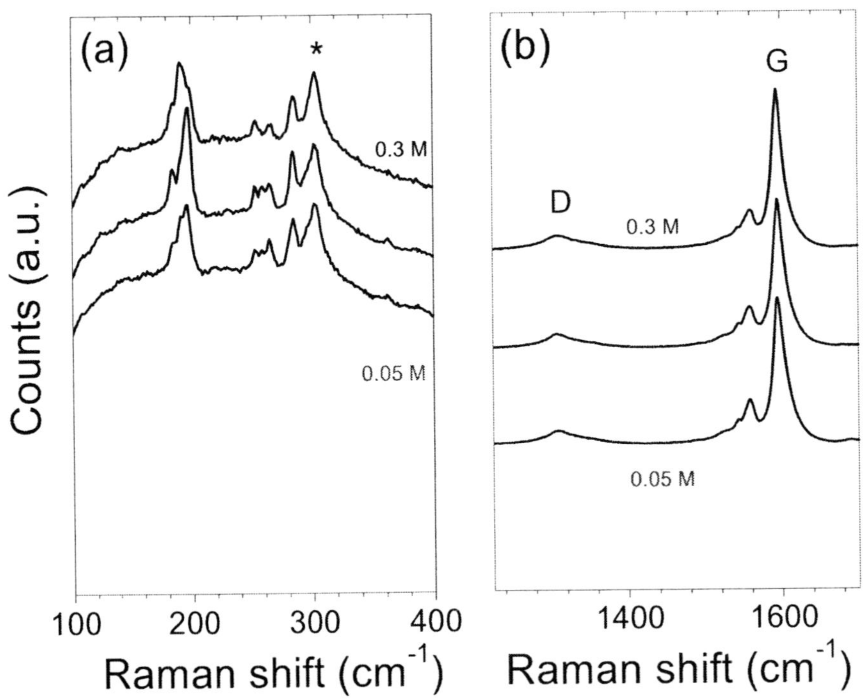

Figure 4.4. Raman spectra of SWNTs from 633 nm laser excitation. (a) Radial breathing modes. (b) Tangential modes. Raman spectra were obtained from 3 different positions of each specimen which were grown with 0.05 M (thin lines) and 0.3 M (thick lines) Cat-PR, respectively. All Raman spectra were calibrated with a peak at 303 cm^{-1} denoted with an asterisk due to the SiO_2/Si substrate [72].

The diameters of nanotubes grown on SiO_2/Si substrates were determined by Raman spectroscopy using 633 nm (Figure 4.4) and 785 nm (Figure 4.5) laser excitations. Figure 4.4 shows the typical Raman spectra of SWNTs, as characterized by the strong G band (tangential stretch mode) and the presence of the sharp radial breathing mode (RBM) in the low frequency region (100 – 300 cm^{-1}) [78]. The diameter (d_t, nm) of SWNTs can be calculated from the frequency (ω_{RBM}, cm^{-1}) of the RBM by the equation, $\omega_{RBM} = A/d_t, + B$ where A and B are experimentally found to be 223.5 $cm^{-1}nm$ and 12.5 cm^{-1} for SWNTs. In the RBM mode from 633 nm laser excitation (Figure 4.5(a)), several peaks appear in a wide range of 170 – 290 cm^{-1} with the main peaks in the frequency range of 170 – 210 cm^{-1}, regardless of the concentration of the Cat-PR or the laser-exposed position. These intensities are comparable to the substrate peak at 303 cm^{-1}. The corresponding diameter of the main peaks is 1.13 – 1.42 nm. However, the intensities of the minor peaks increase with the increase in concentration of the Cat-PR.

A weak peak at 1310 cm^{-1} of the D-band in Figure 4.4(b) is related to disordered graphite and amorphous carbon. The intensity ratios of D-band relative to G-band (I_D/I_G) are 0.38 ± 0.06 and 0.21 ± 0.02 in the CNTs grown with 0.05 M and 0.3 M Cat-PR solution, respectively. The higher frequency component (ϖ_G^+) at 1594 cm^{-1} and the lower frequency component (ϖ_G^-) at 1560 cm^{-1} of the G-band (Figure 4.4(b)) are attributed to vibrations along the direction of the nanotube axis and along the circumferential direction, respectively. The sharp Lorentzian features of the peaks indicate that the grown SWNTs are mainly semiconducting. For metallic SWNTs, the lower frequency component generally shows a very broad Breit-Wigner-Fano line [60]. The G band components, ϖ_G^+ and ϖ_G^-, are related to the equation, $\varpi_G^- = \varpi_G^+ - c/d_t^2$ where $c = 47.7$ $cm^{-1}nm^2$ for semiconducting nanotubes and $c =$ 79.5 $cm^{-1}nm^2$ for metallic nanotubes. Substituting the diameter of 1.21 nm attributed to the strongest RBM peak at 197 cm^{-1} in the equation, the calculated (ϖ_G^-) value for the semiconducting nanotube is 1561 cm^{-1}, which agrees well with the observed value (1560 cm^{-1}). If the grown nanotubes have random chirality distribution, the ratio of semiconducting to metallic nanotubes should be 2:1. The position (1560 cm^{-1}) and the sharp Lorentzian features of the lower frequency component (ϖ_G^-) of the G-band also reveal that nanotubes are mainly semiconducting. However, the strongest peak at 197 cm^{-1} seems to be attributed to an E_{11} transition of metallic nanotubes, according to the Kataura plot (transition energies versus diameter) [66].

In recent studies on the transition energies of nanotubes as a function of diameter [79], it was reported that the first metallic band in the visible can barely be distinguished from the higher-lying band. Therefore, we used 785-nm-laser to characterize the grown nanotubes. The overall tendency of the Raman spectra from 785 nm is similar to that from 633 nm. However, the main peaks of the RBM mode from 785 nm (Figure 4.5(b)) are located in the higher frequency range of 210 – 240 cm^{-1} than those from 633 nm. The intensities are stronger than those of the substrate peak due to the strong coupling between electrons and phonons of nanotubes with 785 nm laser excitation. In Figure 4.5(a), the empirical Kataura plot reported by Weisman et al. was shown as a function of Raman shift.[25] Supposing a resonant window by 785-nm-laser is 1.58 ± 0.1 eV as designated by horizontal lines in Figure 4.5(a) [80], most RBM peaks in Figure 4.5(b) are attributed to E_{22} transitions of semiconducting nanotubes of which frequency range is designated by vertical lines. The strongest peak at 233 cm^{-1} is due to (11, 3) nanotubes, and the diameter is 1.01 nm. This supports that the grown SWNTs are mainly semiconducting, although it is not clear if the semiconducting portion is higher than the ratio of 2:1 from random chirality distribution. The corresponding diameter of the main peaks (210 – 240 cm^{-1}) is 0.98 – 1.13 nm. Considering the frequency range of the main peaks from 633 and 785 nm laser excitation, the diameter of the grown SWNTs is roughly 1.0 – 1.5 nm which is much smaller than the diameters (3 ~ 10 nm) of nanoparticles deposited from the Cat-PR by the burning process.

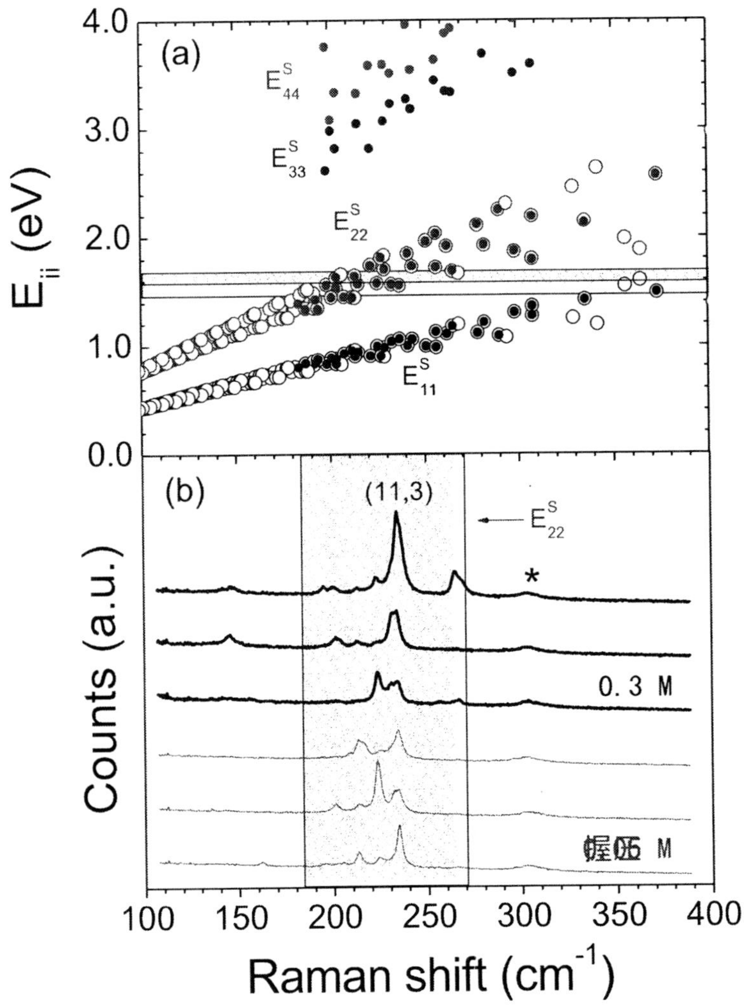

Figure 4.5. (a) Optical transition energy versus Raman shift for semiconducting SWNTs which was re-plotted from the Ref 25. Solid symbols are experimental data; open symbols are from the empirical fitting functions. The horizontal lines designate the resonant window by 785-nm-laser (1.58 eV). (b) Radial breathing modes of SWNTs from 785 nm laser excitation. Raman spectra were obtained from 3 different positions of each specimen which were grown with 0.05 M (thin lines) and 0.3 M (thick lines) Cat-PR, respectively. Vertical lines designate the frequency range due to the E_{22} transition of semiconducting nanotubes. All Raman spectra were calibrated with a peak at 303 cm^{-1} denoted with an asterisk due to the SiO_2/Si substrate.

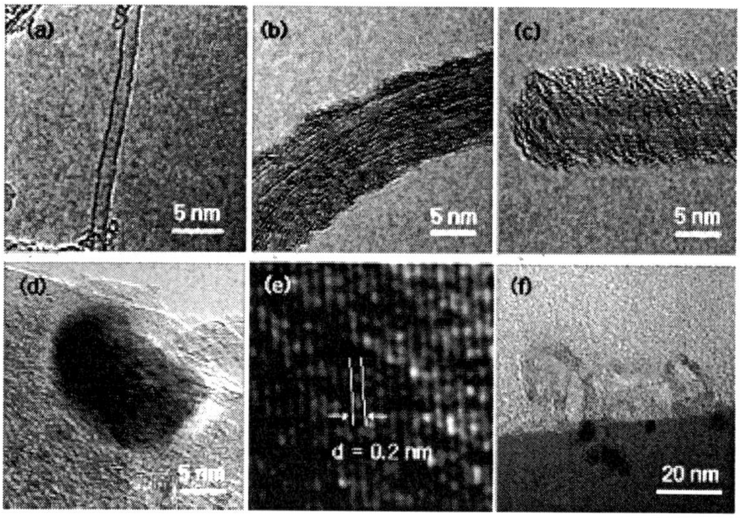

Figure 4.6. High and low resolution TEM images of CNTs and catalysts. (a) An individually separated SWNT with a diameter of 1.3 nm. (b) A bundle of SWNTs. (c) A MWNT with a closing cap. (d) A catalyst nanoparticle after the CNT growth. (e) A magnified image of the catalyst in Figure 7d. (f) Cross-sectional TEM image of nanotubes and nanoparticles after the growth.

The structure and the diameter of nanotubes were confirmed by high resolution TEM. We observed SWNTs as expected from Raman spectroscopy. However, the grown nanotubes were a mixture of individual SWNTs, bundles of SWNTs and multi-walled carbon nanotubes as shown in Figure 4.6. The bundles were most frequently observed and SWNTs were less frequently observed than MWNTs from individually separated nanotubes in the TEM specimen. Figure 4.6(d) shows a catalytic nanoparticle surrounded by graphene sheets after nanotube growth. From a magnified image in Figure 4.7(e), we determined the interplanar spacing to be ~ 0.2 nm. The catalyst particles after nanotube growth were qualitatively analyzed by energy-dispersive X-ray (EDX) spectroscopy, in which iron and carbon were observed without oxygen. This supports the oxidized form of iron is reduced during the growth. However, it is not clear whether the reduced form is metallic iron or iron carbide [81-83].

It should be noted that catalytic nanoparticles are partially or fully embedded in the underlying oxide layer as shown in Figure 4.2(a) and 4.6(f). The depth of the nanoparticles embedded in SiO_2 layer is much deeper after the CNT growth (Figure 4.6(f)) than after the burning process (Figure 2(a)). The embedment may decrease the effective size of catalyst nanoparticles. On the other hand, after the nanotube growth (Figure 4.6(d)), nanoparticles larger than 5 nm are more frequently observed than after the burning process, possibly due to coalescence of small particles. Therefore, the effective nanoparticle size will be influenced by both the embedment and the coalescence of nanoparticles. For the controlled growth of SWNTs, the effective size of catalytic particles should be further optimized by adjustment of the growth temperature and the Cat-PR concentration, since the diameters of nanotubes are determined by the diameters of catalytic nanoparticles [84, 85].

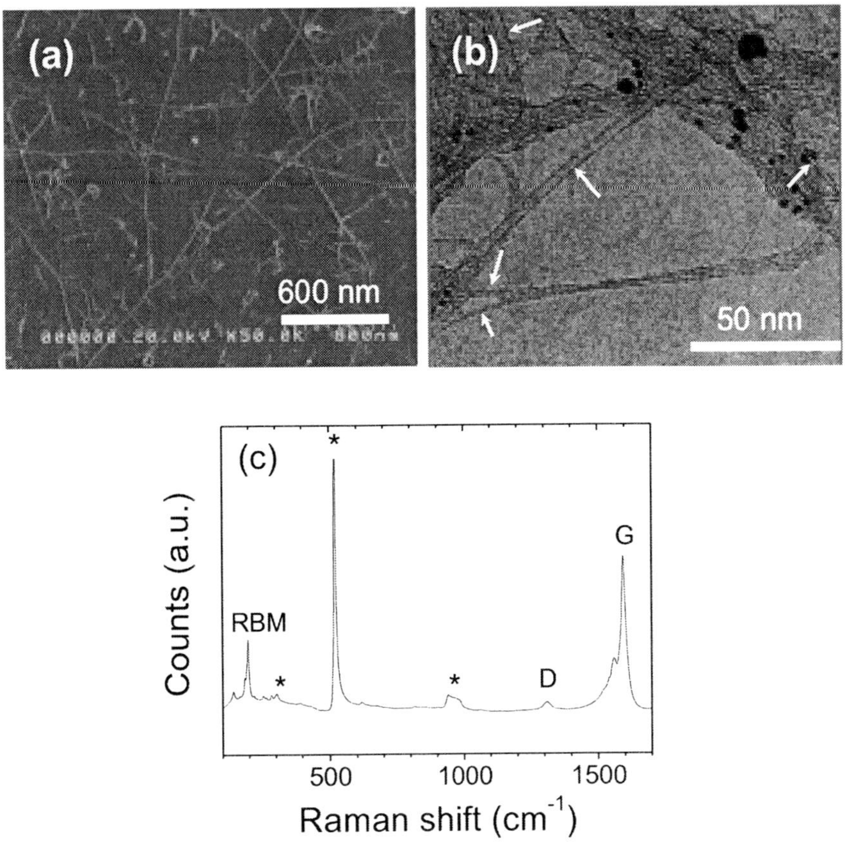

Figure 4.7. SWNTs grown on catalytic nanoparticles (0.1 M Cat-PR) by supplying water vapor during the CVD. (a) SEM image. (b) TEM. Arrows indicates SWNTs with various diameters. (c) Raman spectra (633 nm).

The burning process to remove organics from the Cat-PR pattern should be carefully performed in direct photolithographic methods, because some residues on catalytic particles poison catalytic activity. In contrast, Ishida et al. obtained RBM peaks from only three points among one hundred different scans over the patterned area [71], Raman scan on nanotubes grown with the Cat-PR shows RBM peaks from most points over all area. We believe that the difference originates from the carbonaceous residue remained on catalytic nanoparticles after the removal of organics.

Recently, Hata, et al. reported a thermal CVD synthesis of impurity-free SWNTs by using water vapor where the activity and lifetime of the catalyst nanoparticles were enhanced by water, since water plays a role of a mild oxidant to remove carbon impurities on the catalyst surface. In order to improve the activity of catalytic nanoparticles, CNT growth by using water vapor were also carried out on the catalyst nanoparticles deposited from 0.1 M Cat-PR. The nanotubes grown with water show better quality and higher density than those grown without water, as shown in SEM and TEM images of Figure 4.7(a-b). The high quality of SWNTs by water is verified by much lower I_D/I_G ratio (~ 0.06) in Raman spectra than that (0.2 ~ 0.4) of nanotube grown without water. Due to the higher density of SWNTs, the RBM peaks in Figure 4.7(c) are much stronger than those in Figure 4.4(a) as comparable to the substrate peak at 520 cm^{-1}. The improvement of the quality and the density of SWNTs by water also supports that the activation of catalyst nanoparticles is critical to obtain SWNTs in this direct lithographic method.

Suspended SWNTs. Fabrication of suspended nanotubes between three-dimensional structures is obviously a suitable way for fundamental characterization or electronic device applications of CNT, since nanotubes can be isolated from the adjacent CNTs and the interfaces with substrates [86]. The standard etching process, which is performed with conventional PRs, can be also applied to the Cat-PR. This facilitates the fabrication of the suspended nanotubes, because the Cat-PR pattern can be used as a etch mask during the etching process. A square shape pattern was formed on a SiO_2 layer with Cat-PR (0.05 M) as described above. The oxide layer was subsequently etched with a conventional oxide etching solution (buffered oxide etch 6:1, J.T. Baker Co). After burning the organics from the pattern and growing CNTs, we successfully obtained suspended-CNTs on the SiO_2 stands. Figure 4.8(a) and 4.8(b) show suspended nanotubes grown over four stands and two stands, respectively. Nanotubes grown from the spacing between stands are not observed due to the absence of catalytic particles by etching. Most of suspended nanotubes were connected to the nearest-neighbor stands. It should be further investigated that the direction of the growth can be controlled by changing dimensions of stands and spacing between stands.

The direct photolithographic method for the selective growth of SWNTs by using a simple mixture of ferrocene and conventional photoresist is a promising and practical pathway for large-scale applications of ICs with nanotube-based electronic elements such as transistors, resistors, and interconnects.

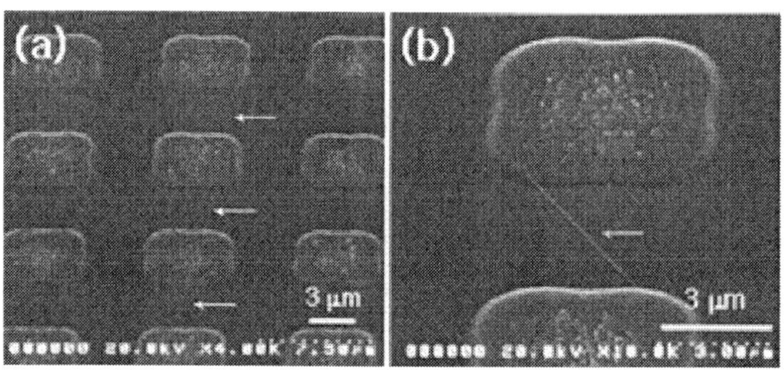

Figure 4.8. Suspended nanotubes obtained by using the Cat-PR pattern as a etch mask. (a) Over four stands. (b) Over two stands [72].

4.3. Sulfidative Purification of Carbon Nanotubes Integrated in Transistors

To obtain any intrinsic properties from CNTs, they should be pure without carbonaceous impurities such as amorphous carbons on their surface. Unfortunately, the growth of CNTs is inevitably accompanied by carbonaceous impurities [87]. Regardless of the preparation method of CNTs, which is mainly arc-discharge, laser ablation, chemical vapor deposition (CVD) or high pressure CO (HiPco) process, the carbonaceous materials on their surface are the most common impurities formed during the growth [88]. Therefore, the separation of pure nanotubes is essential for fundamental research and for device applications.

Since the carbonaceous particles are more susceptible to oxidation, in contrast to the perfect nanotube walls [89], CNTs have been purified by oxidative treatments in gas phase or liquid phase [87, 90-93]. Nowadays, nanotubes for device applications are directly grown on pre-patterned catalyst by CVD due to its compatibility with conventional integration processes and capability on large areas. However, the oxidative treatments are not suitable for the purification of CNTs integrated in a device because CNTs can be lost and deformed during the oxidation process. Therefore, CNTs grown by CVD are generally used without the subsequent purification process in order to avoid damage from the oxidative treatment. Recently, several groups have grown CNTs without surface carbon impurities by using water vapor, ammonia, or hydrogen gas or by using rapid growth process. Nevertheless the appearance of carbonaceous impurities is inevitable during CVD. Although many purification methods have been intensively studied, to our knowledge, there is no report regarding the removal of carbonaceous impurities from *CNTs integrated in a device.*

Recently, we demonstrated a simple purification method of CNTs integrated in field-effect transistors (FET) by using the sulfidation reaction of carbon, $C + 2S \rightarrow CS_2$, to selectively remove carbonaceous impurities. The sulfidative purification of CNTs integrated in FETs results in a dramatic improvement of switching characteristics due to removal of carbonaceous impurities.

It is well known that the sulfidation reaction of charcoal gives carbon disulfide. Until about 1950, the primary industrial production method of carbon disulfide was by heating charcoal to 750-900 °C in the presence of vaporized sulfur [94]. We chose sulfidation reaction to selectively remove carbonaceous impurities from CNTs because sulfidation is less reactive to carbon than oxidation, considering that molar Gibbs free energies of formation of gaseous CO_2 and CS_2 are –394.4 and 67.1 kJ/mol at room temperature, respectively [95]. Furthermore, sulfur shows high volatility in elemental form and the reaction product (CS_2) is easily removable. Carbon disulfide is a colorless liquid, which evaporates at room temperature.

The sulfidation of carbonaceous impurities was performed at 300 °C to enhance the reactivity difference between carbonaceous impurities and CNTs. At first, the occurrence of sulfidation reaction at lower temperatures was proven with simultaneous thermogravimetry-differential scanning calorimetry-mass spectrometry (TG-DSC-MS) analysis. A mixture of single-walled carbon nanotubes (NanoCarb Inc., arc-discharge, purity 60 %, diameter of 1.2.1.4 nm) and elemental sulfur (Aldrich, 99.998 %) was loaded in an alumina crucible and heated to 300 °C with a heating rate of 20 °C/min. The temperature was maintained at 300 °C for 30 min in vacuum (~ 0.01 Torr). During the heating (see the temperature profile in Figure 4.9(a)), TG-DSC-MS was simultaneously analyzed with a Netzsch STA-MS-Skimmer instrument. In the DSC curve (data not shown), two endothermic peaks owing to the melting and vaporization of sulfur are observed at 120 °C and 250 °C, respectively, and an exothermic peak from the sulfidation of carbonaceous impurities appears at 290 °C. The open circles in Figure 4.9(a) show that the mass intensity of carbon disulfide (76 amu) synchronously increases from 150 °C to 290 °C with the exothermic peak, and subsequently reaches its maximum intensity at 306 °C. It then decreases due to the exhaustion of sulfur. This reveals that sulfur reacts to carbonaceous impurities on the surface of CNTs by the sulfidation reaction, $C + 2S \rightarrow CS_2$, at 300 °C in vacuum (~0.01 Torr).

The selectivity of the sulfidation of the carbonaceous impurities is evaluated with Raman spectroscopy (514 nm laser) by comparing the spectra before and after the sulfur treatment as shown in Figure 4.9(b). Spectra were obtained at ten positions and averaged. The overall tendency of the Raman spectra of the sulfur treated CNTs are nearly same with that of the pristine. However, the ratio (I_D/I_G) of the disorder-induced D-band to the tangential G-band (derived from the graphite-like in-plane mode) decreases from 0.05 to 0.035 after the sulfur treatment [60]. This means that the nanotube walls are not sulfidized and not deformed, but carbonaceous impurities are selectively removed by the sulfur treatment at 300 °C. It is also proven by transmission electron microscopic (TEM) images before (Figure 4.10(c)) and after (Figure 4.10(b)) the treatment that the carbonaceous impurities were removed by sulfidation.

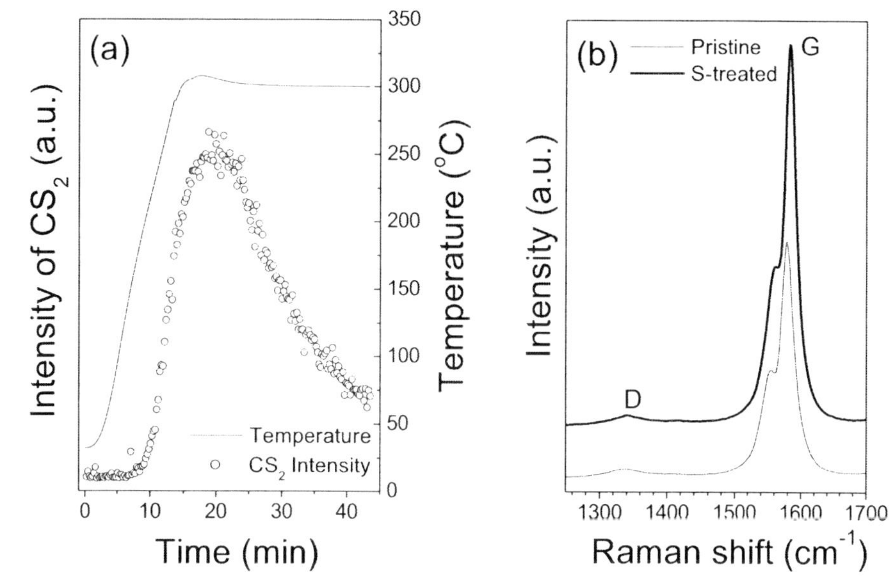

Figure 4.9. Sulfidation of carbon nanotubes with carbonaceous impurities. (a) Mass intensity variation of CS_2 (76 amu) evolved during the heating of a mixture of nanotubes and sulfur. (b) Raman spectra of nanotubes before and after the sulfidation [96].

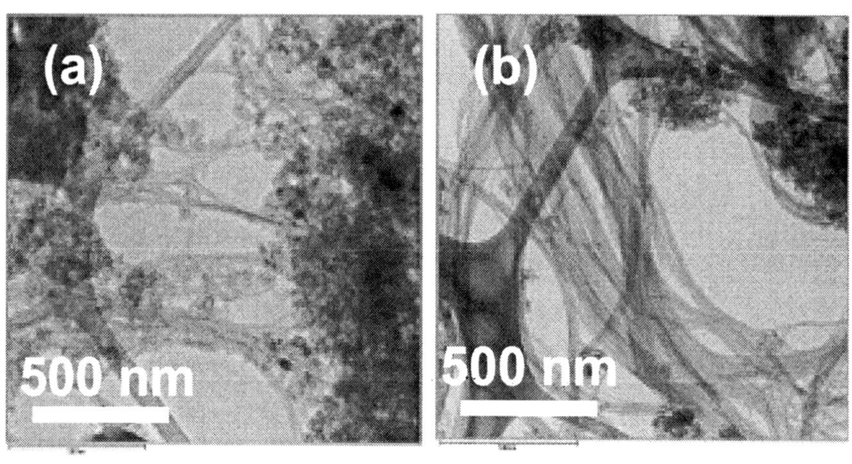

Figure 4.10. TEM images of nanotubes (c) before and (d) after the sulfidation [96].

The purification method by sulfidation can be applied to *CNTs integrated in devices* since nanotube walls are impervious to the sulfidation as proven with Raman spectroscopy. We compared electrical properties of CNTs integrated in a field effect transistor (FET) before and after the sulfur treatment. Figure 4.11(a) shows a typical atomic force microscopic (AFM) image of CNT-FET studied in this work (see Supporting Information for the detailed procedure of FET fabrication).

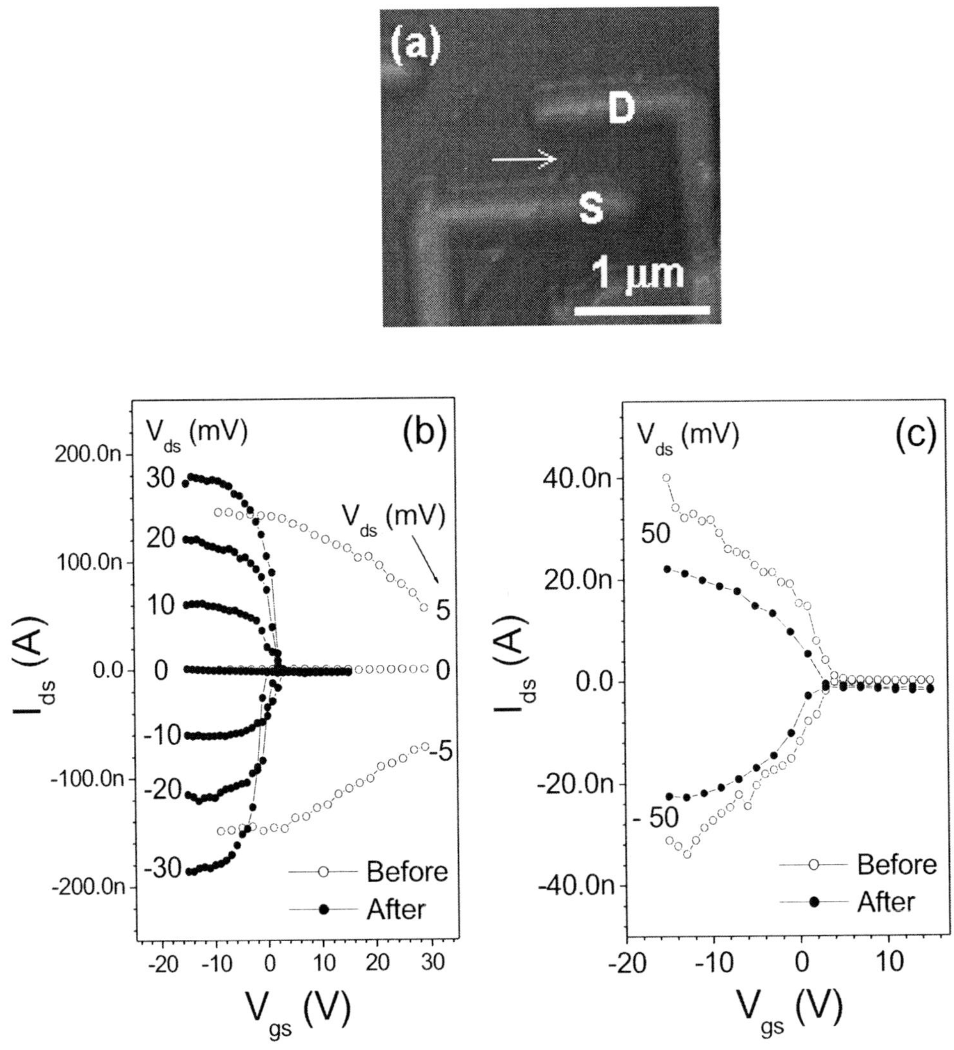

Figure 4.11 (a) AFM image of a back-gated CNT-FET. S and D denote source and drain, respectively, and CNT is designated by the arrow. (b-c) Drain-source current (I_{ds}) versus gate-source (V_{gs}) voltage characteristics of CNT-FETs before and after sulfur treatment. (b) The transistor with high off-state current before the sulfur treatment turns on and off at 3V after treatment. (c) The normally operating transistor before the sulfur treatment shows the same threshold voltage at 3 V after the treatment [96].

Among the CNT-FETs fabricated, we selected a transistor, which shows poor switching characteristics as shown in the drain-source current (I_{ds}) versus gate-source voltage (V_{gs})

curve of Figure 4.11(b) (open circles). The transistor with a resistance of 40 kΩ does not show a clear threshold voltage (V_{th}), which indicates the transition voltage from off-state to on-state. There are two possibilities for the transistor to show such poor switching performance. Firstly, if the CNT integrated in the transistor is metallic or semiconducting with narrow band gap, high off-state current can appear [67]. Secondly, the carbonaceous impurities on the outer surface of the CNT may act as a conduction path between source and drain, consequently increasing the off-state current in the transistor.

Sulfur treatment for the nanotube integrated in the FET was as follows. The CNT-FET specimen was placed on a vacuum chamber with a base pressure of 11.7 mTorr, together with sulfur contained in an Al crucible. The specimen was then heated at 300 °C for 30 min without any carrier gas. During the sulfur treatment, working pressure gradually decreased from 15.3 mTorr to the base pressure as sulfur was evaporated and exhausted. After cooling down, the specimen was unloaded from the chamber.

The solid circles in Figure 4.11(b) show I_{ds}-V_{gs} characteristics after sulfur treatment. Different from the device before the sulfidation, the sulfur treated CNT-FET shows clear on/off switching with a threshold voltage (V_{th}) of 3V and with an on-/off-state current ratio of two orders (see Supporting Information). For comparison, we investigated another CNT-FET, which had normally operated as a transistor before sulfur treatment. As shown with open circles in Figure 4.11(c), the FET shows a typical p-FET I_{ds}-V_{gs} curve with a threshold voltage of ~3V. After sulfur treatment, the FET still operates as a p-FET with the same threshold voltage.

If the sulfur treatment gave rise to the sulfidation of nanotube walls, the threshold voltage should shift due to the widened band gap by the deformation of the nanotube wall. However, as expected from the Raman spectra, there is no shift in V_{th} with sulfur treatment. This reveals that the high off-state current is due to the carbonaceous impurities on the surface of the nanotube, which provide conduction paths between source and drain, and the impurities were selectively removed from the nanotube through the sulfidation reaction [96].

The purification of CNTs by sulfidation can be used to remove carbonaceous impurities that may remain after initial purifying procedures by other methods, and be more useful for nanotubes integrated in devices for fundamental characterizations or industrial applications.

5. Summary

Electronic and transport properties of carbon nanotubes have been discussed theoretically and experimentally for ambipolar behavior on CNT-FET with intrinsic CNT and carrier suppression by molecular adsorption where oxygen molecules act as a main factor for p-type behavior with suppression of electron current. Although the molecular adsorption can be a method for choosing the carrier type of the actual CNT-FET by controlling the CNT electronic structure, as discussed in this article, seeking the more stable methods is the future challenging issue for CNT-FETs to be acceptable for integration and application.

Single wall carbon nanotube with narrow windows of diameter and high purity is essential for the CNT electronics. The sono-chemical route is an efficient method for nanotube synthesis for this point. On the other hand, direct growth of CNT on substrate makes easier way to position the individual CNTs for nanotube arrays. Remote-PECVD technique allows achieving the low temperature growth of single wall carbon nanotube as low as 450 °C so far. Water assisted remote-PECVD is effective for the CNT growth with defect reduction and density increase. The catalytic photo-resist is employed for growing the individual CNT on the pre-defined area with large scale capability. Carbonaceous impurities followed by CVD growth frequently make the undesired electrical leakage paths on the CNT channel. Sulfur based purification is effective to remove this impurities without degradation of CNT electrical characteristics.

ACKNOWLEDGEMENT

The authors would like to thank Dr. H. Kosina for modeling of CNTFET, Dr, N. Park for *ab initio* calculation, Dr. J. Park for HRTEM analysis, and Dr. S. Jeong, Dr. D. Kang and E. Bae for SWNT growth and transport measurements.

This work was supported by TND (Tera-level Nano Devices), a Korean National Research Program

REFERENCES

[1] S. Iijima, *Nature* **354**, 56 (1991).

[2] Javey, H. Kim, M. Brink, Q. Wang, A. Ural, J Guo, P. Mcintyre, P. Mceuen, M. Lundsrtrom, and H. Dai, *Nature Materials* **1**, 241 (2002).

[3] S. Frank, P. Poncharal, Z. Wang, and W. Heer, *Science* **280**, 1744 (1998)..

[4] Dekker, *Phys. Today* **5**, 22 (1999)..

[5] Y. Ando, X. Xhao, T. Sugai, and M. Kumar, *Materials Today* (2004, Oct. 22).

[6] M. Meyyappan, L. Delzeit, A. Cassell, and D. Hash, *Plasma Source Sci. Technol.* **12**, 205 (2003).

[7] R. Krupke and F. Hennrich, *Adv. Func. Mater.* **7**, 111 (2005).

[8] R. Saito, G. Dresselhaus, and M. S. Dresselhaus, *Physical Properties of Carbon Nanotube, (Imperial College Press, London*, 1998).

[9] S. Heinze, J. Tersoff, R. Martel, V. Derycke, J. Appenzeller, and P. Avouris, *Phys. Rev. Lett.* **89**, 106801 (2002).

[10] M. Kim, T. Brintlinger, E. Cobas, H. Zheng, M. Fuhrer, Z.Yu, R. Droopad, J. Ramdani, and K. Eisenbeiser, *Appl.Phys.Lett.* **84**, 1946 (2004).

[11] Javey, J. Guo, Q. Wang, M. Lundstrom, and H. Dai, *Nature* **424**, 654 (2003).

[12] R. Martel, V. Derycke, C. Lavoie, J. Appenzeller, K. Chan, J. Tersoff, and P. Avouris, *Phys. Rev. Lett.* **87**, 256805 (2001).

[13] J. Appenzeller, J. Knoch, V. Derycke, R. Martel, S.Wind, and P. Avouris, *Phys. Rev. Lett.* **89**, 126801 (2002).

[14] M. Radosavljevic, S. Heinze, J. Tersoff, and P. Avouris, *Appl. Phys. Lett.* **83**, 2435 (2003).

[15] S. Heinze, J. Tersoff, and P. Avouris, *Appl. Phys. Lett.* **83**, 5038 (2003).

[16] M. Pourfath, E. Ungersboeck, A. Gehring, B.H. Cheong, W. Park, H. Kosina, and S. Selberherr, *34th European Solid-State Device Research Conference (2004).*

[17] M. Pourfath, A. Gehring, E. Ungersboeck, H. Kosina, S. Selberherr, B. H. Cheong, and W. J. Park, J. *Appl. Phys.* **97**, 106130 (2005).

[18] P. G. Collins, K. Bradley, M. Ishigami, and A. Zettle, *Science* **287**, 1801 (2000).
[19] G. U. Sumanasekera, C. K. W. Adu, S. Fang, and P. C. Eklund, *Phys. Rev. Lett.* **85**, 1096 (2000).
[20] S.-H. Jhi, S. G. Louie, and M. L. Cohen, *Phys. Rev. Lett.* **85**, 1710 (2000).
[21] J. Jhao, A. Buldum, J. Han, and J. P. Lu, *Nanotechnology* **13**, 195 (2002).
[22] Kang, N. Park, E. Bae, J. Ko, J. Kim, and W. Park, *Appl. Phys. Lett.* **86**, 093105 *(2005).*
[23] V. Derycke, R. Martel, J. Appenzeller, and Ph. Avouris, *Appl. Phys. Lett.* **80**, 2773 (2002).
[24] J. Kong, H. T. Soh, A. M. Cassell, C. F. Quate, and H.Dai, *Nature* **395**, 878 (1998).
[25] T. Shimada, T. Sugai, Y. Ohno, K. Kishimoto, T. Mizutani, H. Yoshida, T. Okazaki, and H. Shinohara, *Appl. Phys. Lett.* **84**, 2412 (2004).
[26] W. J. Wildoer, L. G. Venema, A. G. Rinzler, R. E. Smalley, and C. Dekker, *Nature* **391**, 59 (1998).
[27] S. Dag, O. Gulsere, T. Yildirim, and S. Ciraci, *Phys. Rev.* **B 67**, 165424 (2003).
[28] P. Giannozzi, R. Car and G.. Scoles, *J. Chem. Phys.* **118**, 1003 (2003).
[29] P. Hohenberg and W. Kohn, *Phys. Rev.* **136**, B864 (1964).
[30] W. Kohn and L. J. Sham, *Phys. Rev.* **140**, A1133 (1965).
[31] G. Kresse and J. Furthmüller, *Phys. Rev.* **B 54**, 11169 (1996).
[32] J. P. Perdew, K. Burke, and M. Ernzerhof, *Phys. Rev. Lett.* **77**, 3865 (1996).
[33] G. Kresse and D. Joubert, *Phys. Rev.* **B 59**, 1758 (1999).
[34] J. Ihm, A. Zunger, and M. L. Cohen, *J. Phys.* **C 12**, 4409 (1979).
[35] Kang, N. Park, J. Ko, E. Bae, and W. Park, *Nanotechnology* **16**, 1048 (2005)
[36] S. G. Rao, L. Huang, W. Setyawan, and S. Hong, *Nature* **425**, 36 (2003).
[37] Journet, W. K. Maser, P. Bemier, A. Loiseau, M. L. Chapelle, S. Lefrant, P. Deniard, R. Lee, and J. E. Fischer, *Nature* **388**, 756 (1997).
[38] T. Guo, P. Nikolaev, A. Thess, D. T. Colbert, and R. E. Smally, *Chem. Phys. Lett.*, **243**, 49 (1995).
[39] P. Nikolaev, M. J. Bronikowski, R. K. Bradley, F. Rohmund, D. T. Colbert, K. A. Smith, and Smalley, *R.E. Chem. Phys. Lett.* **313**, 91 (1999).
[40] H. Dai, *Acc. Chem. Res.* **35**, 1035 (2002).
[41] J. Kong, H. T. Soh, A. M. Cassell, C. F. Quate, and H.Dai, *Nature* **395**, 878 (1998).
[42] M. Cassell, N. R. Franklin, T. W. Tombler, E. M. Chan, J. Han, and Dai, H., *J. Am. Chem. Soc.* **121**, 7975 (1999).
[43] W. B. Choi, D. S. Chung, J. H. Kang, H. Y. Kim, Y. W. Jin, I. T. Han, Y. H. Lee, J. E. Jung, N. S. Lee, G. S. Park, and J. M. Kim, *Appl. Phys. Lett.* **75**, 3129 (1999).
[44] J. Bae, W. B. Choi, K. S. Jeong, J. W. Chu, G. S. Park, S. Song, and I. K. Yoo, *Adv. Mater.* **14**, 277 (2002).
[45] S. Huang, L. Dai, and A. H. M. Mau, *Adv. Mater.* **14**, 1140 (2002).
[46] K. S. Suslick, MRS *Bull.* **20**, 29 (1995).
[47] S. H. Jeong, J. H. Koh, J. B. Park, and W. Park, *J. Am. Chem. Soc.* **126**, 15982 (2004).
[48] R. Andrews, D. Jacques, A. M. Rao, F. Derbyshire, D. Qian, X. Fan, E. C. Dickey, and C. Chen, J. Chem. *Phys. Lett.* **303**, 467 (1999).
[49] S. J. Doktycz and K. S. Suslick, *Science* **247**, 1067 (1999).
[50] G. Rinzler, J. Liu, H. Dai, P. Nikolaev, C. B. Huffman, F. J. Rodriguez-Macias, P. J. Boul, A. H. Lu, D. Heymann, D. T. Colbert, R. S. Lee, J. E. Fischer, A. M. Rao, P. C. Eklund, and R. E. Smalley, *Appl. Phys.* **A 67**, 29 (1998).

[51] R. Pfeiffer, H. Kuzmany, C. Kramberger, C. Schaman, T. Pichler, H. Kataura, Y. Achiba, J. Kürti, and V. Zólyomi, *Phys. Rev. Lett.* **90**, 225501 (2003).

[52] S. Lefrant, I. Baltog, M. Baibarac, J. Schreiber, and O. Chauvet, *Phys. Rev.* **B 65,** 235401 (2002).

[53] L. Delzeit, I. MaAninch, B. A. Cruden, D. Hash, B. Chen, J. Han, and M. Meyyappan, J. *Appl. Phys.* **91**, 6027 (2002).

[54] J. I. B. Wilson, N. Scheerbaum, S. Karim, N. Polwart, P. John, Y. Fan, and A. G. Fitzgerald, *Diam. Relat. Mater.* **11**, 918 (2002).

[55] T. Kato, G. H. Jeong, T. Hirata, R. Hatakeyama, K. Tohji, and K. Motomiya, *Chem. Phys. Lett.* **381**, 422 (2003).

[56] Y. Li, D. Mann, M. Rolandi, W. Kim, A. Ural, S. Hung, A. Javey, J. Cao, D. Wang, E. Yenilmez, Q. Wang, J. F. Gibbons, Y. Nishi, and H. Dai, *Nano Lett.* **4**, 317 (2004).

[57] Y. Li, S. Peng, D. Mann, J. Cao, R. Tu, K. J. Cho, and H. Dai, *J. Phys. Chem.* **B 109**, 6968 (2005).

[58] E. J. Bae, Y. S. Min, D. Kang, J. H. Ko, and W. Park, *Chem. Mater.*17, 5141 (2005).

[59] Y. S. Min, E. J. Bae, B. S. Oh, D. Kang, and W. Park, *J. Am Chem. Soc.***127**, 12498 (2005).

[60] M. S. Dresselhaus, G. Dresselhaus, A. Jorio, A. G. Souza Filho, and R. Saito, R., *Carbon* **40**, 2043 (2002).

[61] K. Song, Y. S. Cho, G. S. Choi, J. B. Park, and D. J. Kim, *Diam. Relat. Mater.* **13**, 1210 (2004).

[62] J. Wei, B. Jiang, X. Zhang, H. Zhu, and D. Wu, *Chem. Phys. Lett.* **376**, 753 (2003).

[63] K. Tohji, T. Goto, H. Takahashi, Y. Shinoda, N. Shimizu, B. Jeyadevan, and I. Matsuoka, *Nature* **383**, 679 (1996).

[64] J. M. C. Moreno and M. Yoshimura, *J. Am. Chem.* Soc. **123**, 741 (2001).

[65] Z. Kang, E. Wang, L. Gao, S. Lian, M. Jiang, C. Hu, and L. Xu, *J. Am. Chem. Soc.* **125**, 13652 (2003).

[66] Kataura, Y. Kumazawa, Y. Maniwa, I. Umezu, S. Suzuki, Y. Ohtsuka, and Y. Achiba, *Synth. Met.* **103**, 2555 (1999).

[67] E. S. Snow, J. P Novak, P. M. Campbell, and D. Park, *Appl. Phys. Lett.* 82, 2145 (2003).

[68] W. Q. Deng, X. Xu, and W. A. Goddard, *Nano Lett.* **4**, 2331 (2004).

[69] Ago, K. Murata, M. Yumura, J. Yotani, and S. Uemura, *Appl. Phys. Lett.* **82**, 811 (2003).

[70] Hinderling, Y. Keles, T. Stockli, H. F. Knapp, T. Arcos, P. Oelhafen, I. Korczagin, M. A. Hempenius, J. Vancso, R. Pugin, and H. Heinzelmann, *Adv. Mater.* **16**, 876 (2004).

[71] M. Ishida, H. Hongo, F. Nihey, and Y. Ochiai, Jpn. *J. Appl. Phys.* **43**, L1356 (2004).

[72] Y. S. Min, E. J. Bae, J. B. Park, and W. Park, *Nanotechnology* **17**, 116 (2006).

[73] G. Wilkinson, M. Rosenblum, M. C. Whiting, and R. B. Woodward, *J. Am. Chem. Soc.* **74**, 2125 (1952).

[74] W. Zhu, C. L. Xu, D. H. Wu, B. Q. Wei, R. Vajtai, and P. M. Ajayan, *Science* **296**, 884 (2002).

[75] J Tan, K. J. Klabunde, and P. M. A. Sherwood, *Chem. Mater.* **2**, 186 (1990).

[76] D. Welsh and P. M. A. Sherwood, *Phys. Rev.* **B 40**, 6386 (1989).

[77] N. Shabanova and V. A. Trapeznikov, *J. Electron Spectrosc. Relat. Phenom.* **6**, 297 (1975).

[78] S. M. Bachilo, M. S. Strano, C. Kittrell, R. H. Hauge, R. E. Smalley, and R. B. Weisman, *Science* **298**, 2361(2002).
[79] R. B. Weisman and S. M. Bachilo, *Nano Lett.* **3**, 1235 (2003).
[80] Jorio, R. Saito, J. H. Hafner, C. M. Lieber, M. Hunter, T. McClure, G. Dresselhaus, and M. S. Dresselhaus, *Phys. Rev. Lett.* **86**, 1118 (2001).
[81] Y. Homma, Y. Kobayashi, T. Ogino, D. Takagi, R. Ito, Y. Jung, and P. M. Ajayan, J. *Phys. Chem.* **B 107**, 12161 (2003).
[82] H. Hou, A. K. Schaper, Z. Jun, F. Weller, and A. Greiner, *Chem. Mater.* 15, 580 (2003).
[83] N. Y. Jin-Phillipp and M. Ruhle, *Phys. Rev.* **B 70,** 245421 (2004).
[84] Y. J. Jung, B. Wei, R. Vajtai, P. M. Ajayan, Y. Homma, K. Prabhakaran, and T. Ogino, *Nano Lett.* **3**, 561 (2003).
[85] Y. J. Jung, Y. Homma, T. Ogino, Y. Kobayashi, D. Takagi, B. Wei, R. Vajtai, and P. M. Ajayan, *J. Phys. Chem.* **B 107**, 6859 (2003).
[86] B. Farmer and R. G. Gordon, Electrochem. *Solid State Lett.* **8**, G89 (2005).
[87] T. W. Ebbesen, P. M. Ajayan, H. Hiura, and K. Tanigaki, *Nature* **367**, 519 (1994).
[88] T. W. Ebbesen, In Carbon Nanotubes: Preparation and Properties, edited by T. W. Ebbesen (*CRC, Boca Raton,* 1996), *Chapter 4.*
[89] T. Colbert, J. Zhang, S. M. McClure, P. Nikolaev, Z. Chen, J. H. Hafner, D. W. Owens, P. G. Kotula, C. B. Carter, J. H. Weaver, A. G. Rinzler, and R. E. Smalley, *Science* **266**, 1218 (1994).
[90] T. Jeong, W. Y. Kim, and Y. B. Hahn, *Chem. Phys. Lett.* **344**, 18 (2001).
[91] H. Hiura, T. W. Ebbesen, and K. Tanigaki, *Adv. Mater.* **7**, 275 (1995).
[92] M. Bonard, T. Stora, J. P. Salvetat, F. Maier, T. Stockli, C. Duschl, L. Forro, W. A. Heer, and A. Chatelain, *Adv. Mater.* **9,** 827 (1997).
[93] Y. Q. Xu, H. Peng, R. H. Hauge, and R. E. Smalley, *Nano Lett.* **5**, 163 (2005).
[94] A. Lowenheim and M. K. Moran, *Keyes and Clark's Industrial Chemicals (Wiley-Interscience, New York,* 1978).
[95] R. Lide, CRC *Handbook of Chemistry and Physics (CRC, Boca Raton, 2004).*
[96] Y. S. Min, E. J. Bae, and W. Park, *J. Am. Chem. Soc.* **127**, 8300 (2005)

In: Frontal Nanotechnology Research
Editor: Martin V. Berg, pp. 41-59
ISBN: 1-60021-212-3

Chapter 2

APPLICATION OF COORDINATION CHEMISTRY IN THE FABRICATION OF INORGANIC NANOSTRUCTURES

***Yi Xie*†, *Yujie Xiong*‡ *and Zhengquan Li*†**

† School of Chemical and Material Engineering, Southern Yangtze University, Wuxi, Jiangsu 214036, P. R. China

‡ Department of Chemistry, University of Washington, Seattle, Washington 98195, USA

ABSTRACT

Nanostructures of metals and their oxides, chalcogenides have been of extensive interest for their widespread use in catalysis, photography, optics, electronics, optoelectronics, information storage, biological and chemical sensing, and surface-enhanced Raman scattering (SERS). By tailoring the size, shape and/or structure, one can, in principle, tune their intrinsic properties for a range of applications. To this end, coordination structures have been rationally applied as precursors to fabricate diversified nanostructures in solution. In this chapter, we discuss the synthetic strategy and related reaction mechanism through the illustration of some detailed cases developed by our group in the recent years. The simple process, mild synthetic conditions, clean reactions, excellent reproducibility and high product quality make our strategies significant. It is believed that introduction of coordination structures into the synthesis of metal, oxide and chalcogenide nanocrystals may provide a versatile tool to control their growth and assembly into well-defined nanostructures. We also expect that the related synthetic strategy is applicable to extend to chemical vapor transport or laser-assisted catalytic growth.

1. INTRODUCTION

Nanostructures of metals and their oxides, chalcogenides have been of extensive interest for their widespread use in catalysis, photography, optics, electronics, optoelectronics, information storage, biological and chemical sensing, and surface-enhanced Raman scattering (SERS).[1-6] By tailoring the size, shape and/or structure, one can, in principle, tune their intrinsic properties for a range of applications. To this end, diverse strategies, including crystal-structure directed growth, template-directed synthesis (chemical or physical, using solid substrates, porous materials, self-assembled molecular structures or existing nanostructures, such as tubes, wires or rods), electrochemistry, vapor phase growth, capping reagents, laser–assisted catalysis growth, vapor-liquid-solid (VLS), solution-liquid-solid (SLS), and solvothermal methods, have been designed to fabricate diversified nanostructures.[7-14]

Coordination chemistry is an important and expansive branch of chemistry, focusing on transition metals. Transition metals differ from main group metals in several key properties. One of the most interesting aspects of transition metals is their ability to form coordination compounds. The formation of coordination compounds occurs between a metal atom/cation and a molecule with one or more unshared electron pairs that is so-called "ligand". According to the number of their donor atoms, ligands may be classified into two types: monodentate ligand donating a single electron pair to the metal or metal ion, such as NH_3, H_2O, CN^- and SCN^-, bidentate ligand donating two electron pairs, for example, ethylenediamine. Therefore, there may be many coordination forms via diversified connection ways between metal atoms/cations and ligands: simple complexes with single metal-center, linear coordination cluster compounds with multiple metal-centers bridged by bidentate ligands, metal-polymer coordination compounds, coordination polymers with one-, two-, or three-dimensional (1D, 2D or 3D) structures bridged by bidentate ligands. The coordination forms in these compounds are absolutely different, observed from not only their atom numbers (or molecular weight), but also their structural characteristics.

Coordination compounds have applied in the synthesis of nanomaterials long before. For instance, coordination polymers were used as precursors to be decomposed into thin film via a chemical vapor deposition (CVD) route.[15] In general, these techniques could not take advantage of the structural characteristics of coordination compounds, because the reaction condition was extremely rigorous. We believe that intrinsic structural characteristics in specific coordination compounds are advantageous for the oriented growth and simultaneous assembly of nanostructures as they are used as precursors in solution-based reactions, even if the diversified coordination compounds have absolutely different structures. On the other hand, the coordination between metal cations and ligands usually increases their solubility and slows down the hydrolysis in aqueous solution, so that the reactions can proceed more slowly to provide enough time for the crystals to grow, or the ligands can adsorb to specific surfaces to affect the morphology of the crystals. Both the two factors provide useful ways to achieve shape control. Interestingly, Chaudret group also used the tools of organometallic chemistry in preparing 1D nanostructures.[16] The bonding ways of organometallic compounds are very similar to those of coordination structures. The successful synthesis in their related works shows that our strategy is reasonable and general. In this chapter, we review the development of synthesizing inorganic nanostructures from coordination structures by our group. We will describe in detail, six typical strategies of our work on the synthesis of a variety of nanostructures. Each section highlights a unique approach for the fabrication of nanostructures from specific coordination structures: (*i*) simple complex; (*ii*) linear coordination cluster compound; (*iii*) metal-polymer coordination compound; (*iv*) 1D coordination polymer; (*v*) 2D coordination polymer; (*vi*) 3D coordination polymer.

2. SIMPLE COMPLEX (0D)

Simple complex consists of single metal atom/cation center and ligands, in which no extension structure could be found from any dimension. Ethylenediamine is a ligand composing simple complex that has been introduced to synthesize metal chalcogenide 1D nanostructures long before.[17,18] For instance, Qian group prepared CdS, CdSe and CdTe nanorods in ethylenediamine by means of solvothermal reactions.17 The role of ethylenediamine was confirmed by further study of the temporal evolution of CdS nanocrystals in solvothermal reactions.[18] It demonstrated that ethylenediamine molecules adsorbed on the surface of CdS played a critical role in the formation of nanorods.[18] Infrared (IR) spectra of these ethylenediamine molecules show that they are not in chelating (cis) configuration but in trans configuration.[18] Since the ethylenediamine molecules are not in chelating configuration, it is possible to introduce monodentate ligands as shape-controller for the synthesis of metal chalcogenide 1D nanostructures. Qian and Yu employed n-butylamine that has only one anchor atom as ligand and solvent, and successfully obtained CdS, CdSe, ZnSe and PbSe nanorods.[19] It indicates that one anchor atom in a ligand is necessary and sufficient for the formation of nanorods, even though more anchor atoms may be present in a ligand. Reetz et al. also reported that a non-chelating coordination mode of α-hydroxycarboxylates on a metal surface is likely to be the morphology-determining factor in shape-selective preparation.[20] Our group also developed a two-step reaction route to synthesize PbSe nanorods from Pb2+ and Se in the presence of ligand ethylenediamine, in which the ultrasonic radiation and γ-irradiation techniques were combined.[21] The similar route was applicable to be extended to produce PbS rods by refluxing.[22]

As a special case, ammonia could act as structure-directing agent in the formation of NiS layer-rolled structures,23 in which the aqueous solution contains NH3, CS2 and Ni2+. The following reactions were involved in the fabrication process.

$$4NH_3 + CS_2 \rightarrow NH_4SCN + (NH_4)_2S$$

$$Ni^{2+} + (NH_4)_2S \rightarrow NiS + 2NH_4^+$$

It is considered that hydrogen-bonding interactions between $Ni(NH_3)_6^{2+}$ can organize complex ions in space and the linked ones adopt the "least destabilizing" mutual arrangement (Figure 1). Hence the growth of NiS particles is limited in-plane and the surface features of the particles offer the possibility to roll up into tubular structures. This route can also be applied to produce metal sulfide hollow spheres.[24]

Note that the above ligands have no rigid structures, thereby the coordination between metal cations and ligands mainly makes ligands adsorbed on the surface of crystals, affecting the crystal growth, which is so-called "shape-controller". Some other ligands with rigid structures, such as 1, 10-phenanthroline, can coordinate with transition metal cations in chelating configuration. The formed coordination structures have obvious steric effect when chemical reactions occur, due to the rigid characteristics of ligands. For example, it is known that 1,10-phenanthroline can coordinate with Ag^{2+} to form a type of stable complex, $[Ag(phen)_2]^{2+}$.[25] The complex is of d^9 structure, taking a approximate square-planar geometry and having steric inhibition on its *xy* plane. Therefore, if the Ag^{2+} ions in the complex could be reduced to atomic Ag, it would be preferential to orientedly grow along the vertical direction (*z* axis), due to their steric inhibition on the *xy* plane. It was reported that Ag^{+} ions could be oxidized to Ag^{2+} ions by oxygen under conditions of ultrasound.[26] Thus we designed an ultrasound-assisted oxidation-reduction route to prepare silver nanorods with 1,10-phenanthroline as ligand,[27] in which $[Ag(phen)_2]^{2+}$ formed at first and then was reduced to atomic Ag.

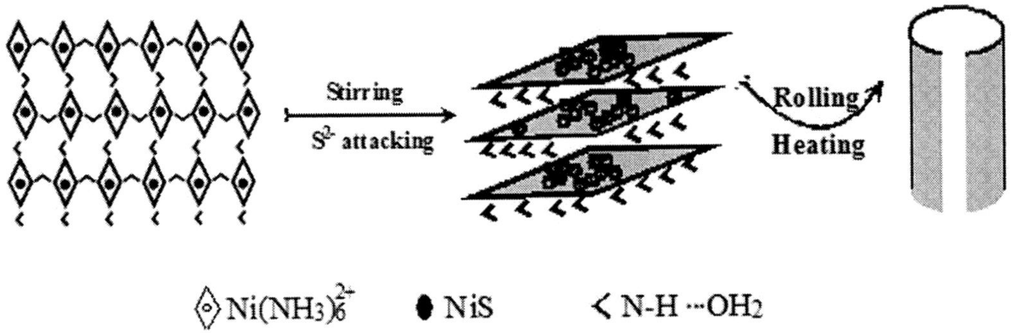

Figure 1. Illustration of the formation of layer-rolled NiS structures. Adapted in part from ref. 23, © 2001 Wiley-VCH. Used with permission.

The coordination situation of iron cations and 1,10-phenanthroline was more complicated. From the structural characteristics, $[Fe(phen)_3]^{2+}$ has a distorted octahedral structure, which can be gradually hydrated to $[Fe(H_2O)_2(phen)_2]^{2+}$, $[Fe(H_2O)_4phen]^{2+}$ and $[Fe(H_2O)_6]^{2+}$. The resulting $[Fe(H_2O)_6]^{2+}$ can be oxidized and hydrolyzed into β-FeOOH in air,[25] which ensures the completion of the hydration from $[Fe(phen)_3]^{2+}$ to $[Fe(H_2O)_6]^{2+}$. It is noteworthy that the ligand of phen can stabilize Fe(II), preventing $[Fe(phen)_3]^{2+}$ from oxidization by air.[25] With the dissociation of phen and the hydration of $[Fe(phen)_3]^{2+}$, the complexes of Fe(II) are more and more inclined to be oxidized by air. As a result, a little $[Fe(H_2O)_2(phen)_2]^{2+}$ can be oxidized to $[Fe(H_2O)_2(phen)_2]^{3+}$ that can also produce β-FeOOH finally. Thus, the first-degree hydration process will most probably form two products: $[Fe(H_2O)_2(phen)_2]^{2+}$ and $[Fe(H_2O)_2(phen)_2]^{3+}$, shown as the Complex (I) and (II) in Figure 2A, respectively. In Complex (I), the Fe^{2+} cation has a d^6 high spin structure and the two ligands of phen lie on an approximate square-plane.[28] Thus there exists planar steric-inhibition for the later oxidization-hydrolysis reactions, leading to the preferential oxidization-hydrolysis along its normal direction. It is advantageous for the growth of β-FeOOH nanowires along [001] axis via oxidization and hydrolysis processes. In Complex (II), the Fe^{2+} cations with d^5 structure has larger splitting energy than Fe^{3+} and the two ligands of phen tend to keep away form a plane,[28] hence the steric-inhibition along the normal direction of the results in the preferential hydrolysis on the plane and hence the formation of β-FeOOH flakes with (001) orientation. Obviously, the oxidization degree from Complex (I) to Complex (II) is relatively low and the first-hydration process is more inclined to produce (I). Thus, it should be more undemanding to produce the nanowires than the flakes, resulting in the production of a mass of nanowires and a small quantity of flakes. Since the axis growth plane of nanowires is consistent with the orientation plane of flakes, the nanowires are facile to grow on flakes directionally. Applied the above strategy, self-supported patterns of oriented alignment of β-FeOOH nanowires (Figure 2B and C) were fabricated from the complex of $[Fe(phen)_3]^{2+}$ at 60 °C.[29] The characterizations of electrochemical performance and UV-vis spectrum show that the product is expected to have potential applications in lithium batteries and semiconductor industry.

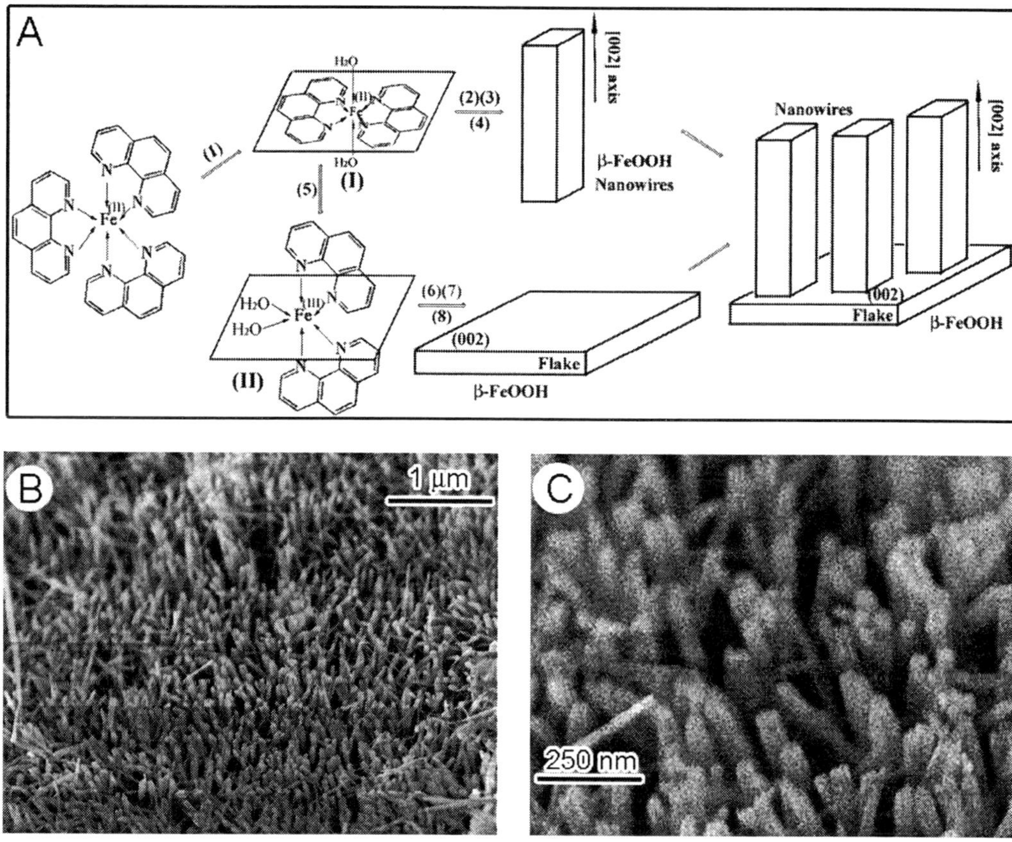

Figure 2. (A) Schematic formation illustration of the oriented alignment of nanowires growing on self-formed flakes. (B), (C) SEM images of the oriented alignment of β-FeOOH nanowires without any treatment. Adapted in part from ref. 29, © 2003 Wiley-VCH. Used with permission.

From the strategy, one can see that a kinetic control is needed to form both the Complex (I) and (II) at the same time, which results in the simultaneous formation of nanowires and flakes. Our experiments indicate that the products prepared at room temperature are all nanowires without flakes supporting. Through the additional reduction or reduction-oxidation process, α-Fe and γ-Fe_2O_3 nanowires were successfully synthesized.[30] It is interesting that there existed some folded ribbon-like nanostructures as well as nanowires in the obtained α-Fe and γ-Fe_2O_3 products, which indicates that this route might provide a new mild approach to metal and metal oxide nanoribbons. Magnetic behavior of as-prepared α-Fe and γ-Fe_2O_3 nanowires was investigated by the temperature dependence of magnetization, showing that the as-prepared α-Fe and γ-Fe_2O_3 nanowires perhaps have more applications in magnetic science and technology, due to their shape anisotropy. Through pyrolysis on β-FeOOH nanowires, α- Fe_2O_3 hollow nanowires could be obtained.[31] The products can catalyze the oxidation of almost 100% carbon monoxide at 320 °C, exhibiting excellent catalytic performances.

The simple coordination structures also have potential applications in the fabrication of micropatterns. In a typical case, novel branch-like CdS micropatterns were synthesized using thiosemicarbazide ($NH_2NHCSNH_2$) as both a sulfur source and a capping ligand in a methanol/water system.[32] The thiosemicarbazide/$CdCl_2$ ratios were found to effectively play crucial roles in the morphologies of CdS crystals. With the ratios decreasing from eight to one, the CdS crystals exhibited different morphologies, from spheres to spindle-like morphologies mediated by cauliflower- and branch-like micropatterns (Figure 3A and B). Control experiments using different sulfur sources showed that the NH_2 groups, which could form hydrogen bonds, were necessary for the formation of the novel branch-like patterns. There involved probably two key processes responsible for the formation of micropatterns, namely "oriented aggregation" and "diffusion-control growth".[33] Figure 3C illustrates the whole growth process of CdS micropatterns.

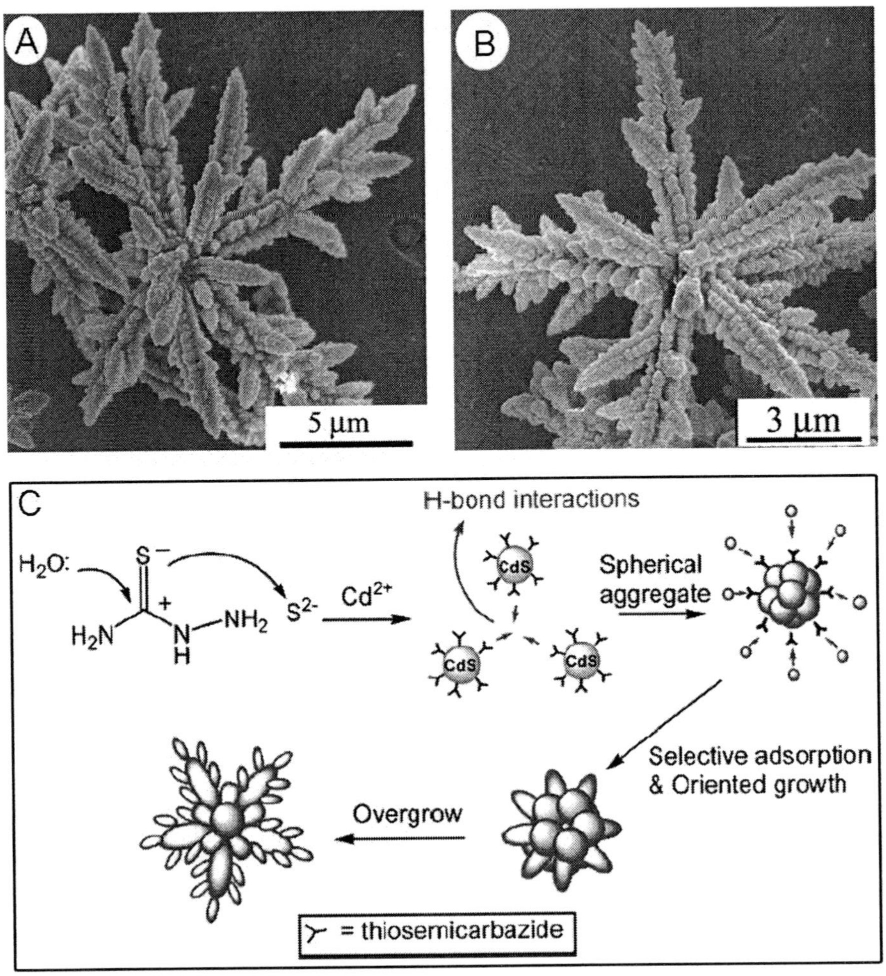

Figure 3. (A), (B) SEM images of branch-like CdS micropatterns. (C) Proposed growth process of the CdS micropatterns. Adapted in part from ref. 32, © 2004 American Chemical Society. Used with permission.

3. LINEAR COORDINATION CLUSTER COMPOUNDS (QUASI-1D)

Distinct from simple complex, linear coordination cluster compounds will directly provide orientation for the growth of 1D nanostructure due to their linear structures. This type of compounds is very familiar to us all, and usually applied in the detection of metal cations in the analytic chemistry. For example, the cluster precursor of cadmium diethyldithiocarbamate, $Cd_2(S_2CNEt_2)_4$, could be treated as an inorganic core $[Cd_2S_2]$ with four capping groups.[34] It was found that nucleophilic attack by ethylenediamine at the thione carbon will strip the capping groups,[35] shown as the following reaction.

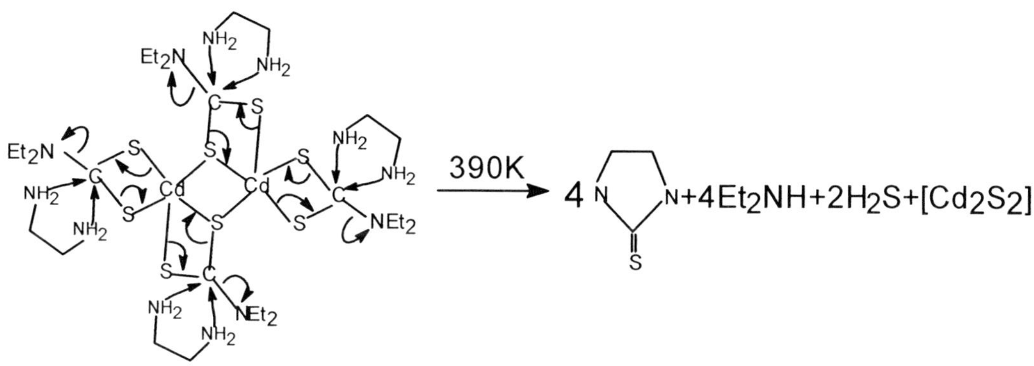

The resulting organic product has been characterized in organic chemistry.[36] The inorganic core $[Cd_2S_2]$ can be combined with each other to form chains after removing the capping groups. Hence under appropriate chemical environment, the $[Cd_2S_2]$ may preferentially grow along the chain direction and result in 1D structure. By using this strategy, quantum-confined CdS nanowires with diameters around 4 nm and lengths ranging from 150 to 250 nm were grown for the first time by our group.[35]

After the successful synthesis of 4 nm CdS nanowires, we did more research works on the application of linear coordination cluster compounds in the preparation of other 1D nanostructure. Unfortunately, no desirable result was obtained. For example, metal diethyldithiocarbamate (M=Pb, Hg, Cu, Mn, Ag) was used as precursors to synthesize metal sulfides, however, some nanoparticles rather than 1D nanostructures were obtained.[37] It should result from the fact that other metal sulfides is not as inclined to grow orientedly as cadmium sulfide. Another important reason is that the coordination cluster is too small to provide orientation for the growth of nanowires.

Thus, the small clusters with several linear-aligned metal cations can most probably used to synthesize 1D nanostructure, but the key problem is how to drive them to connect with each other to form longer chains. Some bridging anions, such as halogen ions, can connect the units to lengthen the linear chains containing metal cations. However, in disorder reaction system, the bridging effect is limited. The previous colloidal research shows that the rod-like micelle can offer a unique microenvironment for chemical reactions and reduce the systematic disorder to small area. It should be an excellent candidate for driving the small linear units' connecting action by bridging anions. Thus, the long chains bridged by bridging anions in rod-like micelles can provide orientation for the growth of nanowires via suitable reactions.

In a detailed case, Cu^{2+} cations can be combined with N and O atoms of dimethylglyoxime (dmgH) to form the trinuclear complex of $Cu_3(dmg)_2Cl_4$ whose 3 copper cations arrange on a line as a unit (Figure 4A).[25] The units composed of 3 copper cations on lines can be bridged by Cl anions, resulting in the formation of longer chains containing copper cations. It is thought that if the copper (II) cations on chains in the complex may be transformed to Cu_2O, the 1D array of copper cations in the precursor will provide orientation for the growth of Cu_2O nanowires. However, the bridging effect of Cl anions is too limited to connect copper cations of every unit to form enough long chains. To solve this problem, an efficient way is to build up rod-like micelles composed of surfactant to confine the linear units of $Cu_3(dmg)_2Cl_4$ in microenvironments and reduce the systematic disorder. Hence the $Cu_3(dmg)_2Cl_4$ units are driven to connect with each other by Cl anions to form $[Cu_3(dmg)_2Cl_2]_n^{2n+}$ (Figure 4B) whose remain Cl^- anions are dissociated near and remain the electric neutrality of reaction system. Note that the rod-like micelles are both hydrophobic and hydrophilic (W/O), and form when the concentration of surfactant is as ten times as critical micelle concentration.[38] As the complex-precursor forms in water phase, it will be enwrapped in rod-like micelles. As a typical result, the copper cations in complex can arrange on long chains of $[Cu_3(dmg)_2Cl_2]_n^{2n+}$ in the rod-like micelles and provide orientation for the growth of Cu_2O nanowires, when the copper (II) cations in the complex are reduced to Cu_2O by glucose.[39] Further studies indicate that the rod-like micelles cannot only drive linear units of $Cu_3(dmg)_2Cl_4$ to connect with each other, but also confine the diameter of nanowires. The characterization results of IR and laser light scattering (LLS) spectra revealed that the linear units of $Cu_3(dmg)_2Cl_4$ actually form longer chains in $[Cu_3(dmg)_2Cl_2]_n^{2n+}$ by the bridge of Cl anions, enwrapped in rod-like SDS micelles. In contrast, without building SDS micelles, only some Cu_2O short rods could be obtained. Thus, the formation of long chains from linear coordination cluster compounds is crucial to the growth of 1D nanostructure. The band gap of Cu_2O nanowires was determined to 2.34 eV by UV-vis absorption spectrum, showing its promising application for the reversible conversion between solar energy and electrical or chemical energy.

4. 1D Metal-Polymer Coordination Chain

An alternative to the precursors of linear coordination cluster compounds is metal-polymer coordination structure with long chains. In a typical case, poly(ethylene glycol) (PEG) as a non-ionic polymer has hydrophilic –O– and hydrophobic $-CH_2-CH_2-$ on the long chains. According to Pearson's results,[40] the –O– of the ether group on the PEG has a strong basic property and possesses a pair of electrons. Its ability to be a donor of oxygen atoms enables it to couple with transition metal cations to form coordination structures. However, the long polymer chain has a high degree of flexibility because its C–O bond is easy to rotate, thereby the chains always incline to be entwined and curled. Therefore, building an appropriate microenvironment for the extension of PEG chains is technically necessary for the application of metal-polymer coordination structures.

The metal-PEG complexes generally form in aqueous media within a narrow range of solvent composition, pH, and ionic strength.[41] Butylacetate is a good solvent for PEG, favoring the full extension of the polymer chain, while cyclohexane is a poor solvent for PEG, favoring the curl and twist of the polymer chain. Thus, Ag^+-PEG coordination chains can extend in the mixture solvent of butylacetate, cyclohexane and water via the curl and extension of PEG. By the strategy, a polymer-surfactant-assisted sandwich-reduction (PSASR) route has been reported to prepare silver nanowires, in which the sandwich structure of reducing agent (η^5-$C_5H_5)_2Fe$ results in the formation of atomic silver along certain axis in *xy* plane and the linear arrangement of silver cations along PEG chains in mixture solvents provide orientation for the 1D growth of silver nanowires (Figure 5).[42] The as-prepared silver nanowires exhibited surface plasma Raman (SPR) peak at ~410 nm. SPR is an important optical property of metal nanostructures, which could lead to applications in colorimetric sensing, nanoscale waveguiding, enhancement of electromagnetic fields, and light transmission. It is noteworthy that the extension of metal-polymer coordination chains is crucial to the successful synthesis of 1D nanostructures. The strategy is applicable to be extended to grow metal oxide 1D nanostructures. For instance, γ-Fe_2O_3 nanorods and ZnO nanowires (Figure 6A) were prepared via a similar process.[43,44]

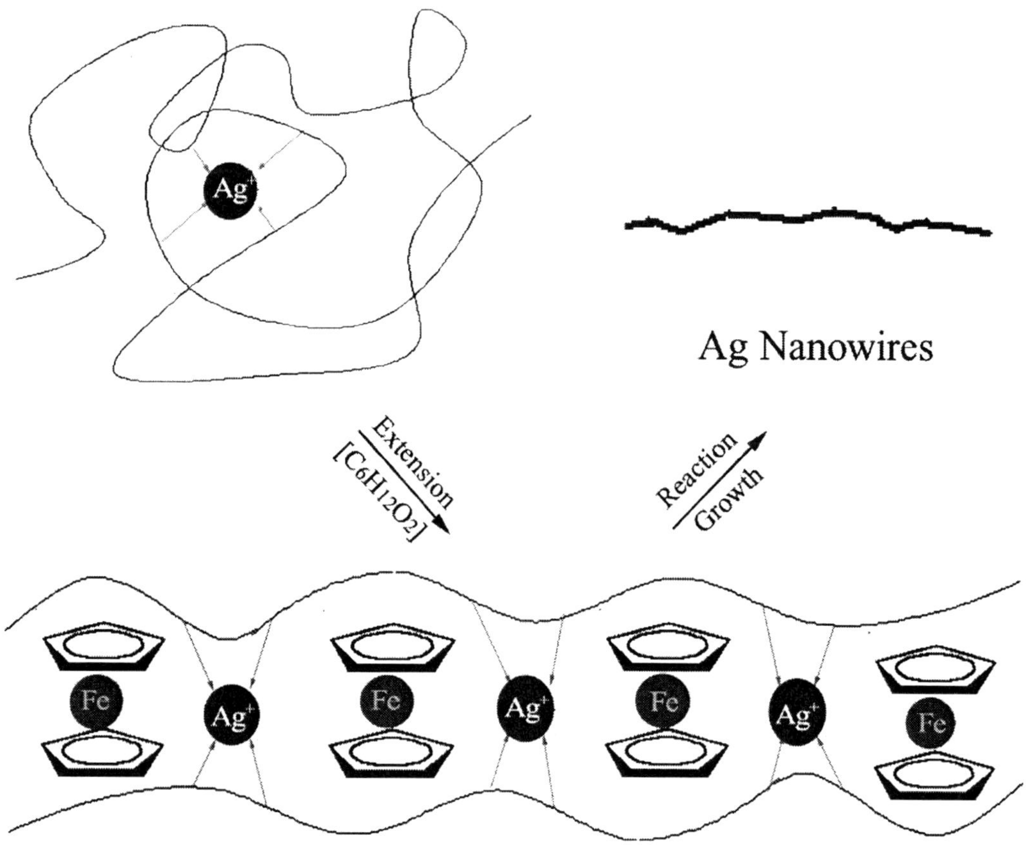

Figure 5. Schematic illustration of preparing silver nanowires from metal-polymer coordination chains. Adapted in part from ref. 42, © 2003 Wiley-VCH. Used with permission.

The polymer of PEG or polyvinylpyrrolidone (PVP) has also been applied to synthesize γ-MnO_2 and α-NiS hierarchical structures.[45,46] In our cases, urchin-like and sisal-like γ-MnO_2 structures (Figure 6B and C), and flower-like and tennis-ball α-NiS structures (Figure 6D) were obtained. The IR spectra of the precipitates composed of metal cations and polymers showed that coordination structures actually formed.[45,46] The formation of assemblies was found to result from the emergence of aggregated particle cores.[45] We found that the key parameter to the formation of these assembly structures was the amount of polymers and solvents.

5. 1D Coordination Polymer

The "1D coordination polymer" means complex with infinite structure in one dimension. For example, a novel supramolecule $[HgCl_2(ptz)]_2{\cdot}HgCl_2$ (ptz = phenothiazine) with uncoordinated inorganic salt $HgCl_2$ presented in a 1D chain was prepared by our group.[47] Single crystal X-ray analysis showed that the 1D chain structure is stabilized by hydrogen bonds between adjacent chains and the coordination mode of the ligand phenothiazine is unusual with large steric inhibition other than the chain directions. From its structure, the preorientation of 1D chain structure in the precursor may facilitate the formation of 1D nanostructure. Hence the submicrometer Hg_2Cl_2 rods could be obtained with $[HgCl_2(ptz)]_2{\cdot}HgCl_2$ as precursor by heating.[47]

Figure 6. SEM images of (A) ZnO nanowires, (B) γ-MnO_2 urchin-like hierarchical structures, (C) γ-MnO_2 sisal-like hierarchical structures, and (D) α-NiS tennis-ball hierarchical structures. Adapted in part from ref. 44 and 45, © 2003 American Chemical Society. Used with permission. Adapted in part from ref. 46, © 2003 Royal Society of Chemistry. Used with permission.

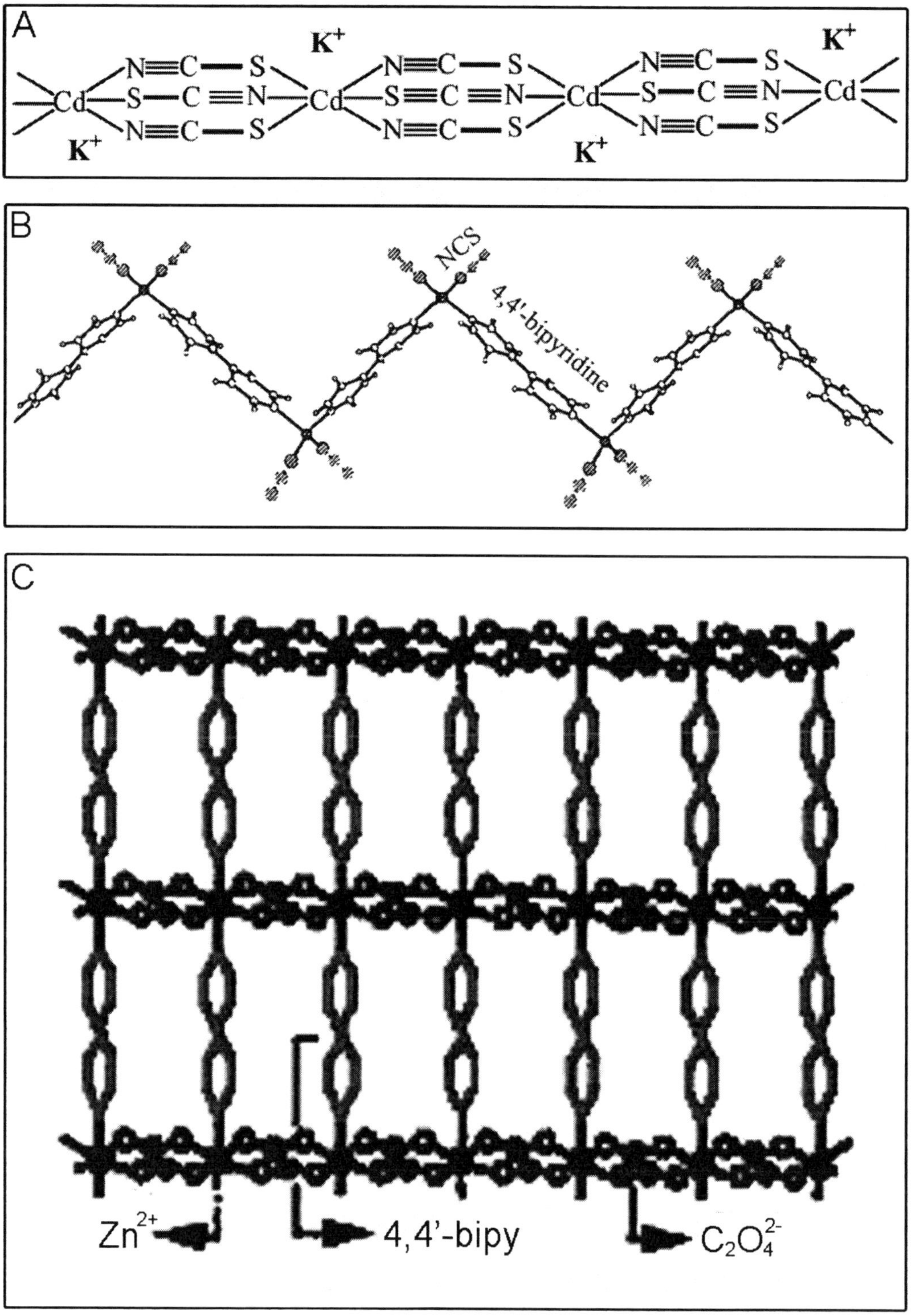

Figure 7. (A) Structural illustration of $KCd(SCN)_3$. Adapted in part from ref. 48, © 2003 Royal Society of Chemistry. Used with permission. (B) Structural illustration of $[Zn(4,4'\text{-bipy})(NCS)_2]_n$. Adapted in part from ref. 51, © 2003 Royal Society of Chemistry. Used with permission. (C) Structural illustration of ${}^{2}_{\infty}[Zn(ox)(4,4'\text{-bipy})]$.

Besides precursor, 1D coordination structure could also be applied as template for the growth of metal sulfide 1D nanostructures. For example, we synthesized CdS and HgS aligned nanowire bundles using 1D-chain-structural $KCd(SCN)_3$ (Figure 7A) as templates.[48] In the reaction process, with the addition of metal cations (Cd^{2+} or Hg^{2+}) and S^{2-}, immediate solution supersaturation of metal sulfide occurred due to its low solubility in the water system.

Most of the metal sulfide nanoparticles produced in the solution then was attached to surrounding negatively charged coordination polymer chains. At the same time, the anionic $Cd(NCS)_3^-$ chain adsorbs selectively to positively charged crystal faces, which directs the aggregation of the MS building blocks along one principal axis to produce the chain-like aggregates. As the $Cd(NCS)_3^-$ template is a 1D linear chain, these interactions are likely to be sufficient to offset the accompanying steric repulsion. The separating surfaces of adjacent particles may have been eliminated after repulsive interactions were overcome as the result of random, Brownian motion-driven particle collisions.[49] Jiggling of nanoparticles by Brownian motion may also allow adjacent particles to rotate to find the low-energy configuration represented by a coherent particle–particle interface. Rotation of particles within aggregates may also be driven by short-range interactions between adjacent surfaces.[50] Diffusion-controlled growth of these metal sulfide nanoparticles then occurs in the aggregates as the amorphous particles are in close proximity and the surface area reduces by particle fusion and structure rearrangement.[33] In aggregation-based growth, reduction in surface free energy is achieved by complete removal of pairs of surfaces.[49] Because the director $Cd(NCS)_3^-$ chains were arranged along one direction and the distances between them were almost equal in this system, which should play a definitive role in the formation of aligned nanowire bundles. The whole process of the formation of MS nanowires can be summarized as follows: primary nucleation → oriented aggregation → particle-particle diffusion → aligned-growth.

Note that the orientation of 1D coordination structure as precursor or template is crucial for the production of 1D nanostructures, therefore, the bonding along the 1D chains is required to be strong enough for reactions. Otherwise, during the reaction process, 1D chains may break down so that they cannot provide orientation for 1D crystal growth. For example, we ever tried to use $[Zn(4,4'\text{-bipy})(NCS)_2]_n$ with zigzag chain (Figure 7B) as precursor in the reaction with ammonia.[51] The NCS^- ligand has a stronger coordination with Zn^{2+} than 4,4'-bipyridine, as seen from their bond lengths [Zn–N (NCS) 1.908 Å, Zn–N (4,4'-bipy) 2.033 Å] in the coordination polymer.[52] Thus, 4,4'-bipyridine could be preferentially removed when attacked by OH^- and the infinite chains of the coordination polymer would gradually be broken. The broken chains with two -OH at either end were likely to aggregate through a dehydration process in the heated basic environment. As the removal of ligands (including 4,4'-bipyridine and –NCS) and the dehydration of –OH were repeated, produced ZnO nanoparticles tended to form hollow spheres, due to their lower surface tension. Therefore, as a typical result, ZnO hollow spheres were prepared from the precursor $[Zn(4,4'\text{-bipy})(NCS)_2]_n$.[51]

6. 2D Coordination Polymer

The coordination polymers with 2D networks have been directly employed as precursors for the growth of metal oxide 1D nanostructures by our group. In a typical case, we used the coordination polymer, $^{2}_{\infty}$[Zn(ox)(4,4′-bipy)], with a 2D coordination network (Figure 7C) to synthesize the radial assemblies of nanoneedles.[53] The experimental results showed that the 2D network of the coordination polymer was crucial to the formation of 1D nanostructures. The coordination polymer $^{2}_{\infty}$[Zn(ox)(4,4′-bipy)] had a 2D coordination network structure constructed by two bridging ligands, oxalate and 4,4′-bipy. All Zn^{2+} ions had an octahedral coordination environment with four oxygen atoms from two oxalate ions and two nitrogen atoms from two 4,4′-bipy molecules along two vertical directions. Zn^{2+} cations were strongly coordinated with both strong ligands that could be removed slowly, thus the newly formed ZnO colloids could not reach monodispersed situation quickly. Therefore, the produced ZnO colloids were inclined to be bound to each other. As ZnO crystal had the anisotropic growth habit, 1D ZnO nanostructures were likely to epitaxially grown from the surface of aggregated colloids during the succedent matured process, such as Ostwald ripenning. It was noteworthy that the role of framework in coordination polymer here was to assemble those produced ZnO crystals that were matured to assembly structures. There is no evidence to show that they acted as template.

7. 3D Coordination Polymer

3D coordination polymers have stable ordered infinite structures in three dimensions. In detail, the metal cations in some 3D coordination polymers bridged by rigid ligands (e. g. 4,4'-bpy) are not only ordered, but also have different connecting fashions. If selecting an appropriate coordination one as precursor, it is possible to make the polymer dissociated and form new metal-oxygen bonds under alkaline hydrothermal conditions. Thus, the formation of the as-desired metal oxide could be achieved by this route, which is significant as there exist many phases of metal oxides. For example, MnO_2 and related manganese (IV) oxide are known to exist in a wide variety of structural forms, such as α-, β-, γ-, δ-type and ramsdellite, etc.

Especially in the process of oxidization reaction, the different coordination abilities of ligands in the coordination polymer will probably lead to a faster oxidization along some direction, thus resulting in an oriented growth of metal oxide nanowires. Therefore, we design a hydrothermal route to synthesize metal oxide nanowires via an alkaline replacement reaction on coordination polymer-precursors. In a specific case, γ-MnO_2 nanowires (Figure 8A) was prepared from the reaction between coordination polymer $[Mn(SO_4)(4,4'\text{-}bpy)(H_2O)_2]_n$ and boiling NaOH solution, in which both the crystal phase and the shape of product are well controlled.[54] The tops of nanowires were found to almost align along one direction (Figure 8B). This ordered alignment of γ-MnO_2 nanowires probably leads to more applications in Lithium batteries.

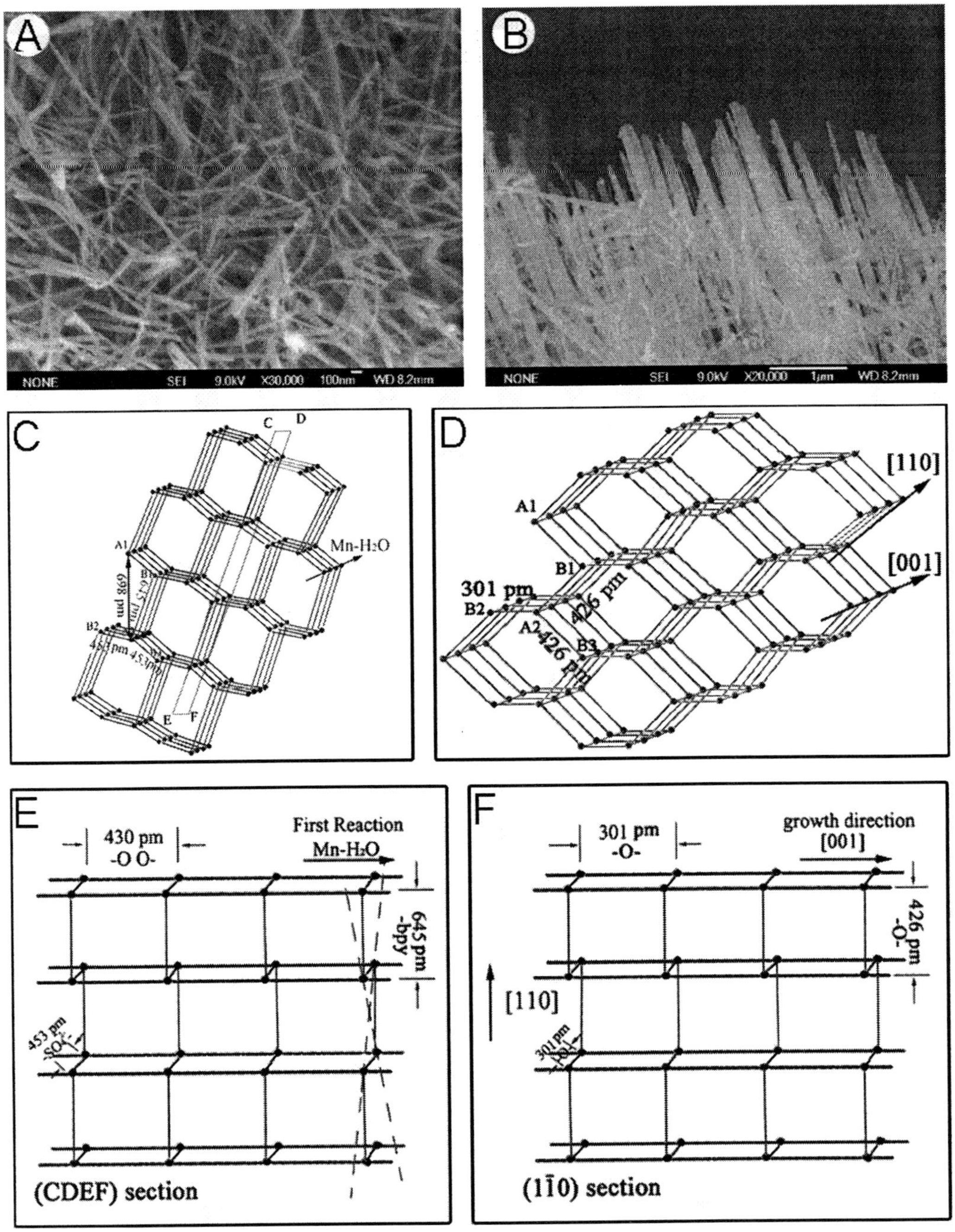

Figure 8. (A), (B) SEM images of γ-MnO_2 nanowires. (C) The Mn cations in $[Mn(SO_4)(4,4'\text{-}bpy)(H_2O)_2]_n$ having more chances to combine with each other and form stable manganese (IV) oxide, when the coordination polymer is dissociated in boiling NaOH solution. (D) The frameworks of Mn cations in γ-MnO_2 crystal lattices. (E) The (CDEF) section in the structure of $[Mn(SO_4)(4,4'\text{-}bpy)(H_2O)_2]_n$ (The dashlines correspond to two directions of –Mn-bpy-Mn-bpy- chains). (F) The (110) section of γ-MnO_2 crystal lattices. The thin lines only represent the distance of Mn cations. Adapted in part from ref. 54, © 2003 Wiley-VCH. Used with permission.

The supplementary experiments revealed that the crystal structure of coordination polymer is crucial to the formation of γ-MnO_2 nanowires. To reveal the reaction process more clearly, the real-time IR spectra were taken, showing that the Mn-water ligands, –Mn-SO_4^{2-}-Mn-SO_4^{2-} and -Mn-bpy-Mn-bpy were dissociated in turn while the new Mn-O of γ-MnO_2 formed, which agrees with the fact that the coordination polymers containing 4,4'-bpy are usually stable. These characterization results reveal that the dissociation of coordination polymer and the formation of γ-MnO_2 occurred at the same time, and the Mn-water ligands were dissociated at first. Therefore, the structural correlation between the polymer and the oxide is important for the formation of product.

The coordination polymer $[Mn(SO_4)(4,4'\text{-bpy})(H_2O)_2]_n$ has 3D infinite layered structure,[55] in which each layer is formed in the *ac* plane by connecting Mn centers via bridging bpy ligands to generate parallel –Mn-bpy-Mn-bpy chains, whereas between adjacent layers the –Mn-bpy-Mn-bpy chains have an angle of 45°. –Mn-bpy-Mn-bpy chains in different layers are connected by zigzag –Mn-SO_4^{2-}-Mn-SO_4^{2-} chains, leading to a 3D polymer framework. The Mn cations having more chances to form Mn-O-Mn through alkaline hydrothermal reaction could be described as distorted hexagonal units in layers (Figure 8C). γ-MnO_2 is considered to be a disordered intergrowth of the β-MnO_2 and ramsdellite structures, consisting of a random arrangement of single and double chains of MnO_6 octahedra.[56] Each Mn cation is connected with other three adjacent ones by oxygen anions to build up distorted Mn hexagonal units in layers, which are composed of Mn cation frameworks shown in Figure 8D. The layers (parallel to *ab* plane) are also connected by sharing oxygen anions. It is found that the structural figure of these Mn cations are similar to that in γ-MnO_2 crystal lattices. More careful studies on the structure of $[Mn(SO_4)(4,4'\text{-bpy})(H_2O)_2]_n$ show that the Mn distorted hexagonal units arrange in layers. These layers deviate from the *ac* plane and each of them is separated by oxygen donors from two water ligands. This characteristic can be clearly shown in the (CDEF) section of the structure of coordination polymer-precursor (Figure 8E). Comparing this section (Figure 8F) with (110) section of γ-MnO_2 crystal lattices, it is found that the two sections are similar in structure and the coordination direction of Mn-water ligands in coordination polymer-precursor can correspond to the [001] axis of γ-MnO_2 crystal lattices. Mn-water ligands perpendicular to the layers were firstly dissociated in boiling NaOH solution and could lead to the oriented growth of γ-MnO_2 crystal lattices to nanowires along [002] axis. The ordered alignment of γ-MnO_2 nanowires might also result from the connecting of –Mn-bpy-Mn-bpy, the last one that was dissociated.

8. Conclusion and Outlook

Coordination chemistry is an ancient subject that has been widely used in many fields. The application of coordination chemistry in the rational synthesis and simultaneous assembly of novel nanostructures provides a new approach to nanomaterials. Especially, it considers more information of chemistry that is so-called "soft-chemistry approach", exhibiting the significance of nano-chemistry. In this chapter, we review the development of synthesizing inorganic nanostructures from coordination structures, including simple complex, linear coordination cluster compound, metal-polymer coordination compound and

1D, 2D, 3D coordination polymers. In detail, simple complex works by steric-inhibition effect and hydrogen bonding; linear coordination cluster compound, metal-polymer coordination compound and 1D coordination polymers directly provide orientation for the growth of 1D nanostructures or act as template; the network of 2D coordination polymer is helpful for the assembly of produced nanostructures; the different coordination ability of ligands in 3D coordination polymer results in the unequal crystal growth rate in three dimensions. A variety of nanostructures have been fabricated via these techniques, which display excellent optical, catalytic, electrochemical, magnetic and SPR properties.

It is believed that these techniques can be readily extended to produce more complex nanostructures covering a wide range of materials by involving more reactions into the fabrication system, incorporating more desired materials into the synthetic scheme, or studying detailed mechanism. It is also expected that the related synthetic strategy is applicable to extend to chemical vapor transport or laser-assisted catalytic growth. As a matter of fact, Lieber group successfully synthesized CdS and ZnS nanowires with cadmium or zinc diethyldithiocarbamate ($Cd_2(S_2CNEt_2)_4$ or $Zn_2(S_2CNEt_2)_4$) as single-source precursors and gold nanoclusters as catalyst by VLS technique recently.[57] In general, by tailoring the size, shape and/or structure, one can, in principle, tune their intrinsic properties in a range of applications. We expect that further research on these techniques can provide more versatile ways to generate desirable nanostructures for potential applications.

ACKNOWLEDGEMENT

This work was supported in part by the National Natural Science Foundation of China.

REFERENCES

[1] Reviews: (a) Hu, J. T.; Odom, T. W.; Lieber, C. M. *Acc. Chem. Res.* 1999, 32, 435. (b) Templeton, A. C.; Wuelfing, W. P.; Murray, R. W. *Acc. Chem. Res.* 2000, 33, 27.

[2] Catalysis and optical: (a) Nicewarner-Peña, S. R. R.; Freeman, G.; Reiss, B. D.; He, L.; Peña, D. J.; Walton, I. D.; Cromer, R.; Keating, C. D.; Natan, M. J. *Science* 2001, 294, 137. (b) Huang, M. H.; Mao, S.; Feick, H.; Yan, H. Q.; Wu, Y. Y.; Kind, H.; Weber, E.; Russo, R.; Yang, P. D. *Science* 2001, 292, 1897. (c) Ahmadi, T. S.; Wang, Z. L.; Green, T. C.; Henglein, A.; El-Sayed, M. A. *Science* 1996, 272, 1924. (d) Peng, X.; Manna, L.; Yang, W.; Wickham, J.; Scher, E.; Kadavanich, A; Alivisatos, A. P. *Nature* 2000, 404, 59. (e) Lewis, L. N. *Chem. Rev.* 1993, 93, 2693.

[3] Electronics and optoelectronics: (a) El-Sayed, M. A. *Acc. Chem. Res.* 2001, 34, 257. (b) Chen, S.; Yang, Y. *J. Am. Chem. Soc.* 2002, 124, 5280. (c) Wu, Y.; Xiang, J.; Yang, C.; Lu, W.; Lieber, C. M. *Nature* 2004, 430, 61.

[4] Information storage: (a) Peyser, L. A.; Vinson, A. E.; Bartko, A. P.; Dickson, R. M. *Science* 2001, 291, 103. (b) Chiba, D.; Yamanouchi, M.; Matsukura, F.; Ohno, H. *Science* 2003, 301, 943.

[5] Sensing and clinical diagnostics: (a) Tkachenko, A. G.; Xie, H.; Coleman, D.; Glomm, W.; Ryan, J.; Anderson, M. F.; Franzen, S.; Feldheim, D. L. *J. Am. Chem. Soc.* 2003, 125, 4700. (b) Taton, T. A.; Mirkin, C. A.; Letsinger, R. L. *Science* 2000, 289, 1757. (c)

Wang, Y.; Jiang, X.; Xia, Y. *J. Am. Chem. Soc.* 2003, 125, 16176. (d) Zhang, X.; Young, M. A.; Lyandres, O.; Van Duyne, R. P. *J. Am. Chem. Soc.* 2005, 127, 4484.

[6] SERS: (a) Cao, Y. C.; Jin, R.; Mirkin, C. A. *Science* 2002, 297, 1536. (b) Nie, S.; Emory, S. R. *Science* 1997, 275, 1102. (c) Tessier, P. M.; Velev, O. D.; Kalambur, A. T.; Rabolt, J. F.; Lenhoff, A. M.; Kaler, E. W. *J. Am. Chem. Soc.* 2000, 122, 9554. (d) Dick, L. A.; McFarland, A. D.; Haynes, C. L.; Van Duyne, R. P. *J. Phys. Chem. B* 2002, 106, 853.

[7] Crystal-structure directed growth: (a) Song, J. H.; Messer, B.; Wu, Y.; Kind, H.; Yang, P. *J. Am. Chem. Soc.* 2001, 123, 9714. (b) Gates, B.; Yin, Y.; Xia, Y. *J. Am. Chem. Soc.* 2000, 122, 12582.(c) Xiong, Y.; Xie, Y.; Du, G.; Su, H. *Inorg. Chem.* 2002, 41, 2953.

[8] Template-directed synthesis: (a) Jones, E. T. T.; Chyan, O. M.; Wrighten, M. S. *J. Am. Chem. Soc.* 1987, 109, 5526. (b) Dai, H.; Wong, E. W.; Lu, Y. Z.; Fan, S.; Lieber, C. M. *Nature* 1995, 375, 769. (c) Song, J. H.; Wu, Y.; Messer, B.; Kind, H.; Yang, P. *J. Am. Chem. Soc.* 2001, 123, 10397.

[9] Electrochemistry: (a) Zhou, Y.; Yu, S. H.; Cui, X. P.; Wang, C. Y.; Chen, Z. Y. *Chem. Mater.* 1999, 11, 545. (b) Zhu, J. J.; Liu, S. W.; Palchik, O.; Koltypin, Y.; Gedanken, A. *Langmuir* 2000, 16, 6396. (c) Xu, D.; Xu, Y.; Chen, D.; Guo, G.; Gui, L.; Tang, Y. *Adv. Mater.* 2000, 12, 520.

[10] Vapor phase growth: (a) Pan, Z. W.; Dai, Z. R.; Wang, Z. L. *Science* 2001, 291, 1947. (b) Jiang, X.; Herricks, T.; Xia, Y. *Nano Lett.* 2002, 2, 1333.

[11] Capping reagents: (a) Peng, Z. A.; Peng, X. G *J. Am. Chem. Soc.* 2001, 123, 1389. (b) Sun, Y.; Gates, B.; Mayers, B.; Xia, Y. *Nano Lett.* 2002, 2, 165.

[12] Laser–assisted catalysis growth and VLS: (a) Morales, A. M.; Lieber, C. M. *Science* 1998, 279, 208. (b) Gudiken, M. S.; Lieber, C. M. *J. Am. Chem. Soc.* 2000, 122, 8801.

[13] SLS: (a) Trentler, T. J.; Hickman, K. M.; Geol, S. C.; Viano, A. M.; Gibbons, P. C.; Buhro, W. E. *Science* 1995, 270, 1791. (b) Yu, H.; Buhro, W. E. *Adv. Mater.* 2003, 15, 416.

[14] Solvothermal method: (a) Wang, X.; Li, Y. *J. Am. Chem. Soc.* 2002, 124, 2880. (b) Xie, Y.; Yan, P.; Lu, J.; Wang, W.; Qian, Y. *Chem. Mater.* 1999, 11, 2619.

[15] Bae, B. J.; Park, J. E.; Kim, B.; Park, J. T. *J. Organomet. Chem.* 2000, 616, 128. (b) Grocholl, L.; Wang, J.; Gillan, E. G. *Chem. Mater.* 2001, 13, 4290.

[16] Soulantica, K.; Maisonnat, A.; Senocq, F.; Fromen, M. C.; Casanove, M. J.; Chaudret, B. *Angew. Chem. Int. Ed.* 2001, 40, 2983.

[17] Li, Y. D.; Liao, H. W.; Ding, Y.; Fan, Y.; Zhang, Y.; Qian, Y. T. *Inorg. Chem.* 1999, 38, 1382.

[18] Yang, J.; Zeng, J. H.; Yu, S. H.; Yang, L.; Zhou, G.; Qian, Y. T. *Chem. Mater.* 2000, 12, 3259.

[19] Yang, J.; Xue, C.; Yu, S. H.; Zeng, J. H.; Qian, Y. T. *Angew. Chem. Int. Ed.* 2002, 41, 4697.

[20] Bradley, J. S.; Tesche, B.; Busser, W.; Maase, M.; Reetz, M. T. *J. Am. Chem. Soc.* 2000, 122, 4631.

[21] Chen, M.; Xie, Y.; Lu, J. C.; Zhu, Y. J.; Qian, Y. T. *J. Mater. Chem.* 2001, 11, 518.

[22] Chen, M.; Xie, Y.; Yao, Z. Y.; Qian, Y. T.; Zhou, G. E. *Mater. Res. Bull.* 2002, 37, 247.

[23] Jiang, X. C.; Xie, Y.; Lu, J.; Zhu, L. Y.; He, W.; Qian, Y. T. *Adv. Mater.* 2001, 13, 1278.

[24] Jiang, X. C.; Xie, Y.; Lu, J.; Zhu, L. Y.; He, W.; Liu, X. M. *Can. J. Chem.* 2002, 80, 263.
[25] Cotton, F. A.; Wilkinson, G. *Advanced Inorganic Chemistry*, John Wilkey and Sons Inc., 1972.
[26] Lorimer, J. P.; Mason, T. J. *Chem. Soc. Rev.* 1987, 16, 275.
[27] Xiong, Y. J.; Xie, Y.; Du, G.; Liu, X. M.; Tian, X. B. *Chem. Lett.* 2002, 98.
[28] König, E. *Coordin. Chem. Rev.* 1968, 3, 471. (b) Madeja, K.; König, E. *J. Inorg. Nucl. Chem.* 1963, 25, 377.
[29] Xiong, Y. J.; Xie, Y.; Chen, S. W.; Li, Z. Q. *Chem.-Eur. J.* 2003, 9, 4991.
[30] Xiong, Y. J.; Xie, Y.; Li, Z. Q.; Zhang, R.; Yang, J. C.; Wu, Z. *New J. Chem.* 2003, 27, 588.
[31] Xiong, Y. J.; Li, Z. Q.; Li, X. X.; Hu, B.; Xie, Y. *Inorg. Chem.* 2004, 43, 6540.
[32] Zhang, X. J.; Zhao, Q. R.; Tian, Y. P.; Xie, Y. *Crys. Growth Des.* 2004, 4, 355.
[33] Manna, L.; Scher, E. C.; Alivisatos, A. P. *J. Am. Chem. Soc.* 2000, 122, 12700. (b) Penn, R. L.; Banfield, J. F. *Am. Mineral.* 1998, 83, 1077. (c) Alivisatos, A. P. *Science* 2000, 289, 736. (d) Chemseddine, A.; Morita, T. *Eur. J. Inorg. Chem.* 1999, 235. (e) Park, S. J.; Kim, S.; Lee, S.; Khim, Z. G.; Char, K.; Hyeon, T. *J. Am. Chem. Soc.* 2000, 122, 8581.
[34] Domenicano, A.; Torelli, L.; Vaciago, A.; Zambonelli, L. *J. Chem. Soc. A* 1968, 1351.
[35] Yang, P.; Xie, Y.; Qian, Y. T.; Liu, X. M. *Chem. Comm.* 1999, 1293.
[36] Ramadas, K.; Janarthanan, N. *J. Chem. Res. S* 1998, 228.
[37] Yan, P.; Xie, Y.; Qian, Y. T. *Chem. Lett.* 1999, 655.
[38] Bourrel, M.; Schechter, R. S. *Microemulsion and Related Systems: Formation, Solvency, and Physical Properties*, Surfactant Science Series Vol. 30, Marcel Dekker Inc., New York and Basel, 1988.
[39] Xiong, Y. J.; Li, Z. Q.; Zhang, R.; Xie, Y.; Yang, J.; Wu, C. Z. *J. Phys. Chem. B* 2003, 107, 3697.
[40] Pearson, P. G. *J. Chem. Educ.* 1968, 45, 581.
[41] Lowman, A. M.; Peppas, N. A. *Polymer* 2000, 41, 73.
[42] Xiong, Y. J.; Xie, Y.; Wu, C. Z.; Yang, J.; Li, Z.; Xu, F. *Adv. Mater.* 2003, 15, 405.
[43] Xu, F.; Zhang, X.; Xie, Y.; Tian, X. B.; Li, Y. Z. *J. Colloid Interface Sci.* 2003, 260, 160.
[44] Li, Z. Q.; Xiong, Y. J.; Xie, Y. *Inorg. Chem.* 2003, 42, 8105.
[45] Wu, C. Z.; Xie, Y.; Wang, D.; Yang, J.; Li, T. W. *J. Phys, Chem. B* 2003, 107, 13583.
[46] Xu, F.; Xie, Y.; Zhang, X.; Wu, C. Z.; Xi, W.; Hong, J.; Tian, X. B. *New. J. Chem.* 2003, 27, 1331.
[47] Zhang, X. J.; Xie, Y.; Yu, W. T.; Zhao, Q. R.; Jiang, M. H.; Tian, Y. P. *Inorg. Chem.* 2003, 42, 3734.
[48] Zhang, X. J.; Xie, Y.; Zhao, Q. R.; Tian, Y. P. *New J. Chem.* 2003, 27, 827.
[49] Banfield, J. F.; Welch, S. A.; Zhang, H.; Ebert, T. T.; Penn, R. L. *Science* 2000, 289, 751.
[50] Zhu, H. L.; Averback, R. S. *Philos. Mag.* 1996, 73, 27.
[51] Li, Z. Q.; Xie, Y.; Xiong, Y. J.; Zhang, R. *New J. Chem.* 2003, 27, 1518.
[52] Pan, L.; Zheng, N. W.; Wu, Y. G.; Jin, X. L.; Huang, X. Y. *J. Coord. Chem.* 1999, 47, 269.
[53] Li, Z. Q.; Xiong, Y. J.; Xie, Y. *unpublished work.*

[54] Xiong, Y. J.; Xie, Y.; Li, Z. Q.; Wu, C. Z. *Chem.-Eur. J.* 2003, 9, 1645.

[55] Hou, H.; Wei, Y.; Fan, Y.; Du, C.; Zhu, Y.; Song, Y.; Niu, Y.; Xin, X. *Inorg. Chim. Acta* 2001, 319, 212.

[56] Hunter, J. C. *J. Solid State Chem.* 1981, 39, 142. (b) Burns, R. G.; Burns, V. M. *Manganese Dioxide Symposium*, I. C. Sample Office, Cleveland, 1970, Vol. 1. (c) Qian, Y. *Introduction to Crystallography*, University of Science and Technology Press, Hefei, 1999 (In Chinese).

[57] Barrelet, C. J.; Wu, Y.; Bell, D. C.; Lieber, C. M. *J. Am. Chem. Soc.* 2003, 125, 11498.

In: Frontal Nanotechnology Research
Editor: Martin V. Berg, pp. 61-88 ISBN: 1-60021-212-3

Chapter 3

1-D ZINC SULFIDE NANOSTRUCTURES: SYNTHESIS AND CHARACTERIZATION

Soumitra Kar[1] and Subhadra Chaudhuri[1]
Department of Materials Science, Indian Association for the Cultivation of Science,
Kolkata- 700 032, India

ABSTRACT

With the growing interest for miniaturization in electronic and optoelectronic devices, scientists are more and more interested in fabricating different materials in nano-scale dimensions with different morphologies with a view to use them as the basic building block for nano-scale devices. One-dimensional (1-D) nanostructures such as nanorods, nanowires, nanobelts or nanoribbons are of special importance because of their improved fundamental, physical and chemical properties, which make them potential candidate for nano-scale devices. Zinc sulfide (ZnS) is a unique luminescent material well known for its photoluminescence, electroluminescence, etc., which enable wide applications in the fields of displays, sensors and lasers. This article reviews the synthesis, structure and properties of 1-D ZnS nanostructures. ZnS 1-D nanostructures have been fabricated by using different physical (vapor-solid, vapor-liquid-solid etc.) and chemical (solvothermal etc.) routes. Optical properties such as optical absorption, photoluminescence and Raman spectra of the 1-D ZnS nanoforms are also discussed here. Wherever possible, the possible applications of these nanostructures are also included.

1. INTRODUCTION

Nanostructures have received tremendous interest among the researchers working in all fields including physics to bioscience, from basic science to technologists for their improved physical and chemical properties and applications superior to their bulk counterparts [1-4]. Since the first discovery of carbon nanotubes, one-dimensional (1-D) semiconductor materials such as nanorods, nanowires, nanotubes and nanobelts/nanoribbons have attracted

[1] E-mail address: mssc2@iacs.res.in (S. C.); kar_mitra@yahoo.co.in (S. K.) Corresponding author. Tel: +91-033-2473-4971, Fax: +91-033-2473-2805

extensive interest because of their fundamental physical, chemical, optical, electrical and magnetic properties, which are distinct from their bulk counterparts, and also due to their potential applications in nano-scale devices [5-16]. It is well known that 1-D nanostructures can play an important role both as interconnect and functional units in fabricating electronic, optoelectronic, electrochemical and electromechanical devices with nanoscale dimension. 1-D nanostructures have been synthesized through a variety of synthesis techniques, such as, template-directed synthesis [17-19], Vapor-solid growth [20-22], vapor-liquid –solid (VLS) growth [23-26], solution-liquid-solid (SLS) growth [27], solvothermal process [28-31] etc. A variety of metal [32-35], semiconducting oxides [22, 36-40], nitrides [41-44] and sulfides [45-55] in different 1-D nanoforms were reported so far.

Out of these materials, zinc sulfide a II-VI semiconductor is one of the most studied material for its wide range of technologically important properties. Zinc Sulfide (ZnS) has a wide band gap of 3.72 eV for cubic phase [56] and 3.77 eV for the hexagonal wurtzite phase [57] at room temperature. It is a key material for ultraviolet light-emitting diodes, injection lasers [58], cathode ray tubes, flat panel displays [59] and IR windows [60]. ZnS is also important for photoluminescence [61], electroluminescence [62], triboluminescence [63], etc. Recently, optical wave confining and lasing has been demonstrated in ZnS nanoribbons [64]. Moreover, zinc sulfide is also widely used in photocatalysis [65-66]. In recent years, nanocrystalline ZnS attracted much attention because properties in nanoforms differ significantly from those of their bulk counterparts [67]. Therefore, much effort has been made to control the size, morphology and crystallinity of the ZnS nanocrystals with a view to tune their physical properties.

Thus keeping in mind the recent trend towards one-dimensional nanostructured materials and applicability of ZnS as an technologically important semiconductor, this article reviews the different aspects of 1-D ZnS nanoforms i.e. synthesis techniques for these nanoforms, their properties and possible applications.

2. ZnS 1-D Nanoforms

The research on the synthesis and characterization of ZnS 1-D nanoforms have progressed much and a variety of 1-D and quasi 1-D nanoforms of ZnS are reported so far. In the following section we will discuss different 1-D nanoforms of ZnS and their synthesis scheme in brief.

2.1. ZnS Nanorods

1-D structures with diameters ranging from several nanometers to few hundred nanometers without having any bending along their lengths are referred as the nanorods. ZnS nanorods were prepared by several routes [68-70, 52]. ZnS nanorods were prepared by ethylenediamine assisted hydrothermal process [68-69] at temperature ~ 200 °C. The fabricated nanorods were found to have diameters within 10-50 nm and lengths ~ few hundred nanometers. These nanorods were single crystalline and no staking faults were revealed from the HRTEM images. Figure 1a-b shows the TEM and HRTEM images of ZnS nanorods synthesized in the author's laboratory via solvothermal route [69]. Interestingly, these nanorods synthesized at ~ 200 °C possessed hexagonal wurtzite structure, which is the meta-stable polymorph of ZnS as wurtzite ZnS is generally stable beyond 1020 °C. ZnS nanorods with cubic zinc blende structure were also reported by chemical routes [70]. The 1-D nanoforms can show quantum confinement effect if their diameter lies well within the limit of 10 nm [71]. ZnS nanorods having quantum size effect have also been synthesized by several groups [72-73] through chemical approach. ZnS nanorods were often appeared as an intermediate product when ZnS nanowires were being prepared by VLS approach as reported in our earlier work [52]. These nanorods further grew in length giving rise to ZnS nanowires.

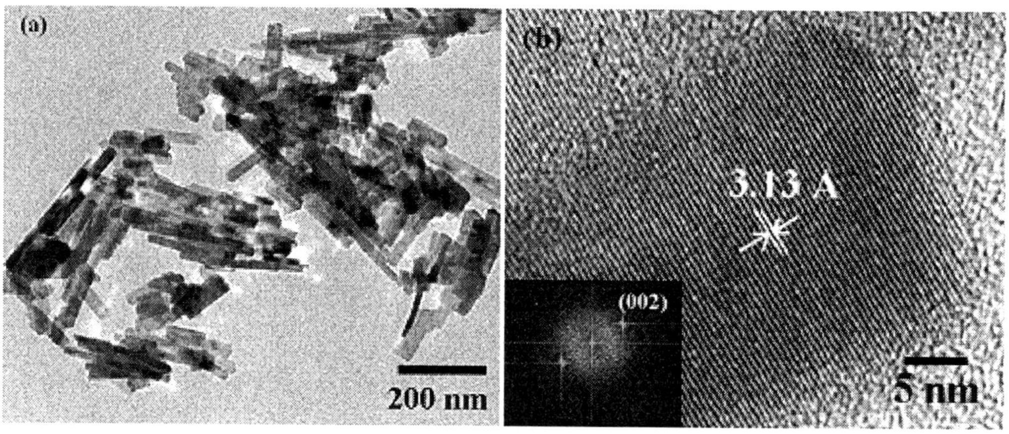

Figure 1. (a) TEM image of the ZnS nanorods synthesized by solvothermal process and (b) HRTEM image of one ZnS nanorod revealing its single crystalline nature.

2.2. ZnS Nanowires

Nanowires are basically elongated version of the nanorods having wire like bending throughout their lengths (Figure 2a-b). The aspect ratio (length/diameter) of the nanowires is generally very high compared to those of the nanorods. Cubic ZnS nanowires with zinc blende structure were prepared by liquid crystal template [74], and miecelle-template [75] based chemical routes. Meng et al. [76] also reported cubic ZnS nanowires by a VLS approach. Unlike the cubic nanowires produced by the chemical approach, the nanowires produced by the VLS approach were found to have better control over their morphology. The nanowires produced by this high temperature (Figure 2c) based approach were smooth and uniform throughout their length and possessed Au nanoparticles at their tips, which were utilized to catalyze the nucleation of the nanowires. Similar VLS approach was also utilized by other groups [52, 77-78] to fabricate ZnS nanowires with hexagonal wurtzite structure. The nanowires produced by these groups were found to have circular cross-section whereas Xiong et al. [79-80] have reported square and hexagonal cross-sectional nanowires by pulsed laser vaporization (PLV) method. Jiang et al. [81] have reported polytype modulated ZnS nanowires via intermittent laser ablation-catalytic growth. The polytype modulated ZnS nanowires are composed of closed packed planes of the wurtzite structure stacking and wurtzite variant half cells of 8H and 10H appearing along the entire wire lengths in turn. Basically the growth mechanism of these ZnS nanowires, produced by laser ablation methods, is VLS mechanism. The nanowires produced by all these groups were non-aligned lying allover the substrate in an irregular manner as revealed by the SEM image (Fig 2a). Freestanding aligned 1-D semiconductor nanostructures are now attracting special attention because of their promising application as building blocks for nano-scale devices and as model system to study the 1-D nano-scale effect [82-83]. Efforts are on to achieve 1-D ZnS aligned nanoforms with a few recent successes [84-85]. Ding et al. [84] and Z. L. Wang's [85] research group have reported synthesis of aligned ZnS nanowires by thermal evaporation of ZnS powder on to aniodic alumina template (AAA) and CdSe crystals respectively. Ding et al [84] have also showed room temperature lasing from the aligned ZnS nanowires.

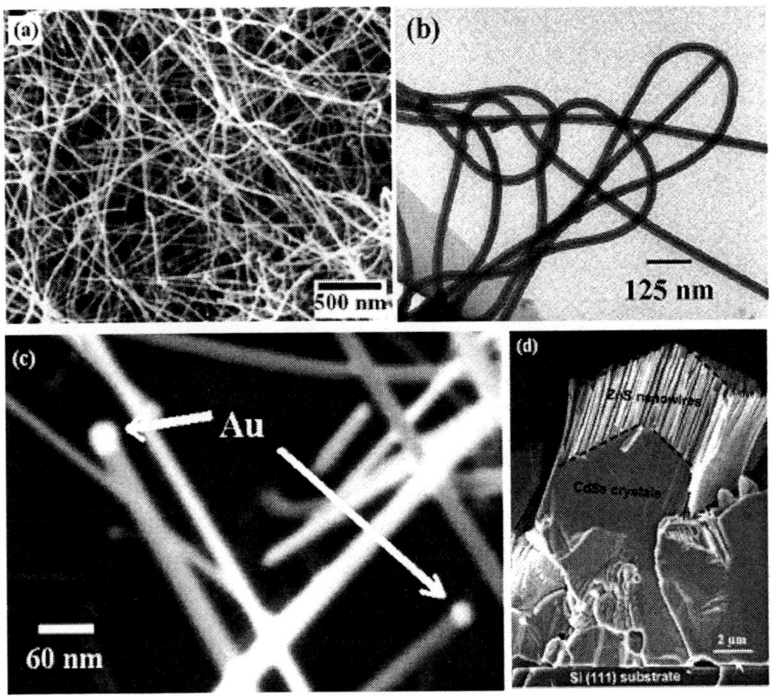

Figure 2. SEM (a) and TEM (b) images of ZnS nanowires. The presence of the Au nanoparticle at the tip of the ZnS nanowires could be seen in SEM image of the ZnS nanowire shown in (c). ZnS nanowire arrays grown on to a lattice matched CdSe crystals is shown in (d) [Figure d reprinted with permission from J. Am. Chem. Soc. 126 (2004) 14372].

2.3. ZnS Nanotubes

Nanotube was the first 1-D nanostructure, the discovery of which by S. Iijima [5] in 1991 triggered the research of 1-D systems. These 1-D systems are basically wire or rod like structures with hollow channel throughout their lengths. Though carbon nanotube is the most popular in this category, other inorganic nanotubes are also attracting interest to explore their physical, chemical properties and technological importance. Dloczik et al. [86] have synthesized ZnS nanotubes by chemical conversion of ZnO nanocolumns with H_2S at 400 °C temperature. ZnS nanotubes were also produced at near room temperature through chemical conversion of ZnS nanobelts in H_2S atmosphere [87]. The wall of these ZnS nanotubes were not single crystalline but, polycrystalline in nature and they had hexagonal wurtzite phase. Single crystalline wurtzite ZnS nanotubes with hexagonal cross-section (Figure 3) was reported by Yin et al [88]. They produced the ZnS nanotubes by thermal evaporation of ZnS powder at 1500-1700 °C in an H_2O-Ar atmosphere. These ZnS nanotubes were few micrometer long having diameter within 100-120 nm and wall thickness ~ 10 nm. The walls of these ZnS nanotubes were perfectly single crystalline with hexagonal crystal structure. These nanotubes showed strong green emission at room temperature.

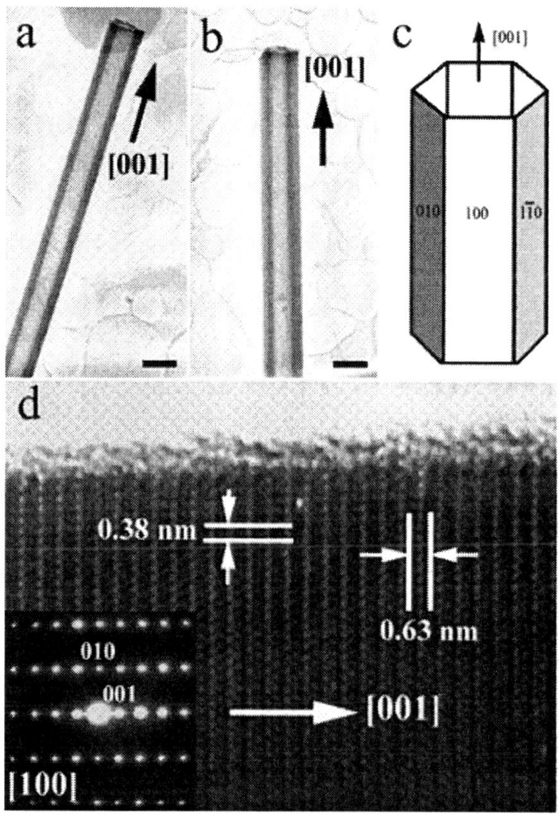

Figure 3. (a, b) TEM images of the ZnS nanotubes, (c) Schematic view of the hexagonal cross-sectional ZnS nanotubes and its growth habit; (d) HRTEM of the wall of the ZnS nanotube with the SAED pattern in its inset. [reprinted with permission from Adv. Mater. 17 (2005) 1972].

2.4. ZnS Nanobelts/Nanoribbons

Semiconductor nanoribbons/nanobelts have been reported as a new subgroup of 1-D nanostructures, which are distinctly different from the nanotubes and nanowires. These nanobelts/nanoribbons gained much interest after Z. L. Wang's group first reported it in 2001 [22]. The nanobelts have specific growth direction and the top, bottom and side surfaces have well-defined crystallographic facets. Whereas, for the nanowires, the side surfaces need not be well defined. For the use of these 1-D nanoforms as a building block for the fabrication of nano-scale devices such as lasers etc. nanoribbons are more suitable candidates than nanowires or nanorods because of their larger widths, which makes the nanobelts/nanoribbons structurally more controllable. ZnS nanobelts/ribbons have been fabricated by several groups [52, 54-55, 64, 89-98]. Almost all the groups have fabricated ZnS nanobelts/nanoribbons by thermal evaporation of ZnS powder at high temperature (900-1600°C) in an inert or reducing atmosphere. For the synthesis, catalytic and non-catalytic both types of approach have been applied. The catalyst-assisted approach is called vapor-liquid-solid (VLS) process whereas; non-catalytic thermal evaporation is termed as vapor-solid (VS) approach. Gong et al. [93] have used elementary Zn and S as the source materials to fabricate ZnS nanobelts by catalyst free thermal evaporation process. The basic advantage of using Zn and S powder rather than using ZnS powder is that ZnS sublimes rather slowly than those of Zn and S resulting in the faster growth of the ZnS nanobelts with Zn and S as the precursors. Recently Yao et al. [94] have fabricated ZnS nanobelts using a low temperature based solvothermal process. Fig 4a shows the SEM image of the ZnS nanobelts/nanoribbons with flat side surfaces fabricated by our group by thermal evaporation of ZnS powder at 1225 °C in flowing Ar atmosphere. Though the side facets of the nanobelts/ribbons are expected to be flat, but, saw-teeth like ZnS ribbons were also reported by Moore et al. [92]. They have suggested that the saw-teeth structure was a direct result from the difference in the surface chemical activities as determined by surface termination. The Zn-terminated surface was chemically active, while the S-terminated surface was relatively inactive, resulting in the growth of saw-teeth like structure. Figure 4b shows the TEM image of such saw-like ZnS nanobelts. These nanobelts were mainly composed of hexagonal wurtzite phase of ZnS but in some cases cubic structured zones could be visible from the HRTEM images. These cubic structured zones were found within two wurtzite regions separated by few staking faults [54-55, 91]. ZnS nanoribbons with a bicrystalline structure was also reported by few groups [54, 97-98]. TEM image of a ZnS bicrystal nanoribbon is displayed in Figure 4c along with the schematic representation of the bi-crystal nanoribbon in the inset. Figure 4d shows the HRTEM image of one representative ZnS bicrystal nanoribbon revealing two identical lattice planes on both sides of a twin boundary. The twin boundary was defined by two stacking faults (marked by drawing two arrows in the image). The SAED pattern recorded on a bicrystal nanoribbon is shown in the inset of Figure 4d revealing two sets of points for the two planes.

3. Synthesis

The basic concept behind the 1-D nanostructures formation is the controlled and restricted crystal growth. Crystal growth means, the evolution of a solid from a vapor, liquid or solid phase resulting into nucleation and subsequently crystal growth. When the concentration of the elementary species or building blocks of a solid becomes sufficient, they assembled together to form a small cluster, which act as the nucleation center or seed. For the successful fabrication of 1-D nanoforms the building blocks needs to be supplied at controlled rate so that the resulting crystal have a homogeneous composition and uniform morphology. The basic need for the growth of the 1-D nanoforms is to prevent the growth of the seeds in two dimensions. This could be achieved by using various templates, use of appropriate capping agents to kinetically control the growth rates of various facets of the seed, use of supersaturation condensation to modify the growth habit of the seed, introducing a liquid-solid interface to reduce the symmetry of a seed etc. [99]. For the synthesis ZnS 1-D nanoforms various synthetic approaches have been adopted and are discussed below.

3.1. Vapor-Solid (VS) Method

The vapor-solid (VS) growth is nothing but an evaporation condensation process. This is one kind of spontaneous growth process, where the growth of 1-D nanoforms is driven by the decrease in Gibbs free energy, which arises from either recrystalization or decrease in supersaturation. The 1-D nanoforms grown by this process are generally single crystalline with very few imperfections. The formation of the 1-D nanostructure via VS process is due to the anisotropic growth. Several mechanisms may be responsible for the anisotropic growth such as, different growth rates at different facets in a crystal, presence of imperfection or defects in a specific crystal directions or preferential accumulations of impurities on specific facets. Sears [20] was the first to explain the growth of mercury whisker by anisotropic growth in 1955.

ZnS nanobelts/nanoribbons are the commonly reported structure grown by VS method. The growth mechanism of the ZnS nanoribbons can be well understood on the basis of the vapor-solid (VS) mechanism [20-22]. In the VS method the source materials vaporize to the molecular level with stoichiometric cation-anion molecules which condense on the substrate and the molecules arranged themselves in such a way that the proper local charge balance and the structural symmetry is maintained resulting in a nucleation center. With further intake of the molecules, the surfaces that have lower energy e.g. side surfaces starts to form and the low energy surfaces tend to be flat. As the molecules have higher mobility at the higher growth temperature, the newly arrived molecules tend to diffuse to the rough growth front resulting in the increase in the surface area, which leads to the formation of the nanoribbons. The rough surface at the growth front leads to the rapid intake of the incoming molecules resulting in the rapid formation of the nanoribbons. The newly arrived molecules can stick to the rough growth front or the side surfaces. But the smooth side surfaces and the high mobilities prevent them from remaining on the surface. The molecules randomly diffuse on the surface and finally accumulate at the lower energy growth sites.

Generally, a horizontal quartz tube furnace fitted with a gas flow system and a vacuum system is used for the synthesis purpose. A simple schematic view of the experimental set up is presented in Figure 5. Jiang et al. [55] have first reported wurtzite ZnS nanobelts by hydrogen assisted thermal evaporation of ZnS powder at 1100 °C for 3-4 h in horizontal quartz tube furnace. Analytical grade ZnS powder was placed in an alumina wafer kept in the central part of the quartz tube, which was then evacuated and Ar gas pre-mixed with 5 % H_2 gas was flown through it at rate of 50 cm^3 /min. ZnS nanobelts were collected after the deposition from the inner wall of the quartz tube from a region having temperature ~ 900 °C. The synthesized products were found to have ribbon like morphology and they possessed wurtzite crystal structure. But, the HRTEM study revealed that the nanoribbons have two kinds of structure zones, one wurtzite and another cubic zone. These two zones were found coexisting in some of the nanoribbons by changing stacking sequences. Zhu et al. [90] also produced wurtzite ZnS nanobelts by thermal evaporation of ZnS powder in presence of graphite and B-N-O powder in an N_2 / H_2O flowing atmosphere. Ma et al. [91] have also reported ZnS nanobelts by simple evaporation of ZnS powder at flowing atmosphere at 1050 °C. But they could not achieved morphological control as other structures like nanocombs, windmills etc were also formed simultaneously. Recently, we have [54] reported controlled growth of ZnS nanoribbons by simple thermal evaporation of ZnS powder in pure Ar flowing condition. Also, the HRTEM observations revealed no cubic structured zone in the ZnS nanoribbons synthesized by us. We ascribed this to the fact that the growth region of the ZnS nanoribbons in our case was situated in a region with temperature ~ 1050 °C which was above the phase transition temperature of ZnS from cubic to wurtzite (1020 °C). Whereas, the growth region of the nanobelts in the previous reports were below this phase transition temperature resulting into some cubic phase formation along with the dominant wurtzite phase.

We have also reported [54] the growth of ZnS bi-crystal nanoribbons, where two single crystal planes were formed on both sides of a twin boundary along the length of the ribbon. Identical growth mechanism was responsible for the formation of single crystal and bicrystal nanoribbons at the higher Ar flow rate. The sword like end was another indication of the vapor-solid growth mechanism. The higher Ar flow rate had carried the ZnS molecular vapor to the distant end with temperature ~ 500 °C. This sudden fall of temperature of the nucleation site relative to the source region (1225 °C) might have introduced the stacking faults, which favors the bicrystal growth. The twin structure observed in the bicrystal nanoribbons (Figure 4d) provided energetically favorable sites onto which the incoming molecules and clusters stack uni-directionally into nanoribbons.

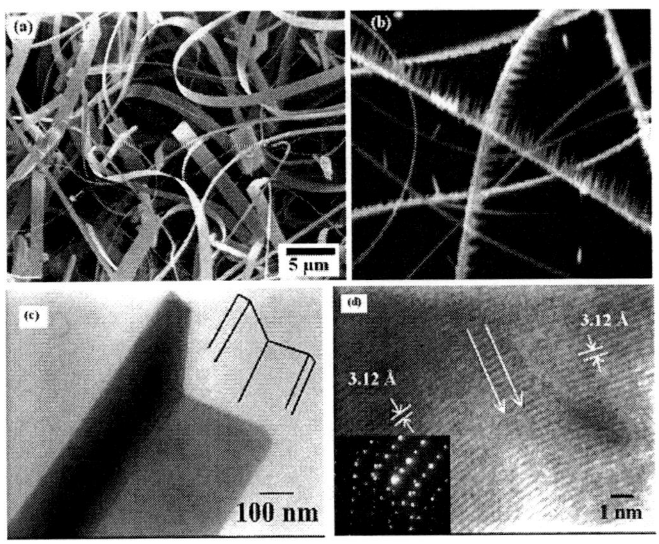

Figure 4. (a) SEM image of the smooth ZnS nanoribbons; (b) SEM image of the ZnS nanosaws; TEM (c) and HRTEM (d) image of ZnS bicrystal nanoribbons and the corresponding inset shows the schematic view of bicrystal ribbon and the SAED pattern showing two different set of points for the bicrystal ribbon respectively [Figure b reprinted with permission from Chem. Phys. Lett. 385 (2004) 8 and Fig c-d reprinted with permission from Chem. Phys. Lett. (2005) Doi:10.1016/j.cplett.2005.08.021].

3.2. Vapor-Liquid-Solid (VLS) Method

For the VLS process similar experimental set up as that of VS process are used. In the VLS process, a second material commonly known as the catalyst is purposefully introduced to direct and confine the crystal growth onto a specific orientation and within a confined area. The catalyst forms a liquid droplet by itself or by alloying with the growth material during the growth, which acts as the trap of growth species. Enriched growth materials in the catalyst droplets subsequently precipitate upon super saturation at the surface of the substrate resulting in the 1-D growth. Wagner et al. [23-26] first proposed the VLS theory in 1964 to explain the experimental results and observations in the growth of Si nanowire. So in the VLS process, the presence of the catalyst is always necessary and a liquid droplet is always found at the tip of the nanowire.

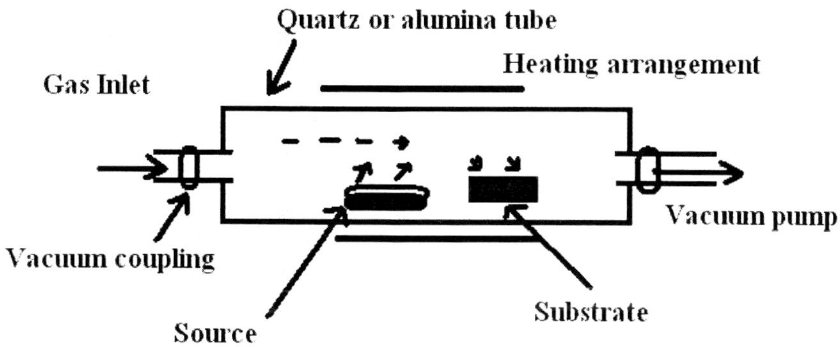

Figure 5. Schematic view of the deposition system for the VS and VLS growth of 1-D ZnS nanoforms.

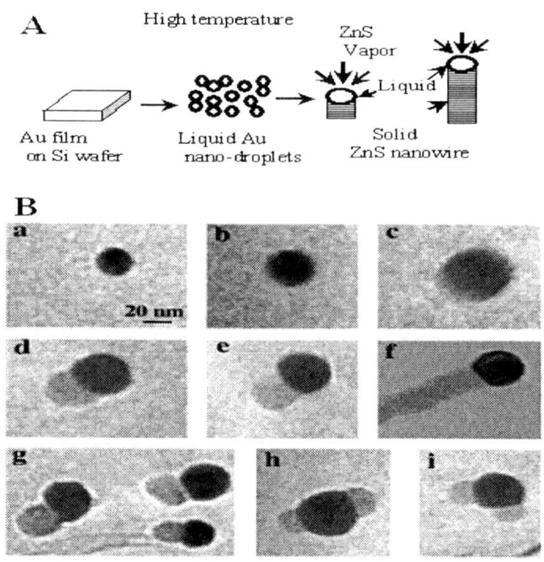

Figure 6. (A) Schematic view of the VLS process, (B) In situ TEM images recorded during the Ge nanowire growth revealing the role of the catalyst particle in the nanowire growth by VLS process [Figure B reprinted with permission from J. Am. Chem. Soc. 123 (2001) 3165].

Schematic representation of the VLS model is presented in Figure 6 on the basis of the in situ TEM observations of the Ge nanowire growth by VLS process [25].

For the growth of ZnS 1-D nanostructures generally gold (Au) was used as the catalyst and ZnS powder as the source material. A thin layer of Au film deposited on suitable substrates broke up at high temperature to form liquid Au nanodroplets on the surface. These liquid Au droplets at high temperature absorbed ZnS vapor and Au-ZnS nanocomposite alloy particles were formed. On super-saturation, pure ZnS nanostructures were originated from these composite particles. Depending on the experimental parameters such as, temperature and flow rate of the Ar gas, these liquid composite particles took different shapes and different morphologies of the ZnS nanostructures were originated from them. There are various reports on the fabrication of ZnS nanowire [52, 77-81] and nanobelt/nanoribbon [52, 89, 97] by VLS process. Generally higher temperature was needed for the fabrication of ZnS nanopbelts/nanoribbons than those of nanowires by the VLS process. We have reported [52] that ZnS nanowires were formed at 850 °C whereas; ZnS nanoribbons were achieved at 950 °C via VLS process. Figure 7 shows the SEM images of ZnS nanowires at 900 °C at various stage of their growth by VLS technique [78].

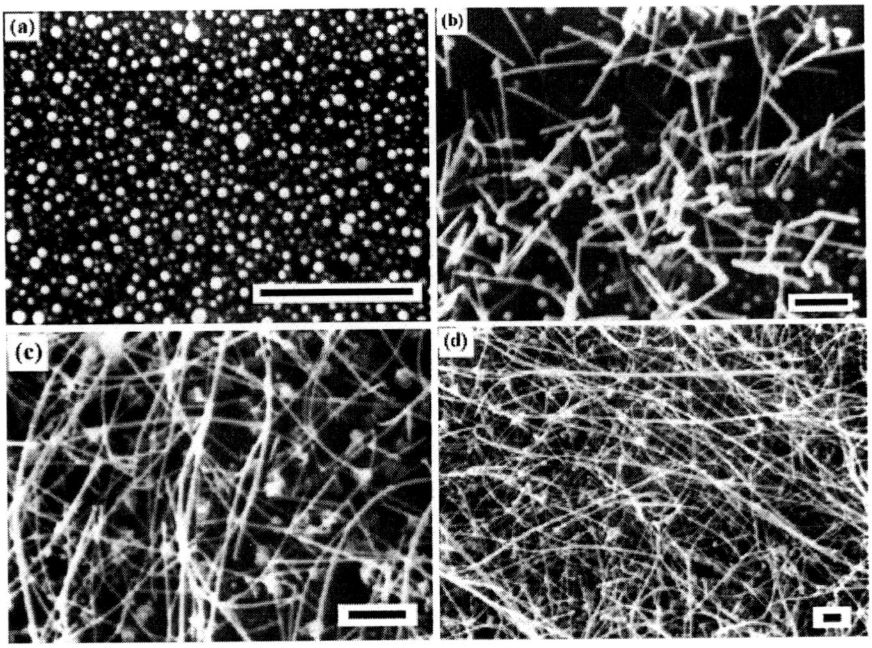

Figure 7. SEM images showing different stages of the ZnS nanowire growth after (a) 5 mins, (b) 15 mins, (c) 20 mins and (d) 30 mins. The scale bars in all these images represent 500 nm [reprinted with permission from Nanotechnology 16 (2005) 737].

We have also [52] shown that the initial shape of the ZnS nanostructures could determine the morphology of the final ZnS nanoforms and we postulated on the basis of experimental observation that in some favorable experimental condition the catalyst particle could be elongated to form cylindrical Au particles, which in turn induces the ZnS nanoribbon growth. In our recent work [53] we have proposed that the formation of ZnS nanobelts/nanoribbons having width greater than 500 nm could not be attributed to the VLS process alone. Instead, we propose that the ZnS nanoribbons with larger width were formed by a combined effect of VLS and VS mechanism. Unlike other groups, instead of using ZnS powder as the source material, Barrelet et al. [100] have used a single source molecular precursor $Zn(S_2CNEt_2)_2$ for the fabrication of ZnS nanowire by the VLS process. Laser assisted catalytic growth process were also reported [79-80, 100-101] for the ZnS nanowire formation, where a target of mixed powder of ZnS and Au was ablated by a laser source. The nanowires obtained by this process were smooth and uniform throughout their length and they were perfectly single crystalline in nature. Both cubic [76] and hexagonal [52, 77-78] nanowire could be achieved by varying the experimental parameters like temperature of the source and the substrate, position of the substrate etc. ZnS nanoribbons/nanobelts were single crystalline, and possessed well-defined side as well as end surfaces. The lengths of the nanowires and nanoribbons could be of the order of few hundred micrometers.

3.3. Template Assisted Growth

Template based synthesis of nanostructures of different materials is a very general method used to fabricate nanorods, nanowires and nanotubes. Various templates with nanosized channels have been explored for the template-assisted growth of 1-D nanoforms. The most commonly used and commercially available templates are anodized alumina membrane, mesoporous silica, carbon nanotube etc. Liquid crystal template [74] and micelle templates [75] were used to fabricate ZnS nanowire at or near room temperature. Porous templates like aniodic alumina were generally used to fabricate nanowire arrays by electrochemical process, sol-gel deposition and other chemical routes. Ding et al. [84] have utilized Au particle filled aniodic alumina membrane as the substrate to synthesize aligned ZnS nanowire arrays by VLS approach at 1100 °C.

3.4. Chemical Conversion Method

1-D nanoforms can also be prepared by using a consumable template i.e. first nanowires or nanoribbons of one phase of the material is formed which are subsequently converted to another phase by chemical reactions. ZnS nanotubes were produced by chemical conversion of ZnO nanocolumns [86] and ZnO nanoribbons [87] with H_2S. Dloczik et al. [86] converted an existing ZnO column arrays by passing H_2S through it at 400 °C, when a thin layer of ZnS was formed on the ZnO core, which was then etched by using H_2SO_4 solution. Being more etch resistant, the ZnS shell remained intact giving rise to a tubular ZnS with ~ 15 nm wall thickness. They have also obtained ZnS nanotubes by passing S vapor at 450 °C. Wang et al. [87] have converted ZnO nanobelts to ZnS nanotubes by adding H_2S solution to an ethanol solution of ZnO nanobelts and subsequently kept it at room temperature for some time for complete conversion. The basic principle behind this kind of conversion can be expressed by the following equation.

$$ZnO + H_2S \rightarrow ZnS + H_2O$$

3.5. Solvothermal Process

Solvothermal synthesis utilizes a solvent under pressure and temperature above its critical point to increase the solubility of the solid reactants and speed up the reactions. In the typical procedure, the precursors and a solvent, possibly an amine capable of regulating or templating the crystal growth to generate 1-D nanoforms. The mixture is placed in an autoclave, which is placed at an elevated temperature and pressure to speed up the reaction to generate 1-D nanoforms. The major advantage of this approach is that most of the materials can be made soluble in a solvent by heating and pressurizing it close to the critical point. The products obtained by this approach are generally in the form of powder. Ethylenediamine (en) was found to promote one-dimensional growth of CdS by solvothermal process due to its chelating property [29-30]. But ZnS nanorods could not be synthesized with ethylenediamine following the same experimental conditions used for the CdS nanorods. Instead, flakes like two-dimensional complex $ZnS.(NH_2CH_2CH_2NH_2)_{0.5}$ were produced [31]. The differences in the coordination ability of ethylenediamine with Zn^{2+} and Cd^{2+} ions and some other kinetic factors related with the solvothermal growth might be the reason behind the difference [31]. However, by annealing this complex precursor in Ar atmosphere at high temperature or by treating the complex hydrothermally, a mixture of pure hexagonal ZnS nanorods and particles could be obtained. It was observed that by mixing water with the ethylenediamine (en), uniform hexagonal ZnS nanorods were produced at 200 °C [68-69]. The hexagonal phase of ZnS is stable at temperature higher than 1020 °C while the cubic phase is stable below this temperature. But the presence of the chelating solvent at high pressure might have brought down the free energy difference between the hexagonal and cubic phases of ZnS resulting in the formation of wurtzite ZnS at much lower temperature. Recently, ZnS nanoribbons were also fabricated by solvothermal process [94] in a mixed solvent of water and diethylenetriamine with a volume ratio 1:2.5.

4. CHARACTERIZATION

ZnS nanoforms were characterized by different techniques to explore their physical and chemical properties. Crystal structure and phase of the products are determined from the powder X-ray diffraction (XRD) studies. Chemical purity and stoichiometry of the ZnS nanoforms can be tested from the energy dispersive analysis of X-ray (EDAX) and X-ray photoelectron spectroscopy (XPS). Morphology of the ZnS nanoforms is determined from the scanning electron microscopy (SEM) and transmission electron microscopy (TEM). The HRTEM of the nanoforms helps to determine the crystal structure of the individual nanoforms. Various chemical characterizations are also performed on the ZnS nanoforms to determine the surface and interior atoms and compounds and their spatial distribution. Optical absorption spectroscopy, photoluminescence spectra (PL), Raman spectroscopy, infrared spectroscopy etc. are the most popular characterization techniques to explore the chemical properties of the ZnS nanoforms.

4.1. Phase and Crystal Structural Identification By XRD

X-ray diffraction XRD is a very important experimental technique that has been used to address all issues related to the crystal structure of samples, including lattice parameters, identification of unknown materials, orientations of single crystals, preferred orientations of polycrystals, defects, stress etc. Also, XRD is used to identify the specimen's crystalline phases. In case of ZnS nanostructures the phase of the as synthesized 1-D nanoforms is identified by XRD i.e. whether they are cubic or hexagonal. Figures 8a-b show the representative XRD patterns of the cubic and hexagonal ZnS. In most of the reports on ZnS 1-D nanoforms XRD was mainly used for the identification of their phases. We have observed some difference in the XRD pattern of the ZnS nanoribbons produced by VLS and VS methods. Figure 8c shows that the relative intensities of the different diffraction peaks obtained from the ZnS nanoribbons produced by VLS and VS approaches are not identical. This indicates that the growth rates of the different crystal planes of the wurtzite ZnS nanoribbons are very much dependent on the experimental condition i.e. the environment at which these nanoforms are formed.

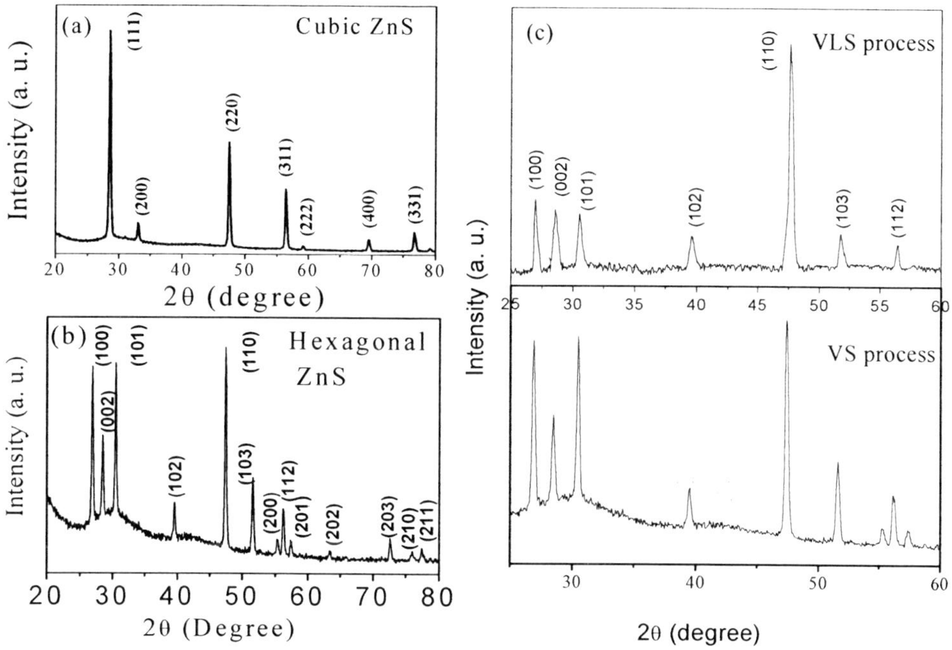

Figure 8. XRD pattern of the different polymorphs of ZnS: (a) cubic ZnS with sphalerite phase and (b) hexagonal ZnS with wurtzite phase and (c) XRD pattern of the wurtzite ZnS nanoribbons synthesized by VLS and VS techniques.

4.2. Chemical Purity and Stoichiometry Test by EDAX

Energy dispersive analysis of X-rays (EDAX) is one kind of electron spectroscopy, which relies on the unique energy levels of the emission of photons (X-rays) or electrons ejected from the atom under investigation. Elementary analysis by EDAX is very important for ZnS nanoforms as the presence of impurity or foreign elements like Mn, Cu etc can change the emission from the ZnS nanoforms drastically. Figure 9a shows the energy dispersive analysis of X-ray (EDAX) pattern recorded over a single ZnS nanoribbon revealing the presence of elementary Zn and S. The atomic ratio of the constituent atoms i.e. the stoichiometry of the sample can also be determined from the compositional analysis of the EDAX spectrum. The EDAX was also utilized to test the identity of the spherical particle obtained at the tip of the 1-D ZnS nanoforms produced by VLS approach. Figure 9b shows the EDAX spectra of the Mn-incorporated ZnS nanorods. Thus the role of the EDAX is not only to test the chemical composition of the ZnS nanoforms but also to investigate the reason behind the visible emission from the ZnS nanoforms. The main drawback of this technique is that it cannot trace the presence of elements having very low concentration.

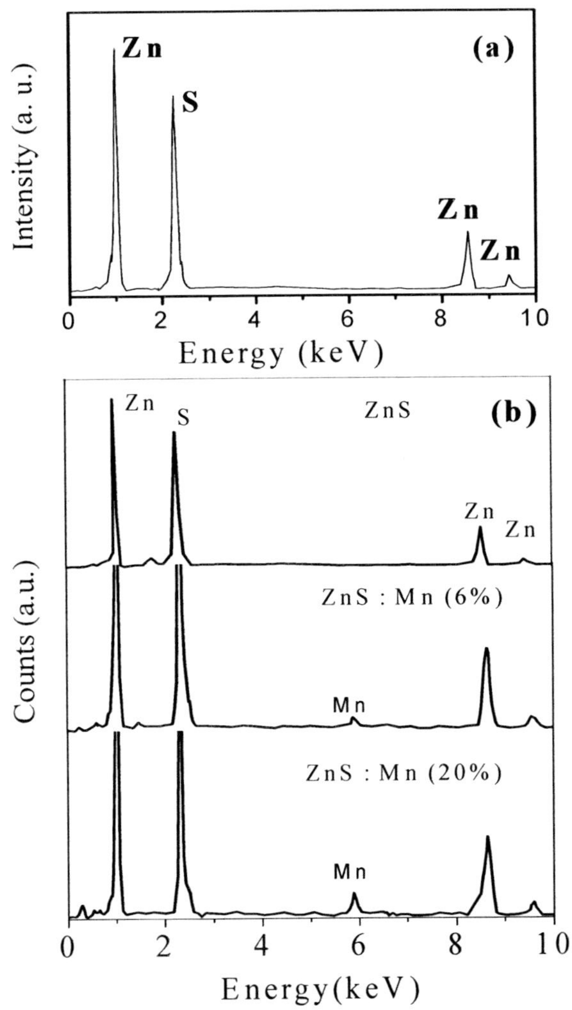

Figure 9. (a) EDAX spectrum of a representative ZnS 1-D nanoforms and (b) EDAX spectra of the $Mn_xZn_{1-x}S$ nanorods.

4.3. XPS Study

Working principle of the XPS spectroscopy is somewhat similar to EDAX. The only difference is that in XPS, relatively low-energy X-rays are used to eject the electrons from the atom via the photoelectric effect. In XPS study the nature of bonding between the elements could also be distinguished. In addition to its ability to test the chemical composition, XPS pattern was successfully utilized to investigate the nature of chemical bonding in the ZnS 1-D nanoforms and their impact on the properties of the ZnS nanoforms. Ye et al. [96] have utilized the XPS to establish the origin of the green emission from their ZnS nanobelts. From the XPS spectrum (Figure 10) they have identified the presence of the elementary S along with the S^{2-} species in the ZnS nanobelts and they have attributed this elementary S as the reason behind the green emission from their sample.

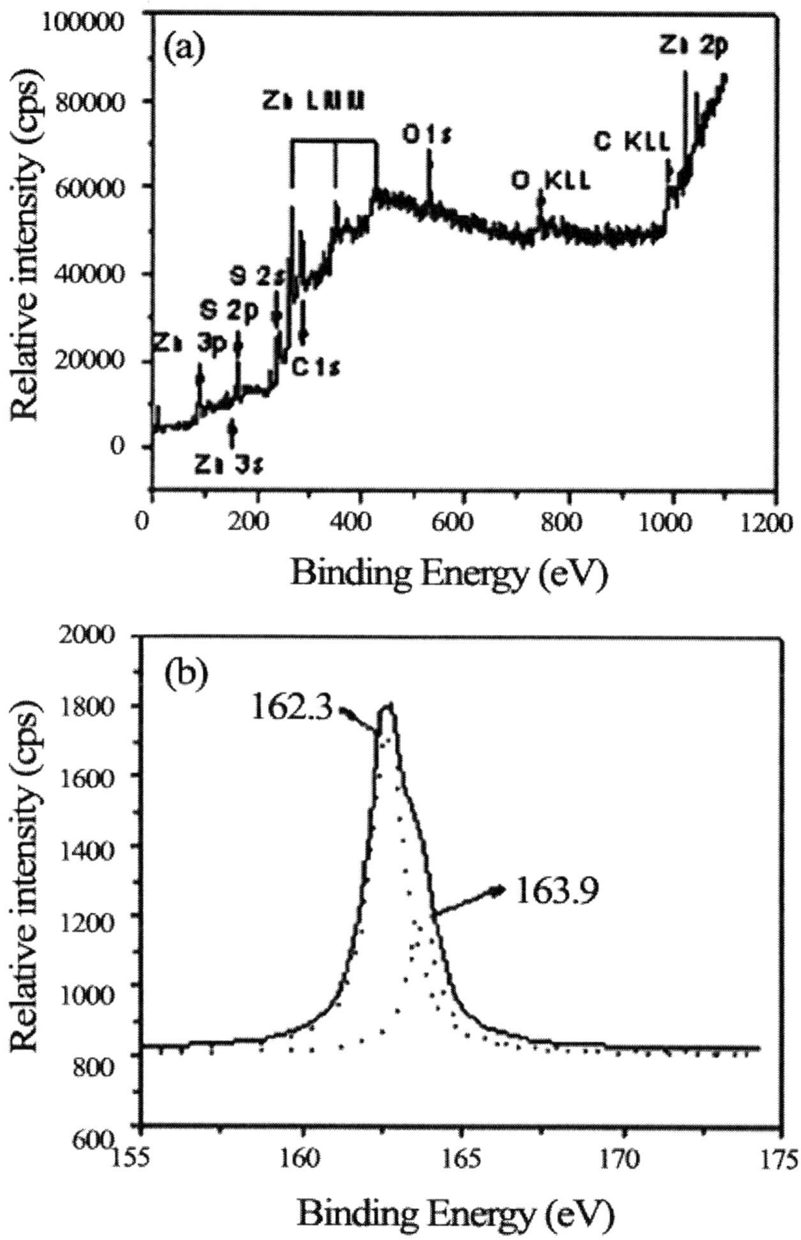

Figure 10. XPS spectra recorded over the ZnS nanobelts: (a) survey spectrum of ZnS nanobelt, and (b) S (2p) spectrum [reprinted with permission from Appl. Phys. Lett. 85 (2004) 3035].

4.4. Morphological Identification by SEM Study

Scanning electron microscopy (SEM) is one of the most widely used techniques to characterize nanomaterials. The resolution of the SEM approaches a few nanometers. The biggest advantage of the SEM is that one can investigate the morphology of the sample without disturbing the original nature of the products. Generally, ZnS 1-D nanoforms are synthesized over a flat surface or in the form of powder. In either case the as grown sample can be transferred to the SEM chamber without disturbing the original nature of the sample and this is most important to investigate the growth nature of the sample. From the SEM image along with the morphological identification of the products one can also deduce the size distribution of the nanoforms. Figures 2a,c-d, 4a and 7 show the SEM image of the different forms of ZnS 1-D nanoforms. The SEM images are widely utilized to distinguish between the nanorods, nanowires, nanobelts/nanoribbons and nanotubes. In addition to its unique ability to identify the morphology of the ZnS nanoforms, the SEM images can also be utilized to monitor the growth of ZnS nanoforms by taking images at different stage of their formation. We have utilized SEM images [78] to investigate the role of VLS process in the formation of ZnS nanowires by recording the images (Figure 7) after different deposition periods.

4.5. Morphology and Crystal Structure Identification by TEM and HRTEM Studies

For the TEM study the sample has to be (after it is scrapped from the substrate if the nanoforms grows over a substrate) dispersed ultrasonically in some volatile solvent and subsequently deposited on a thin layer of carbon coated Cu grid. Figure 11a shows the TEM image of a ZnS nanowire and nanoribbon.

The ripples like contrast observed in the TEM images were originated from the strains in the sample due to the bending of the structures. Selected area electron diffraction (SAED) pattern (inset of Figure 11a) were utilized to identify the crystal structureandthe growth direction of the ZnS 1-D nanoforms. The HRTEM images are more important as they can be utilized to identify the crystallographic structure of the individual nanofrom or even their selected area. Figure 11b and 11c show the HRTEM image of a cubic ZnS nanowire and wurtzite ZnS nanoribbon respectively. Figure 11b also reveals the presence of the catalyst particle at the tip of the ZnS nanowire, which was the essential evidence of the VLS growth mechanism. Also, the presence of defects in the samples could be identified from HRTEM images. From the HRTEM analysis of the ZnS 1-D nanoforms reported by different groups, it was observed that almost all the 1-D ZnS nanoforms were single crystalline in nature. Figure 11d shows the HRTEM image of a ZnS nanoribbon synthesized by us where cubic structured zones are visible within the hexagonal wurtzite zones separated by few staking faults. From the measured spacing of the lattice fringes and the position of the lattice image, growth direction of the 1-D nanoforms as well as their side facets could be identified accurately [102-103].

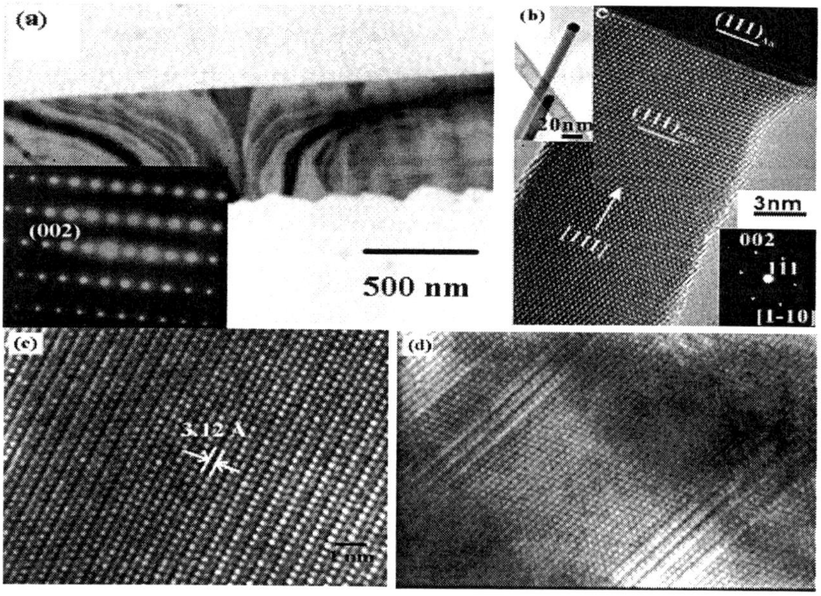

Figure 11. (a) TEM image of a ZNS nanoribbon showing the presence of ripple like contrast originated from the strains developed due to the bending of the ZnS nanoribbons, (b) HRTEM image of a cubic ZnS nanowire showing the presence of the Au catalyst at its tip indicating the role of VLS process, (c) HRTEM image of a ZnS nanoribbons showing the hexagonal crystal structure, and (d) HRTEM image showing the presence of cubic structured zone within wurtzite zone separated by few staking faults. [Figure b reprinted with permission from Chem. Phys. Lett. 382 (2003) 434].

4.6. Optical Absorption/Transmission Spectroscopy

For the optical absorption/transmission spectroscopic measurements ZnS 1-D nanoforms were ultrasonically dispersed in ethanol and subsequently the optical spectra were measured at room temperature [54, 69]. The optical absorption spectrum of the single crystalline ZnS nanoribbon sample is shown in Figure 12a. The absorption study revealed that the nanoribbons were transparent in the visible region. The absorption edge determined from the peak minimum of the first derivative of the absorption plot was at 330 nm (3.76 eV), which matches well with the standard bulk value of band-gap of the wurtzite ZnS. Also, the steep rise of the absorption edge is an indication of the defect free structure of the nanoribbons. If m represents the nature of optical transition in a semiconducting material, the absorption coefficient (α) may be expressed as [104].

$$\alpha = (A/h\nu)(h\nu - E_g)^m \quad (1)$$

Where, A is an optical constant, E_g is the band gap of the semiconductor corresponding to the particular value of m. The parameter m, representing the nature of optical transition may have certain specific values such as, 1/2, 2, 3/2 or 3 corresponding to the allowed direct, allowed indirect, forbidden direct and forbidden indirect transitions respectively.

From equation (1) we have:

$$d(\ln(\alpha h\nu)) / d(h\nu) = m / (h\nu - E_g) \quad (2)$$

So, the plot of d(ln(αhν)) / d(hν) vs. hν will show a discontinuity at $h\nu-E_g=0$. The Figure 12b shows the above plot and the corresponding value of the E_g was 3.75 eV. Now, instead of pre-assuming the nature of optical transition, it may be determined from the ln(αhν) vs. ln(hν-E_g) plot (inset of Figure 5b) with E_g = 3.76 eV. The slope of this plot gives m= 0.51 corresponding to the allowed direct transition.

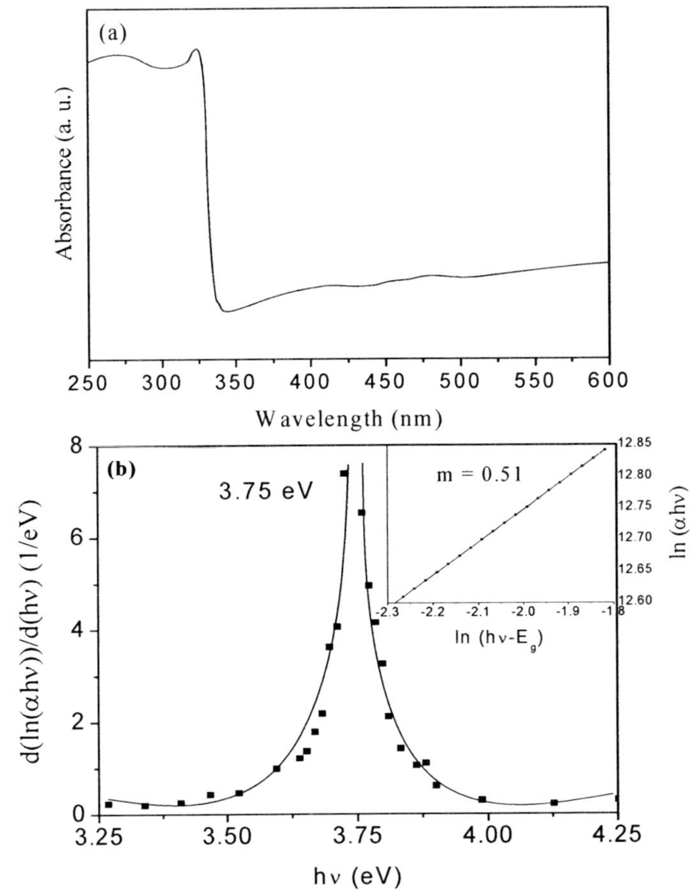

Figure 12. (a) The absorbance spectrum recorded at room temperature, (b) d(ln(αhν)) / d(hν) vs. hν plot with the discontinuity position assigned to the band gap value and inset shows the ln(αhν) vs. ln(hν-E_g) plot for determining the nature of optical transition (m) [reprinted with permission from Chem. Phys. Lett. (2005) Doi:10.1016/j.cplett.2005.08.021].

4.7. Photoluminescence (PL) Study

In PL, a material gains energy by absorbing light at some wavelength by promoting an electron from a low to higher energy level. After a characteristic lifetime in the excited state, electrons will return to the ground state by radiative transition. ZnS is the most popular semiconductor well known for its luminescent properties. Luminescence in the wide range of UV to visible range is observed from the ZnS nanostructures. In this section we will discuss only the luminescence from the 1-D nanostructures. PL study of the ZnS 1-D nanoforms could be carried out either in solid state (for the samples deposited on a substrate) or by dispersing the ZnS nanoforms into a spectroscopic grade alcohol.

Two kinds of emissions are observed from a semiconductor nanoforms - excitonic and trapped luminescence [105-106]. The excitonic emission is sharp and is located near the absorption of the particles, while the trapped luminescence luminescence arising from different kinds of defects or impurity states is broad and stokes-shifted. Only the trapped luminescence is observed in pure ZnS nanoparticles [105-106] and in 1-D nanoforms, reported by most of the groups [52, 54, 77-78, 81, 87-88, 90, 93, 95-97]. Xiong et al. [79] have recently reported band edge emissions from the ZnS nanowires at room temperature. They have correlated the PL and optical absorption spectra (Figure 13) of the ZnS nanowires to explain the origin of the different emissions.

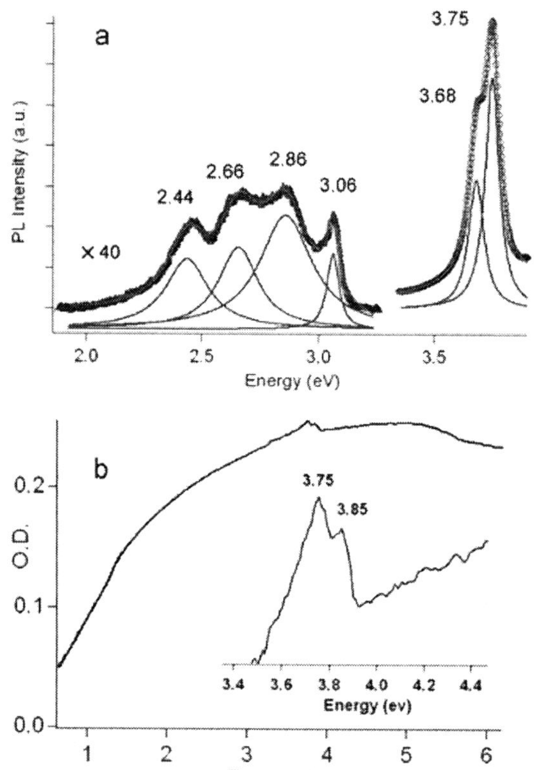

Figure 13. (a) PL spectrum of ZnS nanowires taken at room temperature excited by a 266 nm UV laser. Two strong emission bands near bandedge and four weak defect related luminescence bands were observed. (b) Optical density plot versus photon energy of ZnS nanowires. The inset shows the zoomed view of the vicinity of fundamental absorption edge [reprinted with permission from Nanoletters 4 (204) 1663].

They have observed a strong emission doublet at 3.68 eV and 3.75 eV along with few weaker defect related peaks at 2.44, 2.66, 2.86 and 3.06 eV. In case of wurtzite ZnS with direct bandgap ~ 3.75 eV, the top most valence band is split into three bands due to the crystal field and spin-orbit coupling. Exciton states associated with electron-hole pairs in each case of these valence bands involved are usually denoted as A, B, C excitons [107]. The spin-orbit coupling and crystal-field splitting energy given in energy ref [107] are 89 and 57 meV, respectively. The three features of 3.68, 3.75 and 3.85 eV observed in the PL and OA are assigned to the A, B, C excitons. This kind of bandedge Pl from ZnS was observed for very high quality single crystals [56]. Defect related PL peaks are reported over a wide region ranging from 400 – 600 nm. The blue emission peak at ~ 400 nm was attributed to the sulfur vacancy and interstitial lattice defects in the ZnS nanoforms [52, 78, 95]. The blue emission peaks within the region 440-460 nm are attributed to the trapped luminescence arising from the surface states [52,77-79]. However, various explanations were presented to reveal the origin of the green emission from the ZnS 1-D nanoforms. Few groups [77, 79] believed that the Au used as the catalyst in the VLS growth, might have present in the ZnS lattice as an impurity, which acts as the luminescence center for the green emission from the ZnS nanowires. There are several reports that the self activated defect centers [81], structural defects like point defects [97] or zinc vacancy related defect centers [54, 108] are the origin of the green emission from these 1-D ZnS nanoforms. Ye et al. [96] have reported that the presence of the elementary sulfur species in the ZnS nanobelts could also induces the green emission from the nanoforms. Also the presence of the foreign impurities like Cu^{2+}, Mn^{2+} or rear earth ions even in minute quantity could introduces green emission from the ZnS nanoforms [109-117].

4.8. Raman Spectroscopic Studies

Raman Spectroscopy is very sensitive to lengths, strengths and arrangement of chemical bonds in a material, but less sensitive to chemical composition. Wurtzite ZnS has the space group C_{6v}^4 ($C6_3mc$) [80] with two formula units per primitive cell and with all atoms occupying C_{3v} sites. Group theory predicts the following q=0 phonon symmetries: $\chi_{atom} = 1A_1 + 2B_1 + 1E_1 + 2E_2$, while the A_1 and E_1 symmetry phonons are both Raman- and IR active, the two E_2 pair of modes are only Raman-active, and two B_1 modes are optically silent [118]. There are reports on the Raman scattering study of bulk ZnS [119-121]. But only very recently, Xiong et al. [80] have reported Raman scattering from rectangular cross-sectional ZnS nanowires. They have reported strong first order longitudinal optical (LO), transverse optical (TO), and surface optical (SO) modes and second order combinations and overtones. Figure 14 shows the Raman spectra recorded over the rectangular cross-sectional ZnS nanowires at room temperature. In analogy with the bulk ZnS Raman peaks [118-121] the different Raman peaks were assigned to different phonon modes. The strong peak at 346 cm^{-1} was assigned to the A_1 and E_1 symmetry of LO modes. At lower frequency resolvable doublet at 269 cm^{-1} and 282 cm^{-1} were attributed to (A_1 and E_1) and E_2 TO modes respectively. The broad feature at 216 cm^{-1} was assigned to the LA mode. The continuum scattering within 350-450 cm^{-1} was assigned to the 2-phonon density of states (DOS). The Raman bands between 600-700 cm^{-1} were assigned to the combination and overtones. The peak at 335 cm^{-1} was assigned to the surface optical (SO) mode the line shape of which was sensitive to the real part of the dielectric function of the dielectric medium in contact with the surface.

5. DOPED-ZNS 1-D NANOFORMS

Zinc Sulfide, a II-VI group compound semiconductor is particularly suitable for use as host material for a large variety of dopants because of its wide band gap (3.75 eV).

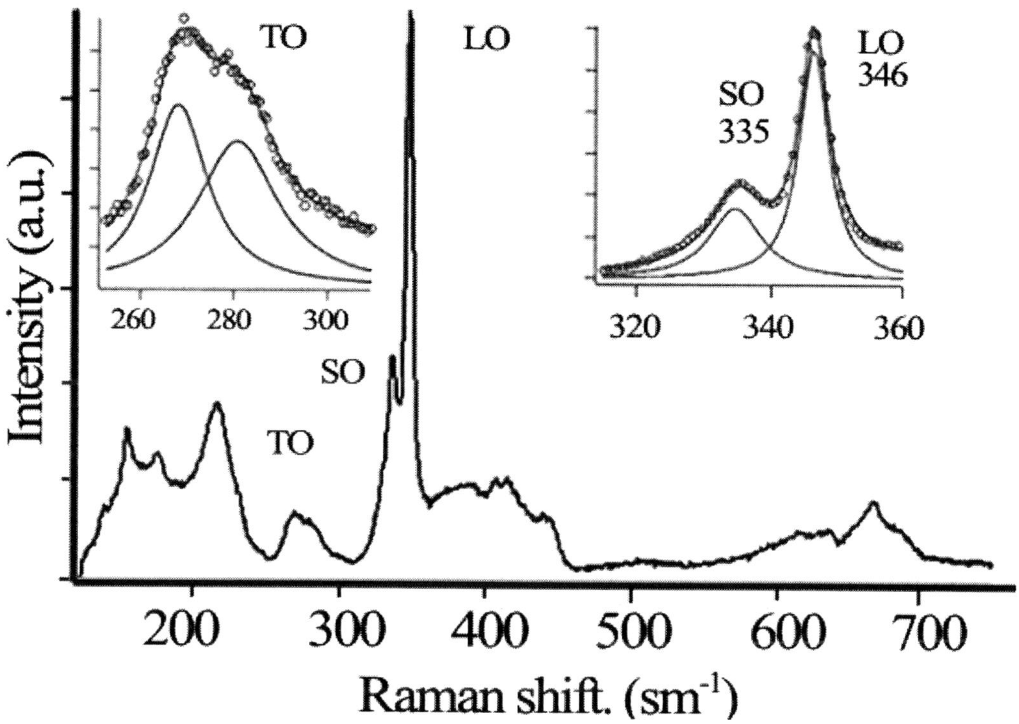

Figure 14. Raman spectrum of ZnS nanowires collected in air. Two insets show Lorentzian line shape analysis of LO (346 cm^{-1}) and SO (335 cm^{-1}) modes and TO doublet (269, 282 cm^{-1}) [reprinted with permission from Nanoletters 4 (2004) 1991].

In fact ZnS has found special importance in electroluminescent devices, lasers, flat panel displays when doped with divalent manganese ions [59, 61-62, 123-125]. Manganese usually occupies substitutionall sites in the Zn lattice as a divalent ion and the excitation and decay of this ion produces an orange luminescence at approximately 590 nm [109, 114]. This emission peak is generally associated with a transition between $^4T_1$ and $^6A_1$ energy levels. Other dopants such as Cu^{2+} [116, 126], Ni^{2+} [127], Cu-Al co-doped [128] and rear earth elements [129] were used to tailor the photoluminescence properties of the ZnS nanocrystals. 1-D ZnS doped with Mn^{2+} [130-133] and Cu^{2+} [131, 133] were also reported. All these processes are based on the thermal evaporation of the precursors at high temperature. Geng et al. [130] have fabricated Mn doped ZnS nanobelts by thermal evaporation of zinc acetylacetonate and manganese acetylacetonate in proper molar ratio in an H_2S atmosphere at 900 C. those doped ZnS nanobelts showed characteristic defect related emission at 450 nm along with the orange emission associated with the Mn^{2+} d-level transition from $^4T_1$ to $^6A_1$. Yuan et al [132] have fabricated Mn-doped ZnS nanowires by thermal evaporation of ZnS and $MnCl_2$ powders onto a Au coated Si substrate at high temperature in an Ar atmosphere. They have reported defect related emission at ~ 503 nm and Mn^{2+} related emissions at 570 nm. The defect related emissions was stronger than that due to ionic level transition at Mn^{2+}. Zhu et al. [131] have synthesized Mn and Cu doped ZnS nanobelt arrays by thermal evaporation of ZnS and MnO_2 or CuO powders in a mixed gas flow (N_2, CO and H_2) at high temperature. These nanobelts arrays showed mainly orange emission for the Mn-incorporated ZnS samles and green emission at 520 nm for the Cu-doped samples. Whereas, the un-doped ZnS samples sowed blue emissions at ~ 450 nm associated with the defect related emissions. Recently, we have synthesized Mn-incorporated ZnS nanorods by a simple solvothermal approach [134]. The sample without Mn corresponds to hexagonal ZnS. Mn^{+2} ion incorporation into the ZnS nanorods resulted into morphological as well as structural changes. The diameter of the ZnS nanorods increased gradually from 12-15 nm to ~ 100 nm for 0 to 20 % Mn incorporations. Mn incorporation introduces a phase transformation from wurtzite to cubic structure. Three different emissions in the blue (400 nm), green (500 nm) and orange (585 nm) regions were visible in the room temperature photoluminescence spectra (Figure 16a).

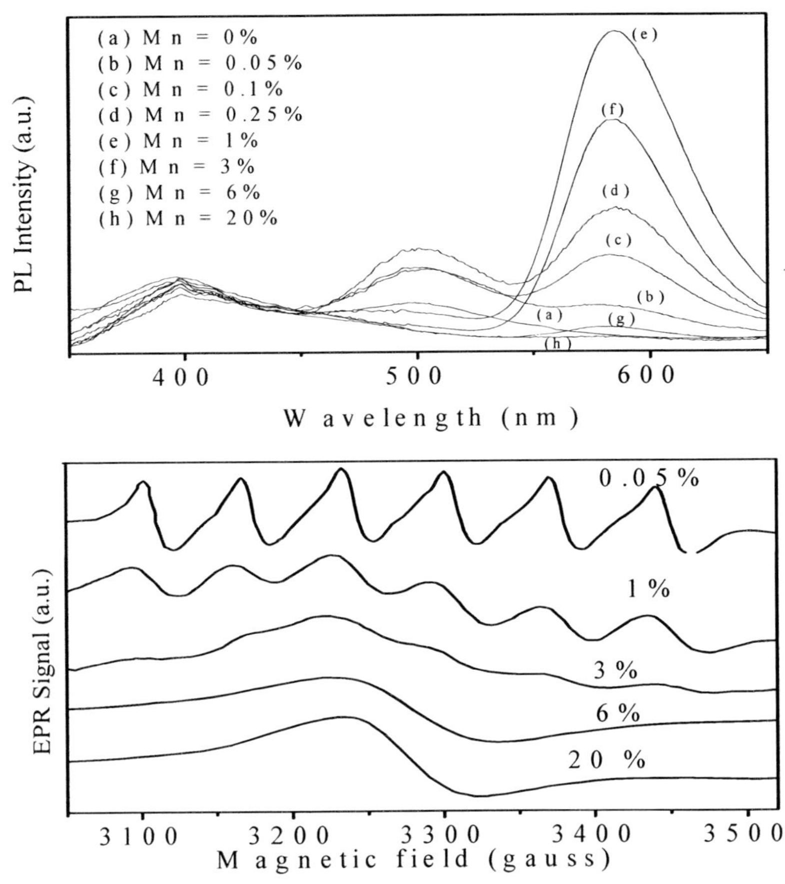

Figure 15. (a) Room-temperature of the $Mn_xZn_{1-x}S$ nanorods and (b) EPR spectra of the $Mn_xZn_{1-x}S$ nanorods [reprinted with permission from J. Phys. Chem. B (2005) DOI: 10.1021/jp053138i].

These blue, green and orange emissions were attributed to the surface defects, zinc vacancy related self activated defect sites and $^4T_1 - {}^6A_1$ transition of the Mn^{2+} ions respectively. The orange emission first increases with the Mn incorporation up to 1 mol % Mn and the decreased gradually for further addition of Mn. Quenching of the orange emission at higher Mn concentrations was attributed to Mn-Mn dipolar interaction. The isolated Mn^{2+} ion and Mn cluster was also detected from the EPR measurement. EPR (Figure 16b) results support the presence of magnetic dipole interaction for the Mn incorporated ZnS nanorods with higher Mn concentration. Also the basic advantage of our approach is that bulk amount of the nanorods can be prepared at low temperature without using any kind of template.

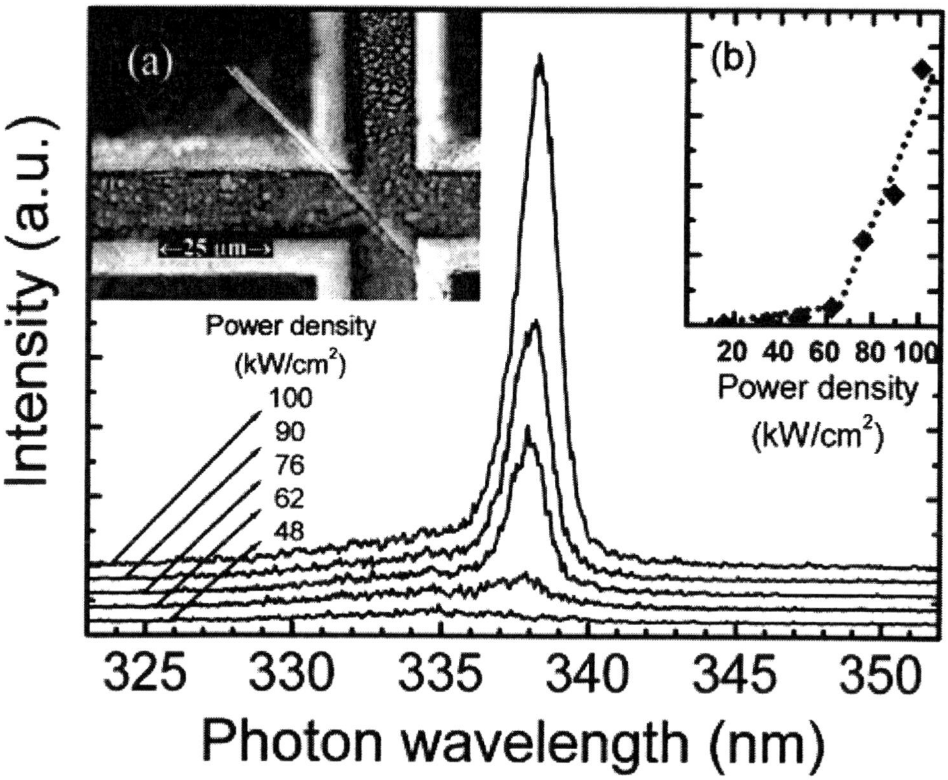

Figure 16. PL of a single ZnS nanoribbon excited by a 266 nm laser with different power densities. (a) The single ZnS nanoribbon dispersed on a TEM grid used for the measurements; the scale is 5 μm. (b) PL intensity versus input power density. Lasing starts at ~60 kW/cm^2 [reprinted with permission from Appl. Phys. Lett. 84 (2004) 1189].

6. APPLICATIONS

Althouhg, ZnS nanocrystals have been utilized for various types of application [56-60], the 1-D ZnS nanostructures are of special importance for laser fabrication. This is because like ZnO, ZnS is also a high band gap (3.75eV) compound semiconductor with exciton binding energy 40 meV, which is much higher than GaN (25 meV) and ZnSe (22 meV). This information indicates that the excitons in ZnS are thermally stable at room temperature. Nanowires and nanoribbons with flat ends are suitable candidate for laser fabrication. Lasing has been observed from the ZnS nanowire arrays [84] and single nanoribbons [64] at room temperature. Figure 15 shows the PL spectra recorded on a single ZnS nanoribbon showing the lasing from the ZnS nanoribbons. The measured threshold pumping power of the room temperature ZnS nanoribbon laser is 45 kW/cm^2, which is comparable to those of the ZnO based lasers. This low threshold and availability of good quality of ZnS nanoribbons could lead to their application in diodes and electrical pumping.

Another application of ZnS 1-D nanoforms is that the ZnS nanowires can be used as a removable template to fabricate other nanoforms. Single-crystalline Si tubular structure was formed by using ZnS nanowires as a template [122].

7. Conclusions

Different aspects of the ZnS 1-D nanoforms such as nanorods, nanowires, nanotubes, nanobelts or nanoribbons are discussed here. Various synthetic procedures ranging from low temperature based chemical approach to high temperature based thermal evaporation have been utilized to achieve different 1-D ZnS nanoforms. The ZnS 1-D nanoforms were found have different crystallographic structures. The synthesized ZnS nanoforms were single crystalline but these nanoforms could be defect free or could contain staking faults depending on the synthesis technique and conditions. These 1-D ZnS nanoforms are excellent visible light emitters and their emission could be tailored by introducing dopants like Mn^{2+} or Cu^{2+} ions in to the ZnS lattice. These emission properties of the ZnS nanoforms could be ideal for the fabrication of various optoelectronic devices like flat panel displays, LED etc. Room temperature UV lasing has been demonstrated from the ZnS nanowires and nanoribbons. So, the further investigation of the 1-D ZnS nanoforms could lead to the fabrication of technologically important nano-scale devices such as LEDs, lasers etc based on these ZnS nanoforms.

References

[1] P. Alvisatos, *Science* **271** (1996) 933.

[2] P.l Alivisatos, P. F. Barbara, A. W. Castleman, J. Chang, D. A. Dixon, M. L. Klein, G. L. McLendon, J. S. Miller, M. A. Ratner, P. J. Rossky, S. I. Stupp, and M. E. Thompson, *Adv. Mater.* **10** (1998) 1297.

[3] Handbook of Nanostructured Materials and Nanotechnology, Edited by H. S. Nalwa, *Academic Press*, New York, 2000.

[4] Nanostructures and Nanomaterials: Synthesis, Properties and Applications, G. Gao, *Imperial College Press*, London, 2004.

[5] S. Iijima, *Nature* 354 *(*1991) 56.

[6] W. Q. Han, S. S. Fan, Q. Q. Li, and Y. D. Hu, *Science* **277** (1997) 1287.

[7] S. S. Wang, E. Joselevich, A. T. Woolley, C. L. Cheung, and C. M. Lieber, *Nature* **394** (1998) 52.

[8] Y. Huang, X. Duan, Y. Cui, L. J. Lauhon, K. Kim, and C. M. Lieber, *Science* **294** (2001) 1313.

[9] J. Lee, T. J. Lee, S. C. Lyu, Y. Zhang, H. Ruh, and H. J. Lee, *Appl. Phys. Lett.* **81** (2002) 3648.

[10] S. Frank, P. Poncharal, Z. L. Wang, and W. A. Heer, *Science* **280** (1998) 1744.

[11] J. Hu, M. Quyang, P. Yang, and C. M. Lieber, *Nature* **399** (1999) 48.

[12] J. C. Johnson, K. P. Knutsen, H. Yan, M. Law, Y. Zhang, P. Yang, and R. J. Saykally, *Nanoletters* **4** (2004) 197.

[13] N. R. Rao, F. L. Deepak, G, Gundiah, and A. Govindaraj, *Prog. Sol. State Chem.* **31** (2003) 5.

[14] M. H. Huang, S. Mao, H. Feick, H. Yan, Y. Wu, H. Kind, E. Weber, R. Russo, and P. Yang, *Science* **292** (2001) 1897.

[15] Z. L. Wang, *Adv. Mater.* **12** (2000) 1295.

[16] J. Hu, T. W. Odom, and C. M. Lieber, *Acc. Chem. Res.* **32** (1999) 435.

[17] Xu, Y. Xu, D. Chen, G. Guo, L. Gui, and Y. Tang, *Adv. Mater.* **12** (2000) 520.

[18] H. Cao, Y. Xu, J. Hong, H. Liu, G. Yin, B. Li, C. Tie, and Z. Xu, *Adv. Mater.* **14** (2001) 1393.

[19] Y. Liang, C. Zhen, D. Zou, and D. Xu, *J. Am. Chem. Soc.* **126** (2004) 16338.

[20] G. W. Sears, *Acta Metal* **3** (1955) 361.

[21] P. Yang, and C. M. Lieber, *J. Mater. Res.* **12** (1997) 2981.

[22] Z. W. Pan, Z. R. Dai, and Z. L. Wang, *Science* **291** (2001) 1947.

[23] R. S. Wagner and W. C. Ellis, *Appl. Phys. Lett.* **4** (1964) 89.

[24] R. S. Wagner, W. C. Ellis, K. A. Jackson, and S.M. Arnold, *J. Appl. Phys.* **35** (1964) 2993.

[25] Y. Wu, and P. J. Yang, *J. Am. Chem. Soc.* **123** (2001) 3165.

[26] A.M. Morales, and C. M. Lieber, *Science* **279** (1998) 208.

[27] T. J. Trentler, K. M. Hickman, S. C. Goel, A. M. Viano, P. C. Gibbons, and W. E. Buhro, *Science* **270** (1995) 1791.

[28] J. R. Heath, and F. K. LeGoues, *Chem. Phys. Lett.* **208** (1993) 263.

[29] Y. D. Li, H. W. Liao, Y. Ding, Y. T. Qian, L. Yang, and G. E. Zhou, *Chem. Mater.* **10** (1998) 2301.

[30] Y. Li, H. Liao, Y. Ding, Y. Fan, Y. Zhang, and Y. Qian, *Inorg. Chem.* 38 (1999) 1382.

[31] Z. X. Deng, C. Wang, X. M. Sun, and Y. D. Li, *Inorg. Chem.* **41** (2002**)** 869.

[32] L. Qu, G. Shi, X. Wu, and B. Fan, *Adv. Mater.* **16** (2004) 1200.

[33] S. R. C. Vivekchand, G. Gundiah, A. Govindaraj, and C. N. R. Rao, *Adv. Mater.* **16** (2004) 1842.

[34] C. Cheng, R. K. Gonela, Q. Gu, and D. T. Haynie, *Nanoletters* **5** (2005) 175.

[35] Y. Wang, L. Zhang, G. Meng, C. Liang, G. Wang, and S. Sun, *Chem. Comm.* (2001) 2632.

[36] X. Wang, C. J. Summers, and Z. L. Wang, *Nanoletters* **3** (2003) 1315.

[37] Banerjee, S. H. Jo, and Z. F. Ren, *Adv. Mater.* **16** (2004) 2028.

[38] C. Liang, G. Meng, Y. Lei, F. Phillipp, and L. Zhang, *Adv. Mater.* **13** (2001) 1330.

[39] P. Guha, S. Kar, and S. Chaudhuria, *Appl. Phys. Lett.* **85** (2004) 3851.

[40] Y. C. Choi, W. S. Kim, Y. S. Park, S. M. Lee, D. J. Bae, Y. H. Lee, G. S. Park, W. B. Choi, N. S. Lee, and J. M. Kim, *Adv. Mater.* **12** (2000) 746.

[41] X. Chen, J. Xu, R. M. Wang, and D. Yu, *Adv. Mater.* **15** (2003) 419.

[42] Y. Huang, X. Duan, Y. Cui, and C. M. Lieber, *Nanoletters* **2** (2002) 101.

[43] Z. Zhong, F. Qian, D. Wang, and C. M. Lieber, *Nanoletters* **3** (2003) 343.

[44] Q. Wu, Z. Hu, X. Wang, Y. Lu, X. Chen, H. Xu, and Y. Chen, J. Am. *Chem. Soc.* **125** (2003) 10176.

[45] S. Kar, and S. Chaudhuri, *Chem. Phys. Lett.* **398** (2004) 22.
[46] M. Nath, A. Choudhury, A. Kundu, and C. N. R. Rao, *Adv. Mater.* **15** (2003) 2098.
[47] Y. Wang, G. Meng, L. Zhang, C. Liang, and J. Zhang, *Chem. Mater.* **14** (2002) 1773.
[48] C. Ye, G. Meng, Y. Wang, Z. Jiang, and L. Zhang, *J. Phys. Chem.* **B 106** (2002) 10338.
[49] L. Dong, J. Jiao, M. Coulter, and L. Love, *Chem. Phys. Lett.* **376** (2003) 653.
[50] X. Duan, C. Niu, V. Sahi, J. Chen, J. W. Parce, S. Empedocles and J. L. Goldman, *Nature* **425** (2003) 274.
[51] R. Agarwal, C. J. Barrelet, and C. M. Lieber, *Nanoletters* **5** (2005) 917.
[52] S. Kar, and S. Chaudhuri, *J. Phys. Chem.* **B 109** (2005) 3298.
[53] S. Kar, S. Biswas, and S. Chaudhuri, *communicated to Nanotechnology*.
[54] S. Kar, and S. Chaudhuri, *Chem. Phys. Lett. In press* (2005).
[55] Y. Jiang, X. M. Meng, J. Liu, Z. Y. Xie, C. S. Lee, and S. T. Lee, *Adv. Mater.* **15** (2003) 323.
[56] T. K.Tran, W. Park, W. Tong, M. M. Kyi, B. K. Wagner, and C. J. Summers, *J. Appl. Phys.* **81** (1997) 2803.
[57] H. C. Ong, and R. P. H. Chang, *Appl. Phys. Lett.* **79** (2001) 3612.
[58] T. Yamamoto, S. Kishimoto, and S. Iida, *Physica* **B 916** (2001) 308310.
[59] M. Bredol, and J. Merikhi, *J. Mater. Sci.* **33** (1998) 471.
[60] P. Calandra, M. Goffredi, and V. Turcoliver, *Colloids Surf.* **A 160** (1999) 9.
[61] C. Falcony, M. Garcia, A. Ortiz, and J. C. Alonso, *J Appl. Phys.* **72** (1992) 1525.
[62] W. Tang, and D. C. Cameron, *Thin Solid Films* **280** (1996) 221.
[63] C. N. Xu, T. Watanabe, M. Akiyama, and X. G. Zhang, *Appl. Phys. Lett.* **74** (1999) 1236.
[64] J. A. Zapien, Y. Jiang, X. M. Meng, W. Chen, F. C. K. Au, Y. Lifshitz, and S. T. Lee, *Appl. Phys. Lett.* **84** (2004) 1189.
[65] Y. Wada, H. Yin, T. Kitamura, and S. Yanagida, *Chem. Comm. (Cambridge)* **24** (1998) 2683.
[66] H. Fujiwara, H. Hosokawa, K. Murakoshi, Y. Wada, and S. Yanagida, *Langmuir* **14** (1998) 5154.
[67] N. A. Dhas, A. Zaban, and A. Gedanken, *Chem. Mater.* **11** (1999) 806.
[68] X. Chen, H. Xu, N. Xu, F. Zhao, W. Lin, G. Lin, Y. Fu, Z. Huang, H. Wang, and M. Wu, *Inorg. Chem.* **42** (2003) 3100.
[69] S. Biswas, S. Kar, and S. Chaudhuri, communicated to Chem. *Phys. Lett.* (2005).
[70] J. H. Yu,J. Joo, H. M. Park, S. Baik, Y. W. Kim, S. C. Kim, and T. Hyeon, J. *Am. Chem. Soc.* **127** (2005) 5662.
[71] X. Wang, Y. Ding, C. J. Summers, and Z. L. Wang, *J. Phys. Chem.* **B. 108** (2004) 8773.
[72] N. Pradhan, and S. Efrima, *J. Phys. Chem.* **B 108** (2004) 11964.
[73] Y. Li, X. Li, C. Yang, and Y. Li, *J. Phys. Chem.* **B 108** (2004) 16002.
[74] X. Jiang, Y. Xie, J. Lu, L. Zhu, W. He, and Y. Qian, *Chem. Mater.* **13** (2001) 1213.
[75] Q. Wu, N. Zheng, Y. Ding, and Y. Li, Inorg. *Chem. Comm.* **5** (2002) 671.
[76] X. M. Meng, J. Liu, Y. Jiang, W. W. Chen, C. S. Lee, I. Bello, and S. T. Lee, Chem. *Phys. Lett.* **382** (2003) 434.
[77] Y. Wang, L. Zhang, C. Liang, G. Wang, and X. Peng, *Chem. Phys. Lett.* **357** (2002) 314.
[78] S. Kar, S. Biswas, and S. Chaudhuri, *Nanotechnology* **16** (2005) 737.

[79] Q. Xiong, G. Chen, J. D. Acord, X. Liu, J. J. Zengel, H. R. Gutierrez, J. M. Redwing, L. C. L. Y. Voon, B. lassen, and P. C. Eklund, *Nanoletters* **4** (2004) 1663.

[80] Q. Xiong, J. Wang, O. Reese, L. C. L. Y. Voon, and P. C. Eklund, *Nanoletters* **4** (2004) 1991.

[81] Y. Jiang, X. M. Meng, J. Liu, Z. R. Ho, C. S. Lee, and S. T. Lee, *Adv. Mater.* **15** (2003) 1195.

[82] D. Appeli, *Nature* **419** (2002) 553.

[83] H. T. Ng, J. Han, T. Yamada, P. Nguyen Y. P. Chen, and M. Meyyappan *Nanoletters* 4 (2004) 1247.

[84] J. X. Ding, J. A. Zapien, W. W. Chen, Y. Lifshitz, S. T. Lee, and X. M. Meng, *Appl. Phys. Lett.* **85** (2004) 2361.

[85] D. F. Moore, Y. Ding, and Z. L. Wang, *J. Am. Chem. Soc.* **126** (2004) 14372.

[86] L. Dloczik, R. Engelhardt, K. Ernst, S. Fiechter, I. Sieber, and R. Ko¨nenkampa, *Appl. Phys. Lett.* **78** (2001) 3687.

[87] X. Wang, P. Gao, J. Li, C. J. Summers, and Z. L. Wang, *Adv. Mater.* **14** (2002) 1732.

[88] L. W. Yin, Y. Bando, J. H. Zhan, M. S. Li, and D. Golberg, *Adv. Mater.* **17** (2005) 1972.

[89] Q. Li, and C. Wang, *Appl. Phys. Lett.* **83** (2003) 359.

[90] Y. C. Zhu, Y. Bando, and D. F. Xue, *Appl. Phys. Lett.* **82** (2003) 1769.

[91] C. Ma, D. Moore, J. Li, and Z. L. Wang, *Adv. Mater.* **15** (2003) 228.

[92] D. Moore, C. Ronning, C. Ma, and Z. L. Wang, *Chem. Phys. Lett.* **385** (2004) 8.

[93] J. Gong, S. Yang, J. Duan, R. Zhang, and Y. Du, *Chem. Comm.* (2005) 351.

[94] W. T. Yao, S. H. Yu, L. Pan, J. Li, Q. S. Wu, L. Zhang, and J. Jiang, *Small* **1** (2005) 320.

[95] P. Hu, Y. Liu, L. Fu, L. Cao, and D. Zhu, *J. Phys. Chem.* **B 108** (2004) 936.

[96] C. Ye, X. Fang, G. Li, and L. Zhang, *Appl. Phys. Lett.* **85** (2004) 3035.

[97] C. Liang, Y. Shimizu, T. Sasaki, H. Umehara, and N. Koshizaki, *J. Phys. Chem.* **B 108** (2004) 9728.

[98] X. M. Meng, Y. Jiang, J. Liu, C. S. Lee, I. Bello, and S. T. Lee, *Appl. Phys. Lett.* **83** (2003) 2244.

[99] Y. Xia, P. Yang, Y. Sun, Y. Wu, B. Mayers, B. Gates, Y. Yin, F. Kim, and H. Yan, *Adv. Mater.* **15** (2003) 353.

[100] C. J. Barrelet, Y. Wu, D. C. Bell, and C. M. Lieber, *J. Am. Chem. Soc.* **125** (2003) 11498.

[101] X. Duan, and C. M. Lieber, *Adv. Mater.* **12** (2002) 298.

[102] D. C. Bell, Y. Wu, C. J. Barrelet, S. Gradecak, J. Xiang, B. P. timko, and C. M. Lieber, *Micros. Res. Tech.* **64** (2004) 373.

[103] Y. Ding, and Z. L. Wang, *J. Phys. Chem.* **B. 108** (2004) 12280.

[104] J. I. Pankove, *Optical Processes in Semiconductors*, Prentice Hall, New Jersey, 1971, p. 34-38.

[105] M. Nell, J. Marohn, and G. Mclendon, *J. Phys. Chem.* 109 (1990) 4359.

[106] L. Spanhel, M. Haase, H. Weller, and A. Henglein, *J. Am. Chem. Soc.* **109** (1987) 5649.

[107] H. Landolt; R. Bornstein, *Numerical data and fundamental relationship in science and technology, New Seriesed*; Springer-Verlag: Berlin, 1999.

[108] H. Samelson, and A. Lempicki, *Phys. Rev.* **125** (1962) 901.

[109] R. N. Bhargava, D. Gaillagher, X. Hong, and A. Nurmikko, *Phys. Rev. Lett.* **72** (1994) 416.
[110] Bol, and A. J. Meijerink, *J. Phys. Chem.* **B 105** (2001) 10197.
[111] Bol, and A. J. Meijerink, *J. Phys. Chem.* **B 105** (2001) 10203.
[112] Yu, and M. Senna, *Appl. Phys. Lett.* **66** (1995) 424.
[113] P. Yang, M. K. Lu, D. Xu, D. L. Yuan, and G. J. Zhou, *Chem. Phys. Lett.* **336** (2001) 76.
[114] S. Sapra, J. Nanda, A. Anand, S. V. Bhat, and D. D. Sarma *J. Nanosci Nanotechnol.* **3** (2003) 392.
[115] P. H. Borse, D. Srinivas, R. E. Shinde, S. K. Date, W. Vogel, and S. K. Kulkarni, *Phys. Rev.* **B 60** (1999) 8659.
[116] W. Que, Y. Zhou, Y. L. lam, Y. C. Chan, C. H. Kam, B. Liu, L. M. Gan, C. H. Chew, G. Q. Xu, S. J. Chua, S. J. Xu, and F. V. C. Mendis, *Appl. Phys. Lett.* **73** (1998) 2727.
[117] J. F. Suyver, S. F. Wuister, J. J. Kelly, and A. Meijerink, *Nanoletters* **1** (2001) 429.
[118] Arguello, D. L. Rousseau, and S. P. S. Porto, *Phys. Rev.* **181** (1969) 1351.
[119] J. Schneider, and R. D. Kirby, *Phys. Rev.* **B 6** (1972) 1290.
[120] Brafman, and S. S. Mitra, *Phys. Rev* **B 171** (1968) 931.
[121] R. L. Schmidt, K. Kunc, M. Cardona, and H. Bilz, *Phys. Rev.* **B 20** (1979) 3345.
[122] J. Hu, Y. Bando, Z. Liu, J. Zhan, D. Golberg, and T. Sekiguchi, *Angew. Chem. Int. Ed.* **43** (2004) 63.
[123] K. Sooklal, B. S. Cullum, S. M. Angel, and C. J. Murphy, *J. Phys. Chem.* **100** (1996) 4551.
[124] F. J. Garlick, and Gibson, *J. Opt. Soc. Am.* **39** (1949) 935.
[125] P. McClean, and C. B. Thomas, *Semicond. Sci. Technol.* **7** (1992) 1394.
[126] Y. Zhang, and Y. Li, *J. Phys. Chem.* **B. 108** (2004) 17805.
[127] P. Yang, M. Lu, D. Xu, D. Yuan, J. Chang, G. Zhou, and M. Pan, *Appl. Phys.* **A 74** (2002) 257.
[128] Ishizumi, C. W. White, and Y. Kanemitsu. *Appl. Phys. Lett.* **84** (2004) 2397.
[129] W. Chen, A. G. Joly, J. O. Malm, and J. O. Bavin, *J. Appl. Phys.* **95** (2004) 667.
[130] Y. Geng, L. D. Zhang, G. Z. Wang, T. Xie, Y. G. Zhang, and G. W. Meng, *Appl. Phys. Lett.* **84** (2004) 2157.
[131] Y. C. Zhu, Y. Bando, D. F. Xue, and D. Golberg, *Adv. Mater.* **16** (2004) 831.
[132] J. Yuan, X. Q. Yan, Z. X. Zhang, D. F. Liu, Z. P. Zhou, L. Cao, J. X. Wang, Y. Gao, L. Song, L.F. Liu, X. W. Zhao, X. Y. Dou, W. Y. Zhou, and S. S. Xie, *J. cryst. Growth,* **271** (2004) 403.
[133] J. P. Ge, J. Wang, H. X. zhang, X. Wang, Q. Peng, and Y. D. Li, *Adv. Funct. Mater.* **15** (2005) 303.
[134] S. Biswas, S. Kar, and S. Chaudhuri, *J. Phys. Chem. B In press* (2005)

In: Frontal Nanotechnology Research
Editor: Martin V. Berg, pp. 89-108
ISBN: 1-60021-212-3

Chapter 4

NANOCRYSTALLINE ZINC OXIDE THIN FILMS PREPARED BY CHEMICAL SOLUTION DEPOSITION FROM A ZINC NAPHTHENATE PRECURSOR

Seung Hwangbo[a] and Kyu-Seog Hwang[b,1]

[a]Major in Photonic Engineering, Division of Electronic & Photonic Engineering, Honam University, 59-1 Seobong-dong, Gwangsan-gu, Gwangju 506-714, Republic of Korea

[b]Department of Applied Optics and Institute of Photoelectronic Technology, Nambu University, 864-1 Wolgye-dong, Gwangsan-gu, Gwangju 506-824, Republic of Korea

ABSTRACT

Nanocrystalline ZnO thin films on silica glass substrates were prepared by spin coating-pyrolysis process from a zinc naphthenate precursor. As-deposited films were heat treated at 500°C ~ 1000°C. The ZnO thin films were analyzed by X-ray diffraction, field emission-scanning electron microscope, scanning probe microscope, and ultra violet-visible-near infrared spectrophotometer. Only (002) peak is observed for all samples on silica glass substrates. With the increase of heat-treatment temperature, the peak intensity of (002) plane increases. No evidently aggregated particles are present. From SPM analyses, three-dimensional grain growth, which is thought to be due to inhomogeneous substrate surface and *c*-axis oriented grain growth of the ZnO phase, is independent on heat treatment temperature. The most homogeneous surface of the highly oriented ZnO film was observed at the heat-treatment temperature of 800°C. All the films exhibit a high transmittance (>80%) in the visible region except film heat-treated at 1000°C and show a sharp fundamental absorption edge at about 0.38 ~ 0.40 μm. The estimated energy band gap for all the films are within the range reported for ZnO films and single crystal. ZnO films, consisting of densely packed grains with smooth surface morphology, were obtained by heat treatment at 600 ~ 800°C, expected to be ideal for practical usage.

[1] E-mail address: khwang@nambu.ac.kr, Phone: +82-62-970-0110, Fax: +82-62-972-0288

INTRODUCTION

High transmittance and good electrical-conductivity thin conducting oxides (TCOs) have potential use in a variety of technological applications, including plasma display panels (PDPs), electroluminescent devices, flat cathode ray tube (CRTs) and solar cells [1-4].

In recent years tin-doped indium oxide [5] (ITO) and tin oxide [6] (SnO_2) have been the most used materials for such applications. Recently, zinc oxide (ZnO) thin films have emerged as most promising oxide materials owing to their optical and electrical properties, together with their high chemical and mechanical stability. ZnO is a wide band gap semiconducting, piezoelectric and photoconducting material, which is transparent in the visible wavelength range and therefore used in a wide variety of applications, ranging from a transparent electrode material to gas sensors and surface acoustic wave (SAW) guides [7, 8]. Furthermore, ZnO has attracted considerable attention because it is a potential suitable candidate for ultra violet (UV) and blue optical applications [9]. ZnO not only has a wide direct band gap (3.37 eV) at room temperature but also shows a large exciton binding energy (60 meV) that is much larger than that of ZnSe (20 meV) and GaN (25 meV) [9]. ZnO generally emits three luminescence bands in the UV, green, and red regions. The UV emission is considered due to exciton-related activities. However, for the visible emissions, various mechanisms have been proposed including the involvement of zinc interstitials, oxygen interstitials, zinc vacancies, and oxygen vacancies [9 – 12].

Several important properties based on zinc oxide materials, such as electroluminescence and photoluminescence (PL), are strongly related to the defect formation. To provide an intuition for improving the PL properties in ZnO thin films, Lin et al [9] reported that the role of the predominant point defects in the as-grown ZnO films on Si substrates.

A variety of methods have been reported for the preparation of ZnO thin films. Physical processes such as metalorganic chemical vapor deposition (MOCVD), sputtering and pulsed laser deposition (PLD) have been adopted by various workers. The high-quality *c*-axis oriented growth of ZnO coatings on 6H-SiC(0001) substrates had been studied by Ashrafi et al [13]. The effect of thermal annealing was investigated on ZnO thin films grown on Si(001) by low-pressure MOCVD. From Zhang's report [14], annealing caused coalescence among different grains, which resulted in the formation of a dense surface and XPS confirmed that ZnO thin films changed from Zn-rich to O-rich due to more zinc atoms being bound to oxygen atoms after annealing. Zhao et al. [15] reported that preparation and optimization of *c*-axis oriented ZnO thin films on single-crystal diamond substrate by MOCVD.

Kim et al. [16] reported that magnetron sputtering growth and characterization of high quality single crystal ZnO thin films on (0001) sapphire substrates. The films showed structural and optical properties comparable with previously reported ZnO thin films grown by MOCVD, when grown at high temperature and O^2-rich atmosphere.

Furthermore, Hirata et al. [17] reported on gallium-doped ZnO transparent-conducting thin films grown on glass at different substrate temperatures by PLD. A widening in the optical band gap and a good gallium-doping efficiency were observed in the films when the substrate temperature was raised from 150°C to 300°C, as determined from optical and electrical measurements.

A chemical solution deposition (CSD) is another attractive technique for obtaining thin films and has the advantages of easy control of the film composition and easy fabrication of a large-area thin film with low cost [18 - 28]. However, the CSD method is relatively new and requires a greater understanding of the relationship between processing and defects in order to optimize film quality. In CSD, although the film showed high orientation or epitaxy to substrate with low fluctuation of alignments, the morphology of the film confirmed by microscope was similar to typical polycrystalline films. Conventional CSD processes utilize a polymeric precursor solution usually prepared from alkoxides, nitrates, or acetates [29, 30] in an organic solvent. A precursor solution is deposited onto a substrate via spin or dip coating. The as-deposited film will sequentially be prefired and heat treated to the desired oxide during heat treatment above 500°C. However, conventional CSD methods have revealed serious problems such as the occurrence of film cracking and pores brought about by stresses arising and volatilization of organics during the thermal conversion of organic-based gel to a ceramic film and instability by starting materials in air.

Thus, recently, interest in finding solutions to overcome the drawbacks has been increasing. In particular ZnO films with preferential orientation along the *c*-axis have been shown to work as SAW devices because of large piezoelectric constant. Ohyama et al. [31] prepared ZnO films on silica glass substrates from zinc acetate sol containing 2-methoxyethanol and monoethanolamine (MEA) with extremely preferred orientation along (002) plane. They described the effects of the heat-treatment conditions, the withdrawal speed of the substrate and the film thickness on the crystal orientation of the ZnO films as well as the relative density and morphology, in order to obtain more detailed insight on the conditions for the formation of highly oriented ZnO films.

Ohya et al. [32] reported transparent, electrically conducting ZnO thin films with polycrystalline structure from zinc acetate and alkoxide. Recently Wessler et al. [33] reported textured ZnO thin films on (0001) sapphire (SAP) by spin-coating a solution of acetate and ethanolamine. As far as we know, however, there is little information on preparing highly oriented ZnO thin films on amorphous substrates by CSD using a metal naphthenate as precursor.

In the present study, the preparation of ZnO thin films with a highly preferred orientation on amorphous silica substrates from a zinc naphthenate precursor is reported. The heat-treatment among many parameters on CSD process had a strong effect on the structural and optical behavior of ZnO thin film. The thin films were characterized by X-ray diffraction (XRD) θ-2θ scans, field emission-scanning electron microscope (FE-SEM), scanning probe microscope (SPM), and ultra violet (UV) – visible - near infrared (NIR) spectrophotometer. The results enabled us to develop guidelines for preparing highly oriented ZnO thin films on silica substrate from a metal naphthenate precursor.

EXPERIMENTAL PROCEDURE

A coating solution was prepared using zinc naphthenate (Nihon Kagaku Sangyo Co., Ltd., Japan) and by diluting the solution with toluene (concentration: 4 wt% metal / 100 mL sol) to achieve an appropriate viscosity for the deposition of smooth and crack-free thin films.

Metal naphthenate which consist of cyclopentanes of cyclohezanes, methylene chains, carboxylates and metals mainly have the following structure (see Fig. 1):

$$[(\text{cyclopentane})\text{-}(CH_2)_n\text{-}COO]^{-}_{m}\text{-}M^{m+}$$

<R-O-M structure> M: metal

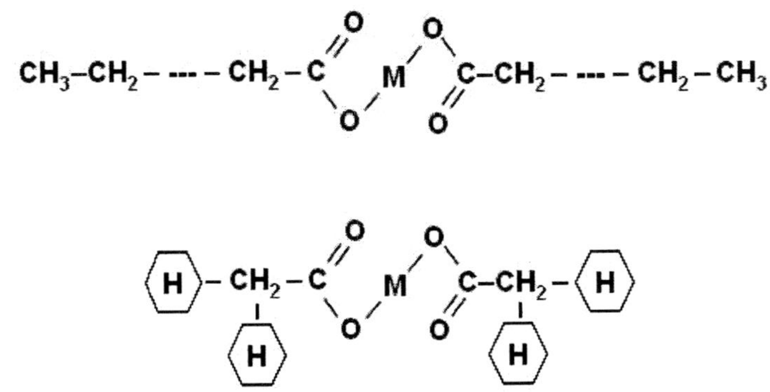

Figure 1. Structure of naphthenic acids.

Metal naphthenates are used in solid form or in organic solution with mineral spirit, toluene, xylene, naphthenic oil or other solvents. The properties mainly depend on the quality of naphthenic acids. Heating decomposition take place above 150 ~ 180°C. In presence of mineral acids are easily decomposed to naphthenic acids, solvent and inorganic salts. Alkaline solutions decomposed metal naphthenates to metal hydroxide and alkaline solution of the naphthenic acids. Metal naphthenates are more advantageous than metal alkoxides as starting materials, in terms of cost, stability in air and ease of handling. Preparation of the coating solution using zinc naphthenate was easily performed by the addition of toluene, while complicated procedures were needed for the preparation of sol from zinc alkoxide [32]. Furthermore, it should be noted that vaporization of the additives, such as alcohol, water and catalyst etc, during prefiring and heat treatment might cause unwanted cracks and pores in the coating [34]. On the other hand, in our paper, prefiring is the process for pyrolytic conversion of zinc naphthenate into ZnO, while the final heat treatment is the process for solid-state reaction. In our previous reports [18, 19] using metal naphthenates, it was observed the heat-treated films were smooth and it was difficult to identify cracks and pores, while defects were easily observed at the surface of metal alkoxide-derived films. In opto-electronic applications, the partially deteriorated film may present some disadvantages.

Prior to coating process, silica glass substrates were cleaned in deionized water, immersed in H_2O_2 solution and finally rinsed in acetone. The starting solution was spin-coated onto the cleaned substrate at 1500 rpm for 10 sec in air. The as-deposited film was prefired at 500°C for 10 min in air to vaporize organic components in coating solution. The coating process was repeated five times to prepare a thick coating of ZnO. Then the final annealing was performed in air at 500°C, 600°C, 700°C, 800°C, 900°C, and 1000°C for 30 min by directly inserting the samples into preheated tube-type furnace, which was followed by fast cooling.

The thickness of the heat-treated ZnO thin film was approximately 0.65 ~ 0.7 μm, as determined by observations of fracture-cross-sections using a FE-SEM (S-4700, Hitachi Co., Japan) and F20 (Filmetric Inc., San Diego, CA, U. S. A.) using reflection spectrum. The crystal structure, crystallinity, and orientation of the thin films were investigated by XRD (D-Max-1200, Rigaku Co., Japan) in θ-2θ geometry. A CuK_α (λ = 1.54056 Å) source was used, and the scanning range was between 2θ = 20° and 70°. The surface morphology of the thin films was evaluated using the FE-SEM micrographs. The microstructure of the samples was observed by FE-SEM cross-sectional view. The growth mechanism and surface roughness of the thin films were studied by a SPM (XE-200, PSIA Co., South Korea).

When one is evaluating surface roughness, the power spectral density (PSD) function is used increasingly as a valuable tool for topography description of extremely flat surfaces, such as epitaxial silicon or optical thin films. Average roughness (R_a) is one of the most commonly used roughness statistics. The illustration as shown in Fig. 2 represents two

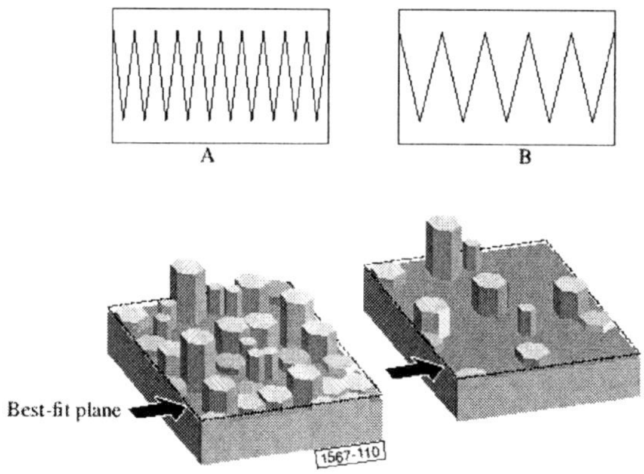

Figure 2. Two surfaces having the same "average roughness"[35].

surfaces having the same "average roughness". Similarly, a number of other roughness values are based upon least-squares calculations [e.g., root mean square (RMS) roughness, or R_q], and their algorithms are more concerned with a best fit of all height points than with the spatial frequency of features. Image "A" surface is represented as having a high frequency profile of features. Image "B" represents a separate surface having the same average feature height, but distributed at wider (lower-frequency) intervals. In terms of average and RMS roughness, both surfaces are "equally rough". What would better distinguish them apart is a method for characterizing each surface's feature frequency or wavelength. For more complex surfaces, moreover, it would be helpful to know which wavelengths occur most often and which impart the greatest influence or "power" to the surface's topography. This is presented in a PSD analysis [35].

Any randomly rough surface can be considered as a Fourier series of sinusoidal waves with different amplitudes, periods, and phases. The grating equation shows that a single grating with spacing d causes scatter into the angle according to $\sin q = l/d$, where l is the wavelength of light. Hence, d can be considered as one spatial wavelength present on the surface, or accordingly, $f = l/d$ as one spatial frequency. For a randomly rough surface such as our thin film component, many different spatial frequencies are present. This is quantitatively expressed by the PSD curve, giving the relative strength of each roughness component of a surface microstructure as a function of the spatial frequency [36], whereas the root mean square (RMS) roughness as a single parameter of surface description reflects the standard deviation of all height values within the considered surface and does not provide any information on the topographical details of the surface.

Consider the following image (see Fig. 3),

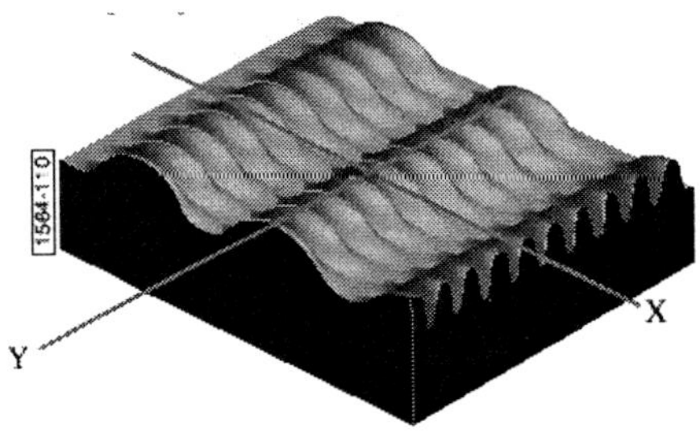

Figure 3. Synthetic surface consists of two dominant wave forms [35].

This synthetic surface consists of two dominant wave forms: a long period waveform along the X-axis, and a shorter period waveform along the Y-axis. A 2-dimentional PSD plot will consist of two dominant spikes (one or each dominant wavelength), plus some number of extra wavelengths inherent within the image. Because of the sinusoidal nature of the composited wave form, a relatively small composite of spectral frequencies suffice to describe the entire surface. By contrast, an image comprised of angular (saw-toothed of square) wave forms will generally yield a much broader spectral plot. This is largely because of the mathematical nature of PSDs, relying as they do upon Fourier transforms. Because PSDs transform the surface's real spatial elements (X-Y-Z dimensions) into an array of spatial frequency constructs (sine waves of differing spatial frequencies), a trend in surface features emerges: surfaces consisting of regular, rounded features yield PSD plots comprised of relatively fewer composite frequencies; surfaces with sharp, angular features yield PSD plots encompassing relatively more composite frequencies [35].

As shown in Fig. 4, this surface is comprised of "terraces" formed from the material's natural lattice structure; each terrace is one atomic monolayer thick and is spaced at fairly regular intervals. This degree of flatness is handily evaluated with PSD, as the terraces will produce a spectral spike corresponding to their spacing wavelength. A PSD plot for this type of surface generally takes the form shown in Fig. 5. This tapered PSD plot is characteristic of flat, isotropic surfaces. Longer wavelengths are present up to the scan width, and are accompanied by uniformly decreasing powers of shorter wavelengths down to 2 pixels. On the plot shown in Fig. 5 a spike stands out, corresponding to the wavelength spacing of the terraced features. Depending upon the qualitatively standards of the person evaluating such a plot, this spike may exceed a threshold standard of flatness [35].

PSD curves used throughout this work were calculated from selected atomic force microscope (AFM, NanoScope Multimode^{TM} SPM, Digital Instruments, U. S. A.) scans with NanoScope off-line analysis software. In order to cover an extended range of spatial frequencies, we combined the PSD functions from 1×1 μm^2 and 10×10 μm^2 scans. All the SPM and AFM measurements were performed in air using the tapping mode with silicon probes having ~ 250 kHz resonant frequency.

Transmittance in the visible wavelength range was measured by a UV – visible - NIR spectrophotometer (CARY 500 Scan, VARIAN Co., Australia). The transmittance was automatically calibrated against a bare glass substrate as a reference sample, and the absorption coefficient was obtained from the transmittance curve.

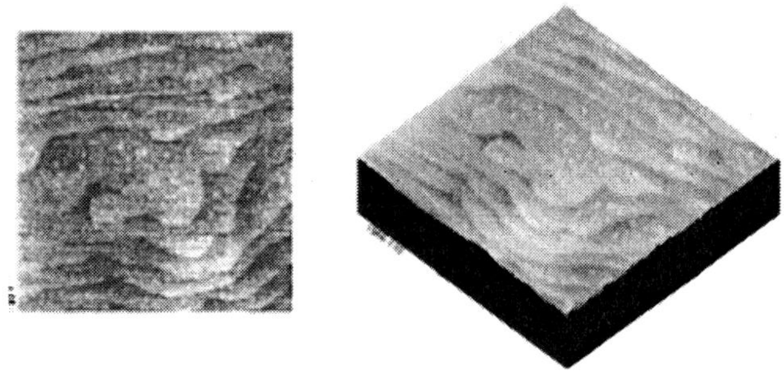

Figure 4. AFM Image of epitaxial gallium arsenide [35].

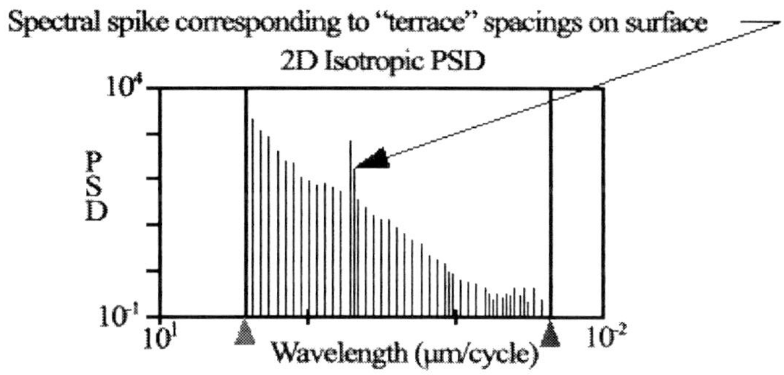

Figure 5. 2D isotropic PSD of epitaxial gallium arsenide [35].

RESULTS AND DISCUSSION

Figure 6 shows the XRD curves of ZnO thin films deposited on the silica glass substrates. Only the (002) ZnO peak at $2\theta \sim 34°$ is observed for all samples. This indicates the ZnO thin films prepared by CSD from a zinc naphthenate show a good *c*-axis orientation, i.e., the vertical growth to the substrate surface. The *c*-axis orientation in ZnO films by physical dry method can be understood by the 'grain boundary movement' model proposed by Loder et al. [37]. According to that model, at the very first stage of film growth, certain grains with particular orientation start to grow. By thermodynamical coalescence of crystallites during film growth, orientational growth (*c*-axis orientation for ZnO film) is achieved.

From Ohyama's work [34], for ZnO films prepared by CSD using zinc acetate – 2-methoxyethanol - MEA solution, removal of the solvent and organic substances produced by acetate decomposition prior to crystallization may be one of the key factors that provide oriented crystal growth. The vaporization of the solvents, the decomposition of the zinc acetate and the crystallization of the zinc oxide may occur almost simultaneously when the heating rate is high. Since the structural relaxation of the gel film, which is induced by the solvent vaporization and acetate decomposition, can take place only before the crystallization, the simultaneous vaporization, decomposition and crystallization may give the film less chance to be structurally relaxed [34].

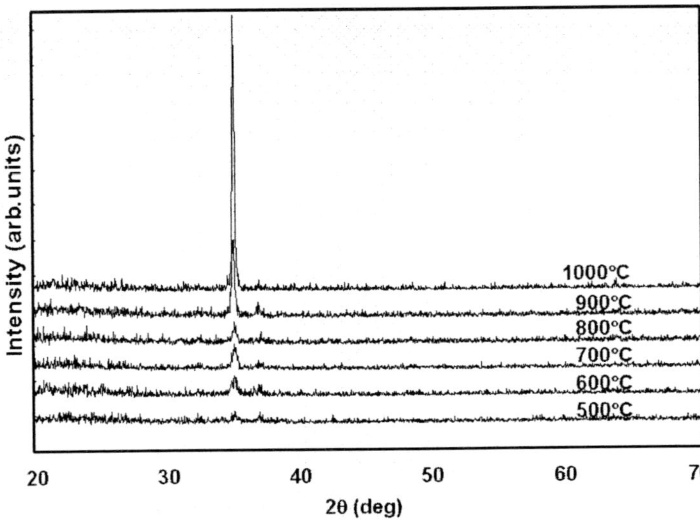

Figure 6. XRD spectra of ZnO films on silica glass substrates as a function of heat treatment temperature.

When the heating rate is low, the gel film is given an enough time to be structurally relaxed before crystallization, resulting in denser ceramic films. Moreover, when the prefiring temperatures too high (>300°C), the vaporization of the solvents, and the thermal decomposition of the zinc acetate may take place abruptly and simultaneously with the crystallization, disturbing the unidirectional crystal growth. On the contrary, when the preheating temperature is too low (<300°C), the complete vaporization and the thermal decomposition of the zinc acetate do not occur at the preheating step but do occur at the post heating step at post heating temperatures over 500°C. In such a case, abrupt solvent vaporization and acetate decomposition occur at the post-heating step, which may also disturb the unidirectional crystal growth. Thus, the preheating temperature seems to be the most important factor for obtaining ZnO films with preferred orientation, affecting the solvent vaporization, zinc acetate decomposition and zinc oxide crystal growth. Removal of the solvent and organic substances produced by acetate decomposition prior to crystallization may be one of the key factors that provide oriented crystal growth [34].

However, in this work, it should be noted here that highly *c*-axis oriented ZnO thin films were obtained by prefiring at high temperature, 500°C, and final heat-treatment with a rapid thermal annealing by directly inserting the samples into preheated furnace. This result suggests that a necessary condition using in Ohyama's work to obtain strongly preferred oriented ZnO thin film such as, usage of solvents having relatively high boiling points of about 200°C, heat-treatment of the film at lower rates and pyrolysis at lower temperatures of 200°C ~ 300°C, is unnecessitated in this work. The boiling point of toluene in our work is much lower, 110.6°C, than those of 2-methoxyethanol (125°C) and MEA (170°C) used in Ohyama's work [34]. Conclusively, the use of toluene, solvent of lower boiling point, resulted in strongly preferred orientation of ZnO crystals as in the present work.

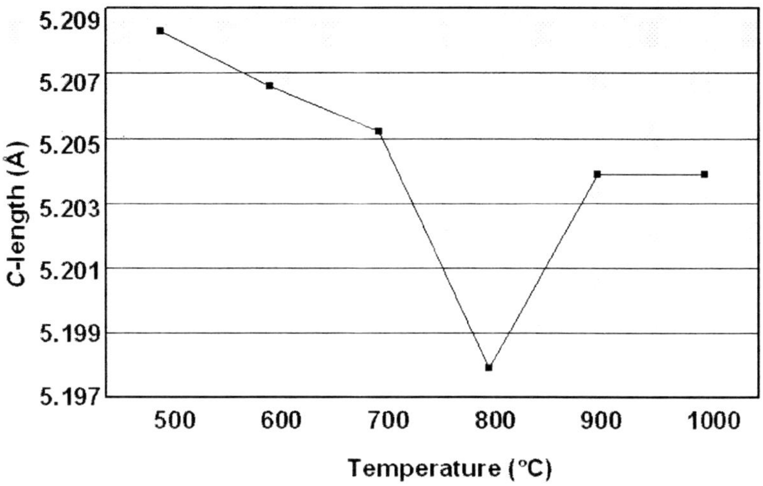

Figure 7. Variation of *c*-length of ZnO films as a function of heat treatment temperature.

It is also observed that with an increase in heat-treatment temperature, the peak intensity of (002) plane increases, as shown in Fig. 6. The lattice parameter, full width at half maximum (FWHM), and peak intensity of (002) reflection as shown in Figs. 7, 8, and 9. The diffraction angle of 2θ ranges from 34.41° to 34.49°. The value of the FWHM ranges from 0.194° to 0.830°. The diffraction 2θ peak position increase from 34.41° at 500°C up to 34.49° at 800°C and then decreases to 34.44° at 900°C and 1000°C. The FWHM decreases from 0.83° at 600°C to 0.194° at 1000°C. The FWHM at >700°C was comparable to those of highly-textured ZnO thin films prepared by the physical coating process [1]. The results indicate that the ZnO thin film using zinc naphthenate heat treated at >700°C is expected to have the best crystallinity.

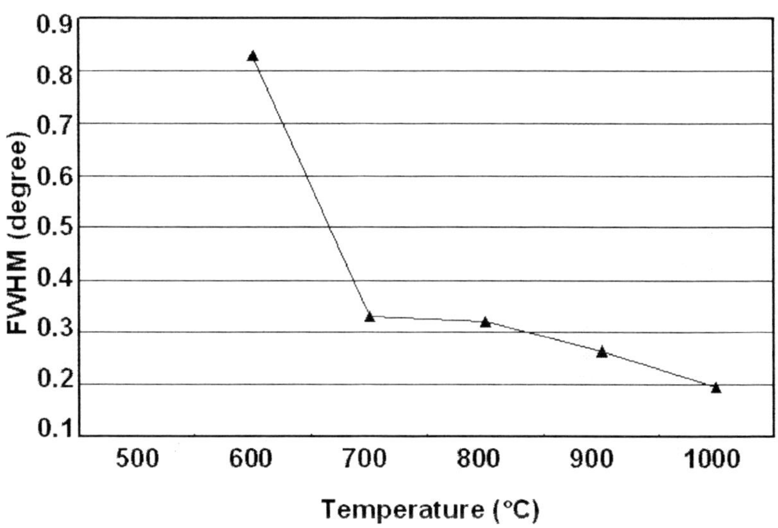

Figure 8. Variation of FWHM of ZnO films as a function of heat treatment temperature.

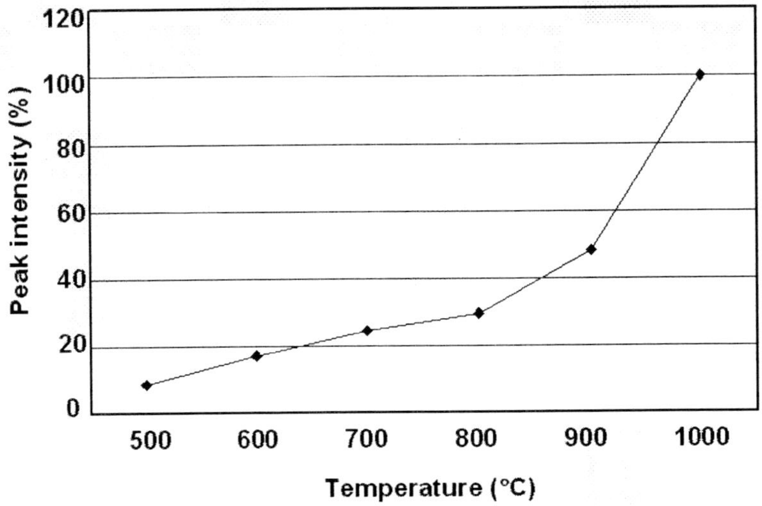

Figure 9. Peak intensity of (002) plane of ZnO films as a function of heat treatment temperature.

On the basis of XRD data, the lattice *c* parameter has been estimated to be in the range of 5.1979 to 5.2083Å, as shown in Fig. 7, which is similar to the JCPDS value of 5.2066 Å for bulk ZnO (JCPDS File 36-1451). The larger value of lattice constant for the film heat treated at 500°C compared to the standard powder value shows that the unit cell is elongated along the *c*-axis, and compressive forces act in the plane of the ZnO film. These compressive forces disappear as the heat treatment temperatures increases.

In order to more exactly investigate the structural properties of ZnO film, we calculated the stress in the film. The calculation of the film stress is based on the biaxial strain model [38].

The strain $\varepsilon = [(c_{bulk} - c_{film}) / c_{bulk}]$ in the direction of the *c*-axis, *i.e.*, perpendicular to the substrate surface, was measured by XRD. To derive the film stress of film parallel to the film surface, the following formula has been used [38], which is valid for a hexagonal lattice;

$$\sigma_{film} = [2c_{13}^2/2c_{13} - c_{33}(c_{11}+c_{12})/2c_{13}][(c_{film}-c_{bulk})/c_{bulk}] \quad (1)$$

The elastic constant c_{ij} of single crystalline ZnO have been used: $c_{11} = 208.8$, $c_{33} = 213.8$, $c_{12} = 119.7$, and $c_{13} = 104.2$ GPa [39]. Stress in the film can be estimated using Eq. (1) and are plotted as a function of heat treatment temperature in Fig. 10. The negative sign for the film heat treated at 500°C indicates that the lattice constant *c* is elongated as compared to unstressed powder and therefore film is state of elongation. The film becomes completely stress free at about 600°C annealing temperature. After further heat treatment (>600°C) a film goes into state of compression as compared to the normal lattice state of bulk ZnO. Previous works [40] observed a state of compression in their films prepared by *rf* sputtering and annealed at 400°C. The stress observed in sputtered ZnO films by previous workers has been associated with interstitial oxygen. Their results can be qualitatively explained by assuming that, after annealing, the interstitial oxygen is probably removed and the defects in the lattice are annealed out. We assume that high stress relief temperatures (SRT), >600°C in this work, could be related with the volatilization of organics during heat treatment. Thus, in this work, the pyrolysis of organic components during annealing caused more defects in ZnO lattice, thus increasing the SRT of the film.

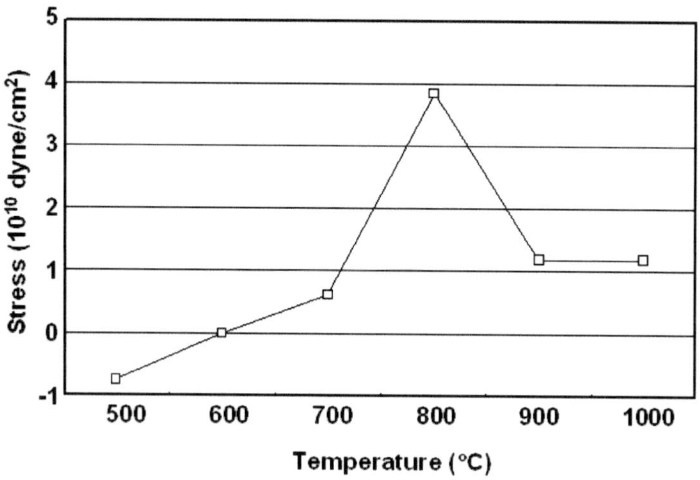

Figure 10. Variation of induced stress on the ZnO films as a function of heat treatment temperature.

Figure 11 shows the FE-SEM photographs of the ZnO thin films after heat treatment at 500°C (a), 600°C (b), 700°C (c), 800°C (d), 900°C (e), and 1000°C (f). Particulate structure is evident in all films except films heat treated at < 600°C. The particle size increases with increasing heat treatment temperature, and grain out-growth of the particles is observed in the films heat treated at 800°C and 900°C. It is noted that no evidently aggregated particles are present and that nano-sized particles were obtained in all the films except the film heat treated at 1000°C. Generally, in the CSD method, pores and cracks are easily recognized in the product films. On the contrary, the surface of the thin film produced in this work was relatively flat and smooth below 800°C.

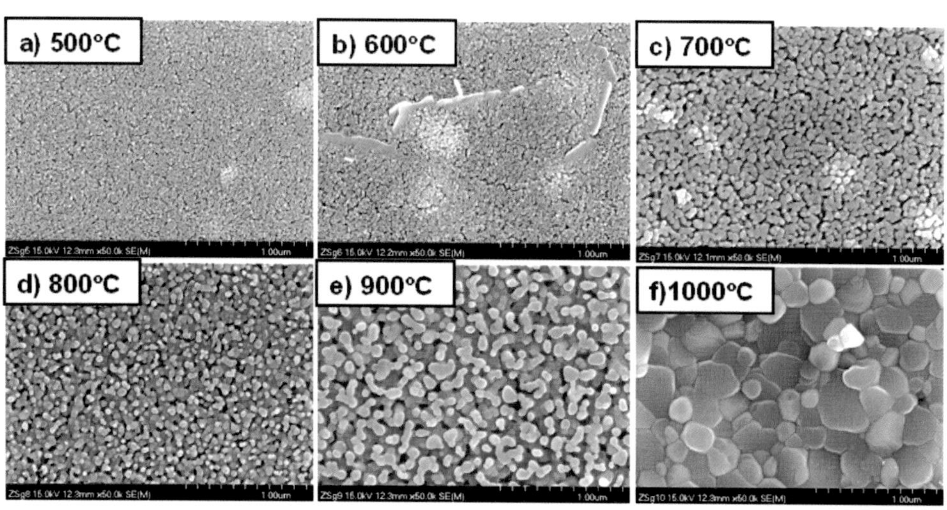

Figure 11. FE-SEM images of the ZnO films after heat treatment at various temperatures.

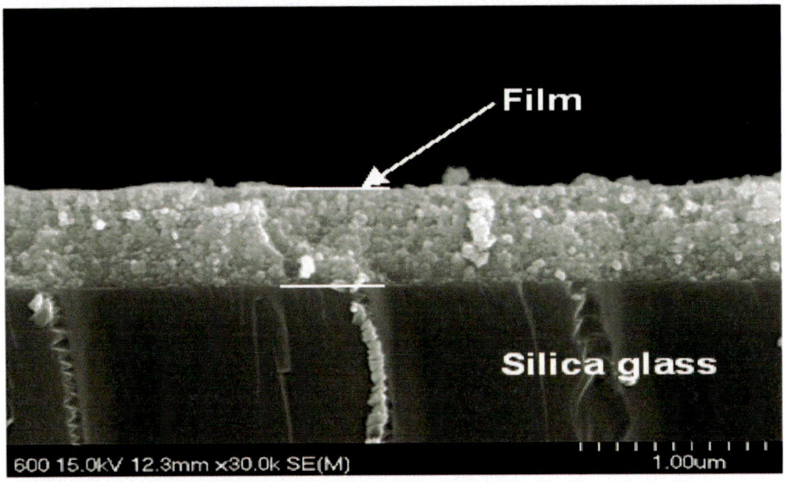

Figure 12. Cross-sectional image of the ZnO film heat treated at 600°C.

Figure 12 shows the cross-sectional morphology of the ZnO films hate treated at 600°C. The fractured cross-section of the ZnO thin film with an approximately 0.67 ~ 0.7 μm thickness appears dense and uniform. This suggests that the bonding strength between the substrate and the film is sufficiently strong.

To evaluate the surface roughness and morphology of the films according to the heat-treatment temperatures, SPM analysis was performed. Figure 13 shows the SPM images (5 × 5 μm^2) of ZnO films heat treated at various temperatures.

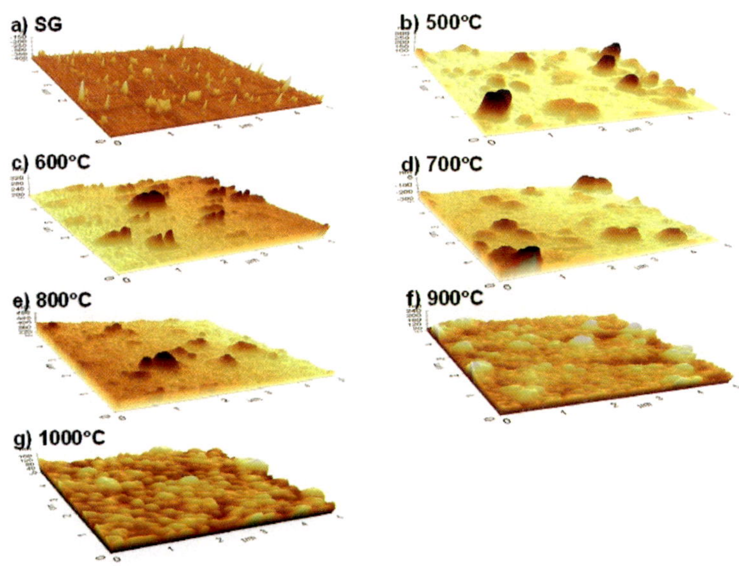

Figure 13. SPM photographs (5 × 5 μm^2) of the ZnO films after heat treatment at various temperatures.

The surface of polished silica glass substrate has a relatively high-surface roughness with some needle-like grain outgrowth. This surface is not smooth enough for automatically textured ZnO film. Due to a needle-like grain outgrowth on substrate surface, the film has three-dimensional island-like features, as shown in Fig. 13. It is observed that with increasing heat-treatment temperature to 800°C, the grain gradually converted to round-shaped peak. Above 900°C, it is easy to identify large grain growth, such as island growth on the two-dimensionally grown layer, although the grain size uniformly increased with increasing heat treatment temperatures to 900°C and 1000°C. In all the samples, three-dimensional grains were observed, although the previous XRD crystallographic results indicate a high orientation. The above results imply that the growth of three-dimensional grains identified in all the samples, which are thought to be due to the silica glass substrate naturally containing some needle-like grain outgrowth, depends on the heat treatment temperature.

Figures 14 and 15 show SPM top-view images and surface roughness of ZnO thin films as a function of heat-treatment temperatures. Generally, in previous works using a SPM, most surface roughness were obtained from line profiles. In this work, however, in order to obtain more exact roughness data, we performed line profiles and area scanning (5 × 5 μm^2) for the surfaces of the ZnO thin films. ZnO thin films on silica glass substrate having relatively flat surface structure annealed to 500°C showed high RMS roughness. We assume that 3-dimensional needle-like grain growth in bare glass substrate may increase surface roughness of ZnO after heating. With an increase in the annealing temperature to 800°C, the RMS roughness decreased to 12.524 or 8.558 nm. At 900°C and 1000°C, the surface roughnesses of the ZnO films were increased due to larger grain growth, resulting in higher RMS roughnesses as shown in Figs. 14 and 15. The most homogeneous film (RMS= 12.524 or 8.558 nm) having a preferred orientation was observed at a heat-treatment temperature of 800°C.

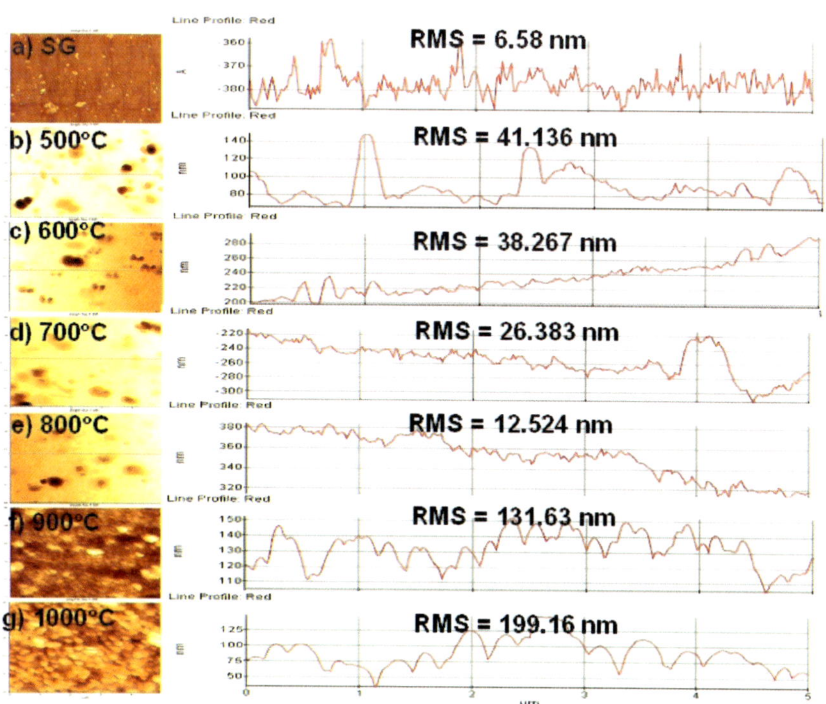

Figure 14. Line profiles (5 × 5 μm^2) for the surfaces of the ZnO films after heat treatment at various temperatures.

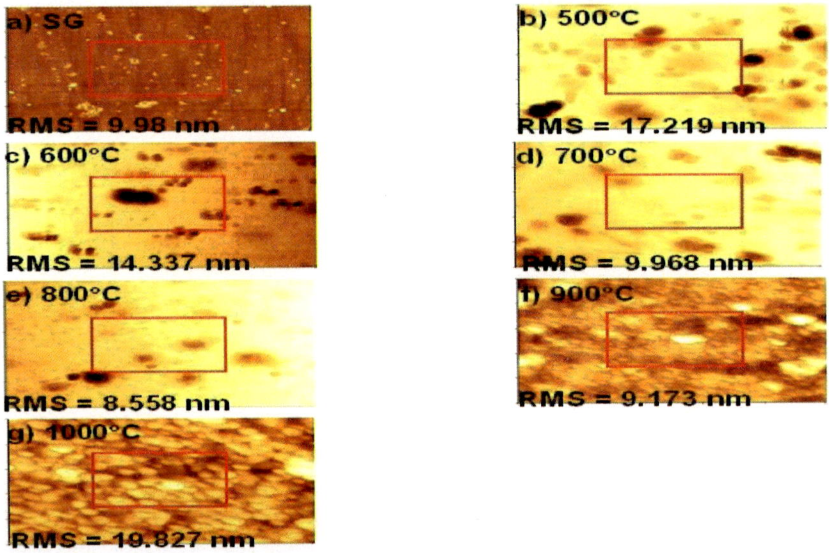

Figure 15. Area scanning (5 × 5 μm^2) for the surfaces of the ZnO films after heat treatment at various temperatures.

Furthermore, we studied the surface roughness characteristics of ZnO thin films as a function of heat treatment temperature by using PSD curves obtained by AFM measurements. In most cases where optical surfaces are to be evaluated, the parameter commonly used for describing the roughness characteristics of a surface of thin film is the RMS roughness. However, when specific roughness properties have to be investigated in detail, spectral roughness analysis should be performed. Because of PSDs transform the surface's real spatial elements (X-Y-Z dimensions) into an array of spatial frequency constructs (sine waves of differing spatial frequencies), a trend in surface features emerges. Determining the PSD provides valuable information not only on the height deviation of the roughness profile, but also on its lateral distribution, and hence it gave a more general description than the RMS roughness alone [18]. PSD is used increasingly as a metrology tool for evaluating extremely flat surfaces, while in the surfaces represented by RMS roughness, peaks and valleys are equidistant from the mean surface centerline.

Figure 16 shows the results of PSD functions for the ZnO thin films according to the heat treatment temperature. Similar to that of SG substrate, the PSD function of the ZnO thin films heat treated at 500°C and 600°C displays no inflated shape generally due to increase in grain size. In addition, PSD curve with a gentle grade indicating that the layer structure has grown uniformly both in height and width. At 800°C, the PSD curve of the ZnO film display inflated shape at spatial frequency around 10 ~ 100 μm^{-1}, which may be due to uniformly formed small grains. For the ZnO film heat treated at 1000°C, however, lower frequency range at a spatial frequency around 0.1 ~ 1 μm^{-1} was increased. At 1000°C, PSD curve of ZnO thin films revealed rough surface morphology, which could be attributed to the grain out growth as previously shown in Figs. 14 and 15.

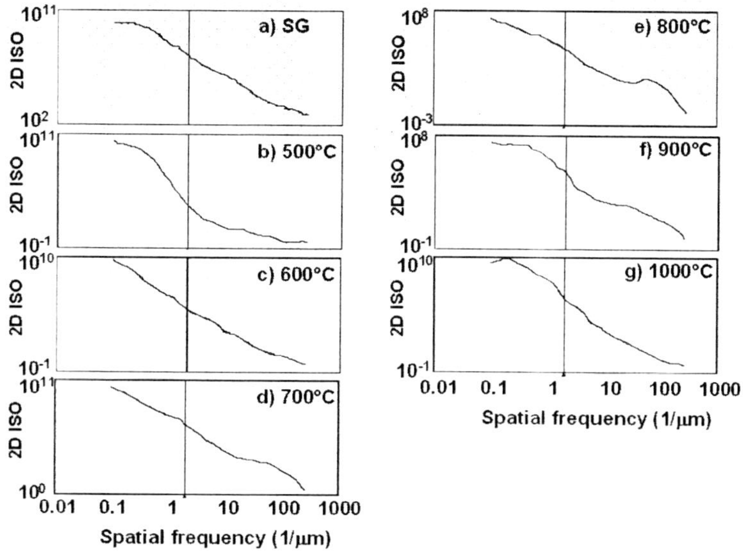

Figure 16. PSD curves of the ZnO thin films as a function of heat treatment temperature.

It is important that the surface of the deposited films should be smooth and flat to exhibit ideal properties for device applications. However, few textured films having excellent surface structure were obtained by the CSD process. Although the films showed epitaxy with the substrate or high orientation with low fluctuation of alignments, the morphology of the film generally exhibited porosity, roughness, and three-dimensional defects. This suggests that further investigation is required for the CSD method to obtain a great understanding of the relation between processing and defects in order to prepare highly *c*-axis oriented ZnO thin film on silica glass substrate without 3-dimensional grain outgrowth.

Figure 17 shows transmission spectra of the ZnO thin films heat treated at different temperatures.

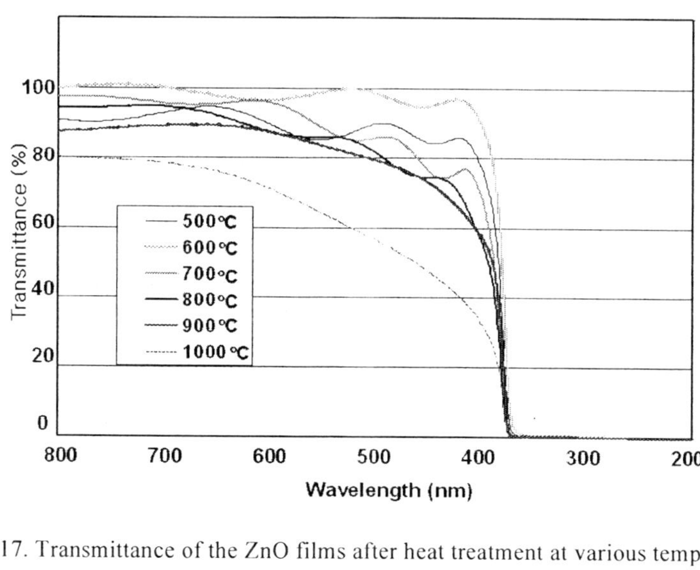

Figure 17. Transmittance of the ZnO films after heat treatment at various temperatures.

All the films exhibit a high transmittance (>80%) in visible region except for the film heat treated at 1000°C, and show a sharp fundamental absorption edge at about 0.38 ~ 0.4 μm, which is very close to the intrinsic band-gap of ZnO (3.2ev). The decrease in transmittance up to 900°C and 1000°C may be attributed to slight increase of surface roughness. For the films annealed below 800°C, the variation of transmittance with wavelength is principally due to the interference phenomenon, and the envelope of fringes starts to shrink after heat treatment the film above 900°C, as show in Fig. 17. The increase in surface roughness may be responsible for such shrinking in the fringes [41]. Briefly, the visible transmittance of ZnO thin film is quite comparable with those of film prepared by physical dry process [42].

The optical absorption coefficient, α, is defined as,

$$I = I_0 e^{-\alpha t} \tag{2}$$

where I is the intensity of transmitted light, I_0 is the intensity of the incident light, and t is the thickness of the ZnO film. As the transmittance is defined as I/I_0, we obtain α from Eq. (2). In the direct transition semiconductor, α and optical energy band gap (E_g) are related by [42, 43],

$$\alpha = (hv - E_g)^{1/2} \tag{3}$$

where, h is Plank's constant, and v is the frequency of the incident photon. Figure 18 shows the plot of α^2 *vs.* hv. The linear dependence of α^2 on hv indicates that the ZnO films are direct-transition-type semiconductors. The photon energy at the point where α^2 is zero is E_g. Thus E_g can be determined by using an extrapolation method [42]. The optical band gap, E_g, is in the range between 3.288 eV and 3.315 eV. The estimated values of the band gap for all the films are within the range reported for such films [30, 31].

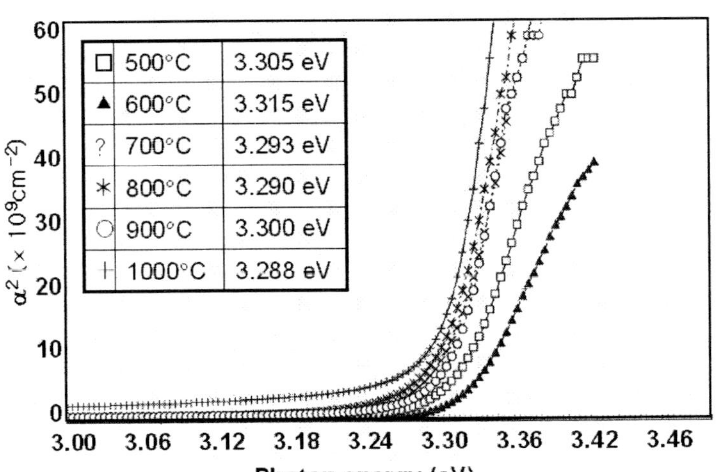

Figure 18. Square of the absorption coefficient as a function of photon energy for the ZnO films after heat treatment at various temperatures.

The refractive index of the ZnO film was calculated by Filmetrics F20, thin film measurement system. The F20 measures thin film characteristics by either reflecting or transmitting light through the sample, and then analyzing this light over a range of wavelengths. Because of its wave-like properties, light reflected from the top and bottom interfaces of a thin film can be in-phase so that reflections add, or out-of-phase so that reflections subtract. Whether the reflections are in- or out-of-phase (or somewhere in between) depends on the wavelength of the light, as well as the thickness and properties of the film [e.g., reflections are in-phase when $\lambda = (2^{nd})/I$, where λ is the wavelength, n is the reflective index, d is the film thickness, and I is an integer.]. The result is characteristic intensity oscillations in the reflectance spectrum. Figure 19 shows refractive index of the ZnO films as a function of heat treatment temperature. The refractive index increases with increasing heat-treatment temperature up to 800°C. At 800°C, the value of refractive index is close to the value of single crystal (n = 2.004). Above 800°C, the refractive index starts decreasing.

The variation of refractive index may be attributed to the change in the packing density with heat treatment temperatures. Tansley et al. [44] assumed a linear dependence of refractive index on packing density and reported a value of 0.95 for packing density for the sputtered ZnO film. Many attempts have been made to correlate the packing density with refractive index of the materials [40, 45]. Macleod [45] reported the relation based on Lorentz-Lorentz polarization theory, as below

$$n_f^2 = \frac{(1-p)(n_s^2+2)n_v^2 + p(n_v^2+2)n_s^2}{(1-p)(n_s^2+2) + p(n_v^2+2)} \quad (4)$$

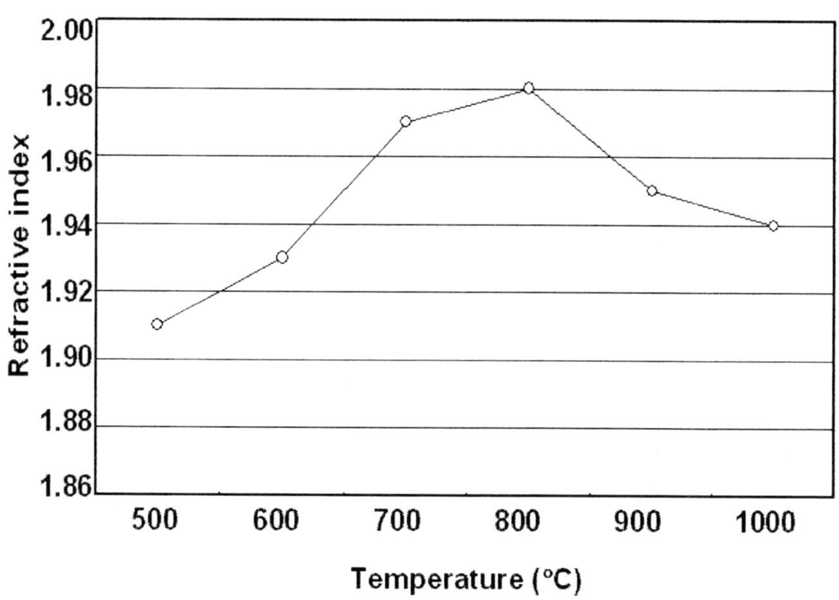

Figure 19. Refractive index of the ZnO films after heat treatment at various temperatures.

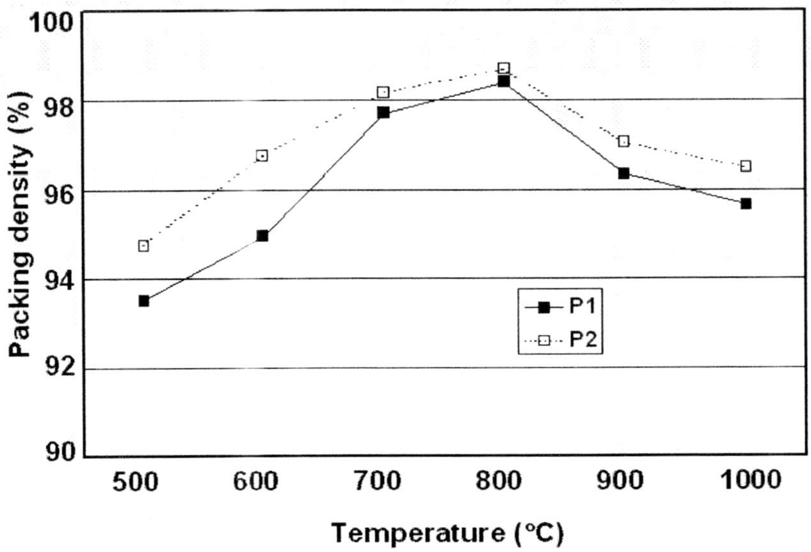

Figure 20. Estimated packing densities from two different formulas as a function of heat treatment temperature.

Where n_f is the refractive index of film (containing voids and columns), n_s is the index of the solid material of the film (single-crystal value), n_v is the index of the voids in the film (equal to one for air), and p is the packing density [40]. Eq. 4 is basically derived by assuming that the grains of the film are of spherical form. This approach is not strictly valid because of the possibility that grains of the film might depart from the spherical form and assuming cylindrical particles expression is [45]

$$n_f^2 = \frac{(1 - p)n_v^4 + (1 + p)n_v^2 n_s^2}{(1 + p)n_v^2 + (1 - p)n_s^2} \tag{5}$$

The values of packing densities of the ZnO film as a function of heat treatment temperature are determined using Eqs. 4 and 5. The variation of packing densities is shown in Fig. 20. The values P_1 and P_2 are the packing densities calculated from the Eqs. 4 and 5, respectively. The maximum packing density was obtained at 800°C. This shows that the film heat treated at 800°C have minimum numbers of voids and pores. On the other hand, at higher temperature the pores owing to larger grain growth, as evident from FE-SEM studies, may be responsible for the decrease in packing density and therefore the refractive index. The FE-SEM images show round-shape grains of uniform size. Therefore Eq. 4, which is based on the assumption of spherical grains, seems to be more appropriate for calculation the packing density of the previous film [40].

Conclusion

Highly *c*-axis oriented ZnO thin films on silica glass substrates were prepared by spin coating-pyrolysis from metal naphthenate precursor. Structural relaxation of the precursor gel before crystallization, by employing solvents having high boiling points about 200°C, by heating the film at low rates or by prefiring at low temperatures of 200°C ~ 300°C, is unessential for obtaining oriented ZnO film from a zinc naphthenate precursor. The most homogeneous film having a preferred orientation was observed at 800°C. All the films exhibit a high transmittance (>80%) in visible region except film at 1000°C and show a sharp fundamental absorption edge at about 0.38 ~ 0.4 μm.

REFERENCES

[1] Kim, K.; Song, J.; Jung, H.; Choi, W.; Park, S.; Song, J.; Lee, J.; *J. Vac. Sci. Tech.* A 2000, 18(6), 2864-2868.

[2] Kim, Y.; Kim, Y.; Yang, H.; Park, J.; Han, J.; Lee, Y.; Kim, H.; *J. Vac. Sci. Tech.* A 1997, 15(3), 1103-1107.

[3] Ohtomo, A.; Kawasaki, M.; Ohkubo, I.; Koinuma, H.; Yasuda, T.; Segawa, Y.; *Appl. Phys. Lett.* 1999, 75(7), 980-982.

[4] Bundesmann, C.; Ashkenov, N.; Schubert, M.; Spemann, D.; Butz, T.; Kaidashev, E. M.; Lorenz, M.; Grundmann, M.; *Appl. Phys. Lett.* 2003, 83(10), 1974-1976.

[5] Shigesato, Y.; Takaki, S.; Haranoh, T.; *J. Appl. Phys.* 1992, 71, 3356-3364.

[6] Dominguez, J. E.; Fu, L.; Pan, X. Q.; *Appl. Phys. Lett.* 2002, 81(27), 5168-5170.

[7] Matsubara, K.; Fons, P.; Yamada, A.; Watanabe, M.; Niki, S.; *Thin Solid Films* 1999, 347, 238-240.

[8] Zhu, S.; Su, C.; Lehoczky, S.; George, M.; Lowndes, D.; *J. Mater. Res.* 2000, 15(5), 1125 - 1130.

[9] Lin, C. C.; Hsiao, C. S.; Chen, S. Y.; Cheng, S. Y.; *J. Electrochem. Soc.* 2004, 151, G285 - G288.

[10] Cho, S.; Ma, J.; Kim, Y.; Wang, G. K. L.; Ketterson, J. B.; *Appl. Phys. Lett.* 1999, 75, 2761 - 2763.

[11] Zhang, Y.; Du, G.; Liu, D.; Wang, X.; Ma, Y.; Wang, J.; Yin, J.; Yang, X.; Hou, X.; Yang, S.; *J. Cryst. Growth* 2002, 243, 439 - 443.

[12] Jin, B. J.; Im, S.; Lee, S. Y.; *Thin Solid Films* 2000, 366, 107 - 110.

[13] Ashrafi, A. B. M. A.; Zhang, B.; Binh, N. T.; Wakatsuki, K.; Segawa, Y.; *Jpn. J. Appl. Phys.* 2004, 43, 1114 - 1117.

[14] Zhang, Y.; Du, G.; Yang, X.; Zhao, B.; Ma, Y.; Yang, T.; Ong, H. C.; Liu, D.; Yang, S.; *Semicond. Sci. Technol.* 2004, 19, 755 - 758.

[15] Zhao, B.; Yang, H.; Du, G.; Fang, X.; Liu, D.; Gao, C.; Liu, X.; Xie, B.; *Semicond. Sci. Technol.* 2004, 19, 770 - 773.

[16] Kim, I. S.; Jeong, S. H.; Kim, S. S.; Lee, B. T.; *Semicond. Sci. Technol.* 2004, 19, L29 - L31.

[17] Hirata, G. A.; McKittrick, J.; Cheeks, T.; Siqueiros, J. M.; Diaz, J. A.; Contreras, O.; Lopez, O. A.; *Thin Solid Films* 1996, 288, 29 - 31.

[18] Hwang, K.; Park, Y.; *J. Mater. Res.* 2001, 16(9), 2519 – 2525.

[19] Hwang, K.; Yun, Y.; Kim, Y.; Ryu, H.; Kang, B.; *Jpn. J. Appl. Phys.* 2002, 41, 795 – 799.

[20] Hwang, K.; Yun, Y.; Kang, B.; Kim, S.; Jang, S.; Kim, C.; Oh, J.; *J. Mater. Sci.: Mater. in Med.* 2003, 14, 761 - 765.

[21] Hwang, K.; Manabe, T.; Yamaguchi, I.; Kumagai, T.; Mizuta, S.; *Jpn. J. Appl. Phys.* 1997, 36, 5221 – 5225.

[22] Hwang, K.; Ahn, J.; Kim, B.; *J. Mater. Sci.* 2006, 41, 3151 – 3153.

[23] Hwang, K.; Jeong, J.; Jeon, Y.; Jeon, K.; Kim, B.; *J. Sol-Gel Sci. & Tech.* 2005, 35, 237 – 243.

[24] Kim, B.; An, J.; Jeon, Y.; Jeong, J.; Kang, B.; Hwang, K.; *J. Mater. Sci.* 2005, 40, 237 – 239.

[25] Hwang, K., Kang, B.; Jeon, Y.; An, J.; Kim, B.; Nishio, K.; Tsuchiya, T.; *Surf. & Coatings Tech.* 2005, 190, 331 – 335.

[26] Ryu, H.; Park, J.; Hong, K.; Hwang, K.; Kang, B.; Oh, J.; *Surf. & Coatings Tech.* 2003, 173, 19 –23.

[27] Kim, B.; An, J.; Jeon, Y.; Kang, B.; Hwang, K.; *J. Kor. Phys. Soc.* 2004, 44(4), 1012 – 1014.

[28] Jeong, J.; Ahn, J.; Kim, B.; Jeon, Y.; Jeon, K.; Hwang, K.; *J. Kor. Phys. Soc.* 2005, 46(2), 559 – 561.

[29] Jiménez-González, A.; Soto Urueta, J.; Suárez-Parra, R.; *J. Crys. Growth* 1998, 192, 430 – 438.

[30] Shimono, D.; Tanaka, S.; Torikai, T.; Watari, T.; Murano, M.; *J. Ceram. Proc. Res.* 2001, 2(4), 184 - 188.

[31] Ohyama, M.; Kozuka, H.; Yoko, T.; *Thin Solid Films* 1997, 306, 78 – 85.

[32] Ohya, Y.; Saiki, H.; Takahashi, Y.; *J. Mater. Sci.* 1994, 29, 4099 – 4103.

[33] Wessler, B.; Lange, F. F.; Mader, W.; *J. Mater. Res.* 2002, 17(7), 1644 – 1650.

[34] Ohyama, M.; Kozuka, H.; Yoko, T.; Sakka, S.; *J. Ceram. Soc. Jpn.* 1996, 104(4), 296 – 300.

[35] *NanoScope Command Reference Manual*, Ver. 4.42 (Digital Instruments, Inc., Santa Barbara, CA, 1999), pp. 12.19 – 12.45.

[36] Ryu, H. W.; Hwang, K. S.; *J. Kor. Phys. Soc.* 2002, 40, 493 - 496.

[37] Loder, J. C.; Wielinga, T.; Worst, J.; *Thin Solid Films* 1983, 101, 61 - 73.

[38] Chen, M.; Pei, Z.; Wang, X.; Sun, C.; Wen, L.; *J. Mater. Res.* 2001, 16(7), 2118 – 2123.

[39] Cebulla, R.; Wendt, R.; Ellmer, K.; *J. Appl. Phys.* 1998, 83, 1087 - 1095.

[40] Gupta, V.; Mansingh, A.; *J. Appl. Phys.* 1996, 80(2), 1063 – 1073.

[41] Swanpoel, R.; *J. Phys. E* 1984, 17, 896 - 903.

[42] Park, K.; Ma, D.; Kim, K.; *Thin Solid Films* 1997, 305, 201 - 209.

[43] Ziegler, E.; Heinrich, A.; Oppermann, H.; Stover, G.; *Phys. Status Solidi A* 1981, 66, 635 - 648.

[44] Tansley, T. L.; Neely, D. F.; Foley, C. P.; *Phys. Status Solidi A* 1983, 77, 491 - 496.

[45] Macleod, H. A., *J. Vac. Sci. Tech. A* 1986, 4, 418 - 422.

In: Frontal Nanotechnology Research
Editor: Martin V. Berg, pp. 109-129
ISBN: 1-60021-212-3

Chapter 5

REFLUXING POLYOL SYNTHESIS OF TRANSITION METAL CHALCOGENIDES NANOCRYSTALLITES: A GENERAL PROCESS

***Guozhen Shen*[a,b1] *and Di Chen*[b]**

[a] Nanoscale Materials Center, National Institute for Materials Science (NIMS), Namiki 1-1, Tsukuba, Ibaraki 305-0044, Japan

[b] Nano-materials and Nano-chemistry, Hefei National laboratory for Physical Sciences at Microscale, University of Science and Technology of China, Hefei 230026, P. R. China

ABSTRACT

Transition metal chalcogenides, including sulfides, selenides, and tellurides are important optoelectronic, luminescent, and semiconducting materials. With regard to all of the applications of chalcogenides, highly crystalline particles with almost monodisperse size distribution and regular morphology are required. The synthesis of chalcogenides nanocrystallites under mild conditions is of great interest in both basic research and engineering applications. In this paper, the synthesis of chalcogenides nanocrystallites, such as nanoparticles, nanoplatelets, nanorods, nanowires, and their assembled patterns, via the simple, rapid and convenient refluxing polyol method was reviewed based on our works. Several techniques including powder X-ray diffraction (XRD), scanning electron microscopy (SEM), transmission electron microscope (TEM), X-ray photoelectron (XPS), Raman, photoluminescence (PL), etc. have been used to characterize the produced nanocrystallites. Possible growth mechanism of the refluxing polyol technique to transition metal chalcogenides nanocrystallites is discussed. Combining with other techniques, the refluxing polyol method could be extended to synthesize other important inorganic materials nanocrystallites. Given to the generality of this simple refluxing polyol method, coating and modification of the surfaces of transition metal chalcogenide nanocrystallites may also be fulfilled.

[1] E-mail address: SHEN.Guozhen@nims.go.jp; gzshen@ustc.edu; Fax: +81-29-851-6280.

1. Introduction

The development of novel materials is a fundamental focal point of chemical research, and this interest is mandated by advancements in all areas of industry and technology. Over the past decade, there has been a dramatic increase of interest in both the preparation and properties of nanocrystalline materials.[1-14] This interest has been fueled by the unique properties that such materials possess when compared to bulk phases, as well as the potential they hold for such varied applications as electronics, catalysis, and biological labeling, etc. Transition metal chalcogenides including sulfides, selenides and tellurides are often found in minerals and have attracted considerable attention in recent decades due to the interesting properties they possess.[15-17] They have a number of commercial applications in pigments, semiconductors, fluorescence devices and even superconductors.[18,19] With regard to all of the applications mentioned, highly crystalline particles with almost monodisperse size distribution and regular morphology are required. In fact, many different methods of preparation, liquid-based as well as gas phase-based, are aiming at the fulfillment of these requirements.[20-32]

The polyol method has been intensively investigated for the synthesis of metals, alloys, and oxides nanocrystallites, in which the reducing properties of a high-boiling alcohol (e.g. glycerol, glycol) towards a suitable metal precursor were utilized.[33-36] In the present work, we briefly review our recent progress in the refluxing polyol preparation of metal chalcogenide nanomaterials. The refluxing polyol process can also be combined with microwave irradiation technique.

2. Apparatus and Characterization

All reagents used for the refluxing polyol synthesis were of analytical purity and purchased from different chemical companies and used without further purification. The experimental setup for the refluxing polyol synthesis can be illustrated in Figure 1, which includes a heating system, a refluxing system, and a cooling system. The source materials are put into the round-bottom flask and heated until the finish of the reaction.

The obtained precipitates were washed with absolute alcohol and distilled water for several times and dried in air.

The phase purity of the products was examined by X-ray powder diffraction (XRD) using a Japan Rigaku D/max γA X-ray diffractometer equipped with graphite monochromatized Cu $K\alpha$ radiation (λ = 1.54178 Å). To obtain further information about stoichiometry of some key compounds, X-ray photoelectron spectra was recorded on a VGESCALAB MKII X-ray photoelectron spectrometer using non-monochromatized Mg $K\alpha$ as the excitation source and choosing C_{1s} (284.60 eV) as the reference line. Scanning electron microscope (SEM) measurements were carried out with a field-emission microscope (JEOL, 7500B) operated at an acceleration voltage of 10 kV. Transmission electron microscopy (TEM) investigation was made on a Hitachi H-800 transmission electron microscope with an accelerating voltage of 200 kV.

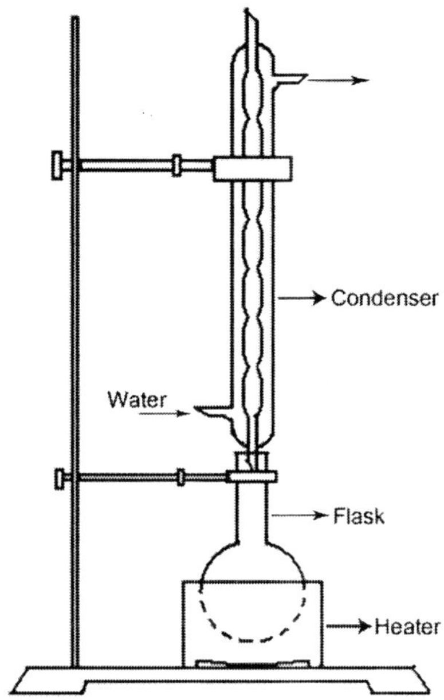

Figure 1. Schematic illustration of the apparatus used in the refluxing polyol process.

3. Results and Discussion

3.1. Preparation of Binary Sulfides Nanocrystallites by the Refluxing Polyol Process

To get transition metal sulfide nanocrystallites, appropriate amount of metal salts and thiourea (Tu) were placed in a round-bottomed flask. Then the flask was filled with glycol solution. Stirred for a few minutes, the solution was heated and refluxed for 0.5-2 h. After the mixture had been cooled, the solid was filtered and washed several times with carbon bisulfide, absolute alcohol and distilled water. Then vacuum dried at 60 ^{o}C for 3 h. Lots of binary sulfides can be obtained by this technique.[37-39] The detail experimental results are shown in Table 1.

Figure 2 shows the SEM image of the synthesized Bi_2S_3 product in the refluxing polyol process. It can be seen that urchin-like patterns consisted of Bi_2S_3 nanorods are synthesized on a large scale. Each urchin-like Bi_2S_3 pattern has a diameter of ~1 μm with acicular crystallites radiating from the center in a uniform size distribution. The nanorods grown from the center have almost uniform diameters of about 50 nm.

The analysis by TEM accompanied with selected area electron diffraction (SAED) gives further information about this product. As shown in Figure 3a, all samples dispersed on the TEM grids show spherical aggregates with diameters of ~1 μm, consistent with the SEM results. Figure 3b is a TEM image of a single Bi_2S_3 nanorod radiating from the center. The corresponding SAED pattern shows its single crystal nature and can be indexed to the reflections of orthorhombic Bi_2S_3 (Figure 3b inset). These unusual urchin-like Bi_2S_3 patterns can be used as self-sacrificing templates for the synthesis of $Bi_2(Se,S)_3$ nanorods and Bi_2Se_3 flakes with similar urchin-like morphology.[40]

Table 1. Results of the products from the reactions between metal salts and thiourea by the simple polyol process

Reactants	Reactive time	Products	Morphology	JCPDS Card No.
$Bi(NO_3)_3{\cdot}5H_2O$+Tu	0.5 h	Bi_2S_3	urchins	17-230
$NiCl_2{\cdot}6H_2O$+Tu	1.5 h	NiS	nanoparticles	75-0613
CuCl+Tu	1 h	$Cu_{1.8}S$	nanoparticles	23-0962
$CuCl_2{\cdot}2H_2O$+Tu	1 h	$CuS+Cu_{1.8}S^a$	nanoparticles	/
$CuSO_4{\cdot}5H_2O$+Tu	1 h	$Cu_{1.8}S+CuS^a$	nanoparticles	/
$PbCl_2$+Tu	0.5 h	PbS	nanoparticles	5-0592
$Pb(NO_3)_2$+Tu	0.5 h	PbS	nanoparticles	5-0592
AgCl+Tu	0.5 h	Ag_2S	nanoparticles	14-0072
$AgNO_3$+Tu	0.5 h	Ag_2S	nanoparticles	14-0072
$InCl_3{\cdot}4H_2O$+Tu	2 h	In_2S_3	nanoflakes	25-390
$FeCl_3{\cdot}6H_2O$+Tu	1.5 h	Fe_3S_4	nanoparticles	16-0713
$SnCl_2$+Tu	1.5 h	SnS	flake-like	39-354
$SnCl_4$+Tu	1.5 h	SnS_2	flake-like	23-677
$AgNO_3+SbCl_3$+Tu	2 h	Ag_3SbS_3	nanoparticles	21-1173
$Cu(NO_3)_2+SbCl_3$+Tu	2 h	Cu_3SbS_3	nanoparticles	75-1574
$CuCl_2{\cdot}2H_2O+InCl_3{\cdot}4H_2O$+Tu	1 h	$CuInS_2$	foam-like	75-106
$AgNO_3+Bi(NO_3)_3{\cdot}5H_2O$+Tu	1.5 h	$AgBiS_2$	nanorods	21-1178

[a] trace product

Figure 2. SEM images of as-prepared product, which shows large-scale urchin-like patterns of three-dimensional aligned Bi_2S_3 nanorods [Reprinted from Ref. 37, Copyright 2003, with permission from Elsevier].

Based on our experimental results, we found that the polyol used has great effect on the final products. For comparison, different polyols are tested. The results reveal that the viscosity and boiling point of polyol play an important role and ethylene glycol is the best one. During the experiment, the ethylene glycol provides a uniform reaction temperature and makes the reactants mixing homogeneously, which is important to the formation of the homogeneously distributed nanoscale crystallines.

Figure 4a is the XRD pattern of the synthesized NiS product by using $NiCl_2 \cdot 6H_2O$ and thiourea as the reactants in ethylene glycol for 1.5 h. All the peaks in this pattern can be indexed to hexagonal phase NiS (JCPDS Card No. 75-0613). No impurities such as Ni, NiO_x or intermediary phase nickel sulfides are detected in the XRD pattern. According to Scherrer's formula, the average size of the prepared NiS is about 25 nm. Figure 4b is the corresponding TEM image of NiS product, which reveals the formation of nanoparticles with average diamters of about 35 nm.

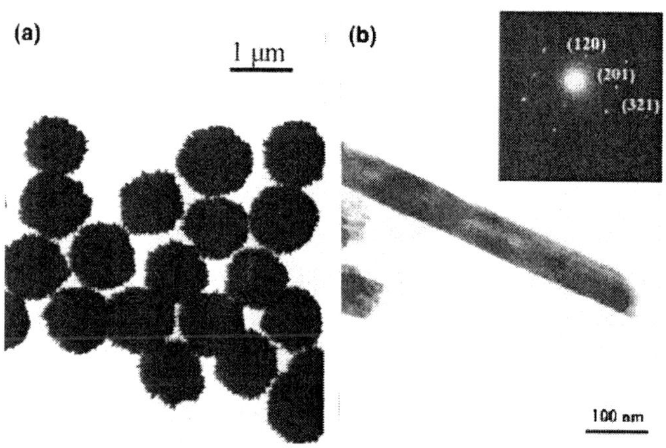

Figure 3. (a) TEM image of Bi_2S_3 urchin-like patterns; (b) TEM image of a single Bi_2S_3 nanorod (inset: the corresponding ED pattern) [Reprinted from Ref. 37, Copyright 2003, with permission from Elsevier].

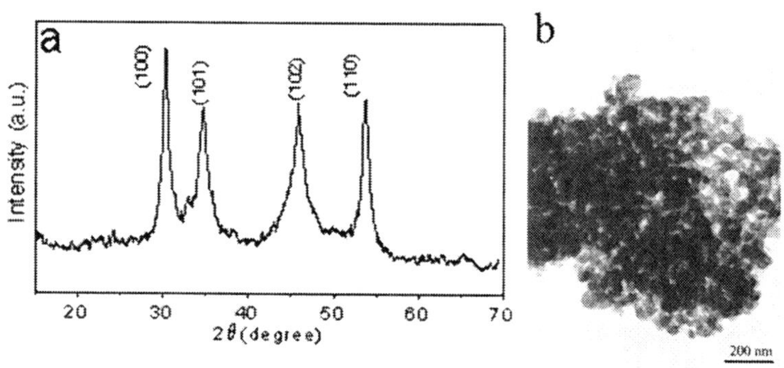

Figure 4. (a) XRD pattern of NiS; (b) TEM image of NiS [Reprinted from Ref. 38, Copyright 2003, with permission from Elsevier].

In fact, in the refluxing polyol process, it was found that the metal salts used have no obviously influence on the morphology and structure of the product. Figure 5 is the XRD patterns of produced PbS by using $PbCl_2$ and $Pb(NO_3)_2$ as the metal source, respectively. All peaks in both patterns can be indexed to PbS with cubic cell. The phases and morphologies of these two samples are the same. TEM observation shows that they are of similar morphology and size, namely, the average size of both PbS products is about 50~120 nm.

Table 2. The relationship between viscosity of different solvents and the morphology of the products [Reprinted from Ref. 37, Copyright 2003, with permission from Elsevier]

Solvent	Viscosity ($mN \cdot s \cdot m^{-2}$)/25°C	Morphology
Distilled water	0.89	Aggregates of nanorods
Mixture of water and EG	0.89~26.09	Nanorods and urchin-like patterns
Ethylene glycol (EG)	26.09	Uniform urchin-like patterns
Polyol ethylene glycol 200	>26.09	Ununiform urchin-like patterns

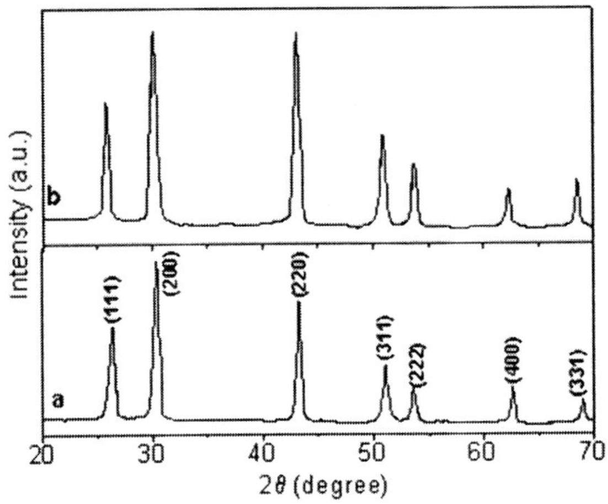

Figure 5. XRD patterns of PbS obtained (a) from $PbCl_2$; (b) $Pb(NO_3)_2$ [Reprinted from Ref. 38, Copyright 2003, with permission from Elsevier].

Besides the analysis of XRD, EDX analysis was also used to determine the composition of these products. Typical EDX pattern of as-obtained PbS sample is shown in Figure 6. The results show that the PbS nanoparticles are composed of the elements Pb and S and the ratio of Pb:S is 1:1.07, in agreement with the expected value (the Cu peaks arise from the copper grid). EDX analyses of other products have the similar results, which are also in good agreement with the expected values of final products.

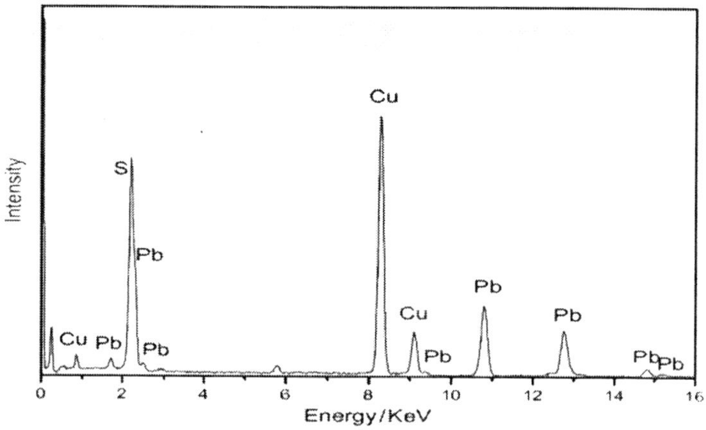

Figure 6. EDX spectrum of the PbS sample [Reprinted from Ref. 38, Copyright 2003. with permission from Elsevier]

3.2. Preparation of Ternary Sulfides Nanocrystallites by the Simple Polyol Process

By carefully controlled the experimental parameters, the refluxing polyol process was found to be very useful for the synthesis of ternary sulfides nanocrystallites.[41-44]

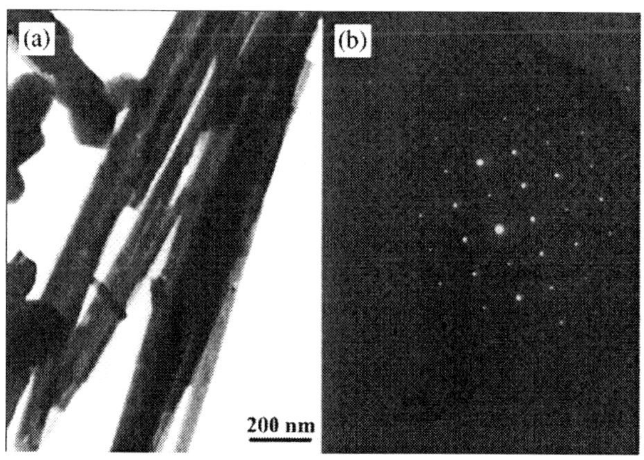

Figure 7. (a) TEM image of the synthesized $AgBiS_2$ nanorods. (b) the corresponding SAED pattern [Reprinted from Ref. 42, Copyright 2003, with permission from Elsevier].

Ternary metal chalcogenide $AgBiS_2$ belongs to the class of I–V–VI semiconducting materials, which are technologically important in linear, nonlinear, optoelectronic, and thermoelectric devices as well as optical recording media.[45] The TEM image (Figure 7a) shows the $AgBiS_2$ samples obtained in the present refluxing polyol process consist of nanowires with ~40 nm in diameter and length up to 10 μm. The corresponding SAED pattern shown in Figure 7b indicates their single-crystal nature.

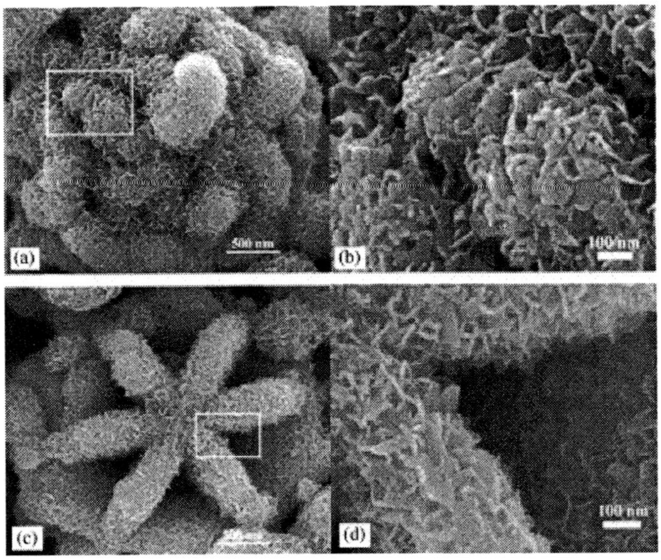

Figure 8. SEM images of the as-prepared $CuInS_2$ product, which clearly shows the porous foam-like microstructures [Reprinted from Ref. 43, Copyright 2003, with permission from Elsevier].

Due to its direct energy gap of 1.5 eV, ternary compound $CuInS_2$ with the chalcopyrite structure has emerged as a leading material for high-efficiency and radiation-hard solar cell applications. In addition, $CuInS_2$ is also a candidate for photochemical applications because its band gap matches well the solar spectrum.[46,47] Through the reaction of $CuCl_2$, $InCl_3$ and thiourea in ethylene glycol at 195 ^oC, porous foam-like $CuInS_2$ nanocrystallines were obtained. The low-magnification SEM in Figure 8a shows the formation of a porous and sponge-like structure with an extended network of features separated by some pores-like structures. The fine texture of the foam features can be clearly seen in Figure 8b. No existence of grainy matter was observed in the product. The structure resembles a network of interconnected flakes forming irregular polygonal cavities with different sizes and the wall thickness is very thin and less than 10 nm. We also observed $CuInS_2$ flower- and hexagram-like crystals composed of foam-like microstructure (Figure 8c, 8d).

The growth process for binary or ternary metal sulfides in the refluxing process can be expressed as follows. When metal salts mixed with thiourea in ethylene glycol, metal-thiourea complexes can form according to literatures.[48,49] In the polyol route, the relatively high solubility of metal salts and thiourea in polyol solution and the change of solution color also reveal the strong complex action between metal ions and thiourea. So the formation of nanoscale metal sulfides crystallines in glycol may be through the following steps: first, metal ions can combine with thiourea molecules to form metal-thiourea complexes in the solution. Second, at a given temperature, the stability of metal-thiourea complexes decreases and undergoes thermal decomposition to form metal sulfides. Meanwhile, it is well known that the thiourea takes part in the following reversible decomposition reaction in solution:

$$NH_2CSNH_2 \leftrightarrow CH_2N_2 + H_2S \quad (1)$$

The formed CH_2N_2 reacts with H_2O at high temperature to produce NH_3 and CO_2:

$$CH_2N_2 + H_2O \leftrightarrow NH_3 + CO_2 \quad (2)$$

So the decomposition of NH_2CSNH_2 can be expressed as:

$$NH_2CSNH_2 + H_2O \leftrightarrow NH_3 + CO_2 + H_2S \quad (3)$$

In this reaction, the trace water (coming from analytical polyol or hydrated metal salts in our experiments) plays a critical role in the formation of metal sulfides crystals, since it can promote the decomposition of NH_2CSNH_2 and the formation of metal sulfides. In the synthesis of ternary metal sulfide nanocrystallites, the molar ratio of metal salts should be well controlled; otherwise, binary metal sulfide would be produced among with the ternary sulfide products.

3.3. Preparation of Selenides Compounds by the Rapid Ethylenediamine-Assisted Polyol Process

Ethylenediamine (EN) is strongly basic and elemental chalcogen is activated by a nucleophilic attack of the amine to form E^{2-} (E=S, Se) in the polyol route. So ethylenediamine-assisted polyol process was designed to synthesize metal selenide nanocrystallites.[50-52] Table 3 shows the products obtained from the ethylenediamine(en)-assisted polyol route.

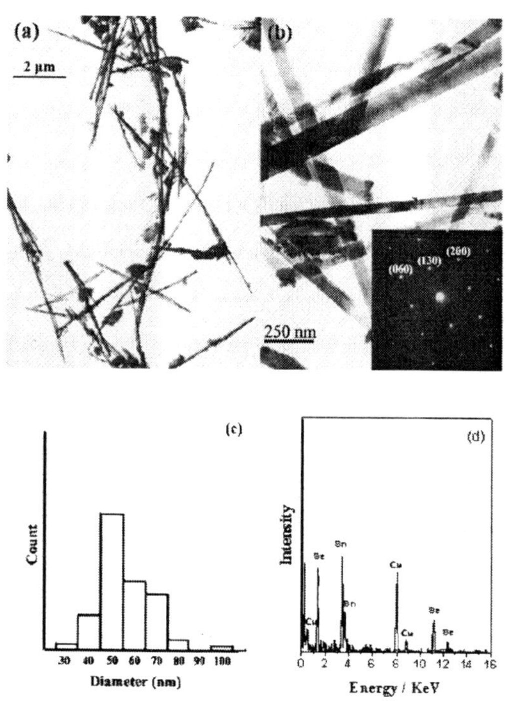

Figure 9. (a,b) TEM images of the as-grown SnSe nanowires. (c) the diameter distribution of SnSe nanowires. (d) EDX spectrum of the SnSe nanowrie [Reprinted from Ref. 51, Copyright 2003, with permission from the Chemical Society of Japan].

Table 3. Results of the products from the rapid ethylenediamine-assisted polyol process

Reactants	Reactive time	Products	Morphology	JCPDS Card No.
$SbCl_3$+S+DEG+EN	1.5 h	Sb_2S_3	nanowires	6-474
$SbCl_3$+Se+DEG+EN	1.5 h	Sb_2Se_3	nanowires	72-1184
$ZnCl_2$+Se+EG+EN	1 h	ZnSe	spheres	80-0021
$CdCl_2$+Se+EG+EN	1 h	CdSe	flowers	77-0021
$SnCl_2 \cdot 2H_2O$+Se+DEG+EN	1.5 h	SnSe	nanowires	32-1382

SnSe belongs to the interesting class of isomorphic materials that are in many ways between two-dimensional systems and three-dimensional crystals. It has been widely used in infrared optoelectronic devices, memory switching devices and holographic recording devices.[53-55] As shown in Figures 9a and 9b, the as-prepared product are SnSe nanowires having a narrow size distribution with a mean diameter of ~50 nm (Figure 9c). The observed average aspect ratio of these SnSe nanowires is about 150. The electron diffraction pattern (Figure 9b inset) taken from the selected area of SnSe nanowires shown in Figure 9b shows that the as-prepared SnSe nanowires are well crystallized.

Energy-dispersive X-ray (EDX) analysis was used to determine the composition of these nanowires. It (Figure 9d) shows that these nanowires are composed of the elements Sn and Se and the ratio of Sn:Se is 1:1.02, in agreement with the expected values (the Cu peaks arise from the copper grid).

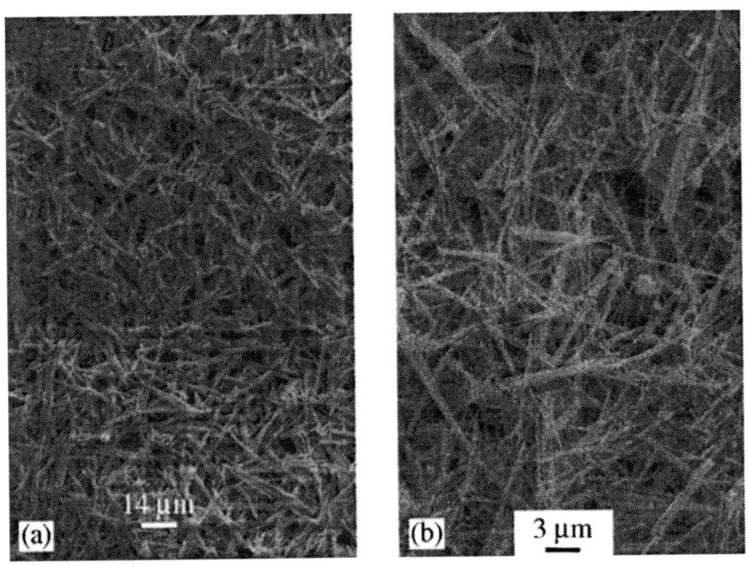

Figure 10. SEM image of (a) Sb_2S_3 and (b) Sb_2Se_3 nanowires obtained in the ethylenediamine(en)-assisted polyol route [Reprinted from Ref. 50, Copyright 2003, with permission from Elsevier].

Group V–VI binary compounds (Sb_2S_3, Sb_2Se_3), are highly anisotropic semiconductor with a layered structure parallel to the growth direction and with orthorhombic phase crystal structure, which is known to adopt a number of packing structures resulting in either trigonal prismatic or octahedral coordination of the metals within a layered matrix of chalcogens.[56,57] The compounds have a great deal of attention due to its switching effects and excellent photovoltaic and thermoelectric power (TEP), which allow possible applications in photo-electrochemical devices, optical and thermoelectric cooling devices, decorative coatings and solar selective.[58] In the ethylenediamine-assisted polyol process, Sb_2S_3, and Sb_2Se_3 nanowires are prepared by the reaction of $SbCl_3$ and S (or Se) powder. The SEM images are shown in Figure 10a and 10b. They showed that the obtained nanowires are with typical lengths in the range of several to several tens of micrometers. While the lengths of as-prepared Sb_2Se_3 nanowires range from 1 to 15 μm.

As a direct band gap material, zinc selenide, ZnSe, can be widely used in many fields. It exhibits great potential for various optoelectronic and high-speed applications like blue–green light emitting diodes, photoluminescent and electroluminescent devices, lasers and thin film solar cells. Lots of methods of synthesizing ZnSe have been developed, including molecular beam epitaxy (MBE), metalorganic chemical vapor deposition (MOCVD), solvothermal or hydrothermal route, organometallic vapor phase epitaxy (OMVPE) and some other methods. Figure 11 is the SEM image of the synthesized ZnSe product. It can be seen that the as-prepared sample composed of relatively uniform spheres crystals with a diameter of about 1 μm on average. Besides the spheres, some nanorods are also detected, which have a diameter of about 50 nm and length up to 10 μm. The ratio of this structure is about 5%. XRD analysis reveals that the ZnSe spheres obtained in the present enthylenediamine-assisted polyol process are of cubic structures.

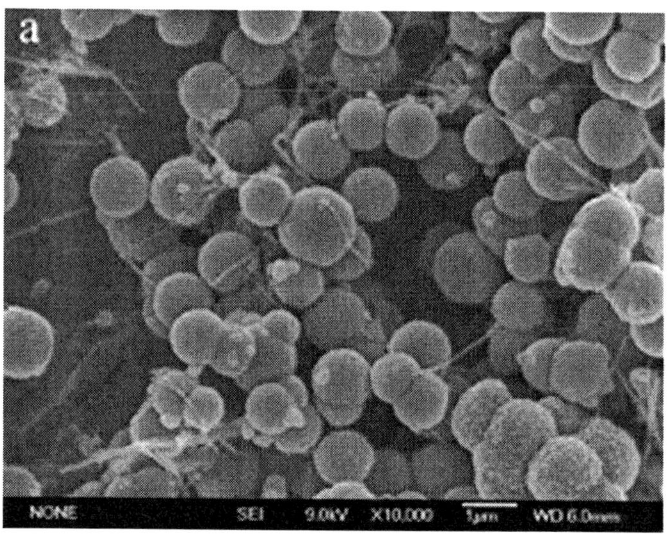

Figure 11. FE-SEM image of as-prepared sample, which shows the formation of ZnSe spheres with a diameter of about 1 μm [Reprinted from Ref. 52, Copyright 2003, with permission from Elsevier].

It is well known that solvent plays an important role in determining the crystal morphology. Solvents with different physical and chemical properties can influence the solubility, reactivity, and diffusion behavior of the reactants, in particular, the polarity and coordinating ability of the solvents can influence the crystal morphology of the final product. EN is strongly basic and elemental E is activated by a nucleophilic attack of the amine to form E^{2-} (E=S, Se) in the polyol route. On the other hand, the N-chelating behavior makes EN a strong donor ligand, and it plays an important role in the formation of one-dimensional nanowires. EN can react with metal ions to form relatively stable antimony complexes, which may serve as molecular templates in control of the crystal growth. We deduced the growth mechanism of metal selenide nanocrystallites as follows. (1) metal salts dissolved in EN to form the chain type structure of metal ions complexes under the template effect of EN; (2) Elemental E is activated by EN to form E^{2-}; (3) At appropriate temperature, E^{2-} attacks the metal ions complexes and the ligands are lost gradually to form the final metal selenide nanocrystallites.

3.4. Preparation of other Chalcogen-Related Compounds by the Simple Refluxing Polyol Process

As important members of the semiconductor family, selenium and tellurium show unique physical and chemical properties, which provide them with the chances to widen the range of applications. The polyol synthesis of Se and Te nanorods is carry out under atmosphere and the chemical reaction we employed can be formulated as the following equations:[59]

$$Na_2SeO_3+(NH_4)_2S_2O_3+2H_2O \rightarrow Se+Na_2SO_4+(NH_4)_2SO_4+2H_2 \quad (1)$$

$$Na_2TeO_3+(NH_4)_2S_2O_3+2H_2O \rightarrow Te+Na_2SO_4+(NH_4)_2SO_4+2H_2 \quad (2)$$

Figure 12. (a) TEM image of submicron Se rod-like crystals; (b) TEM image of a single submicron Se rod-like crystal; (c) SAED pattern of (b); (d) SEM image of submicron Te rod-like crystals [Reprinted from Ref. 59, Copyright 2004, with permission from Elsevier].

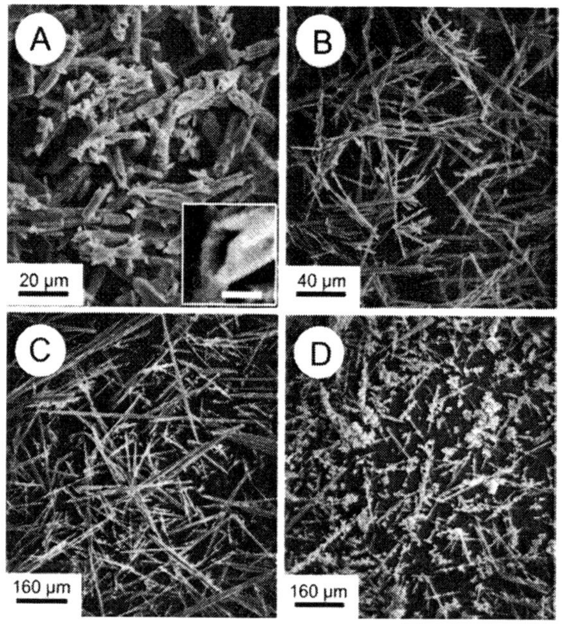

Figure 13. SEM images of the as-obtained Se crystals under different reaction conditions: (a) 150 °C, 0.5 h; (b) 195 °C, 0.5 h; (c) 180 °C, 5 h; (d) 180 °C, without stirring [Reprinted from Ref. 59, Copyright 2003, with permission from Elsevier].

Figure 12a is the SEM image of as-obtained Se nanorods with diameter of ~500 nm and length up to 15 μm. TEM result (Figure 12b) shows the long Se rod like crystal with diameter of about 450 nm, consisting with the SEM results. The corresponding selected area electron diffraction (SAED) pattern shows its single crystal nature (Figure 12c). SEM image of as-obtained Te sample shown in Figure 12d also reveals that the obtained Te sample is consisted of uniform rod-like crystals. The average diameter of these nanorods is ca. 400 nm and length ranges from 2 mm to 10 μm.

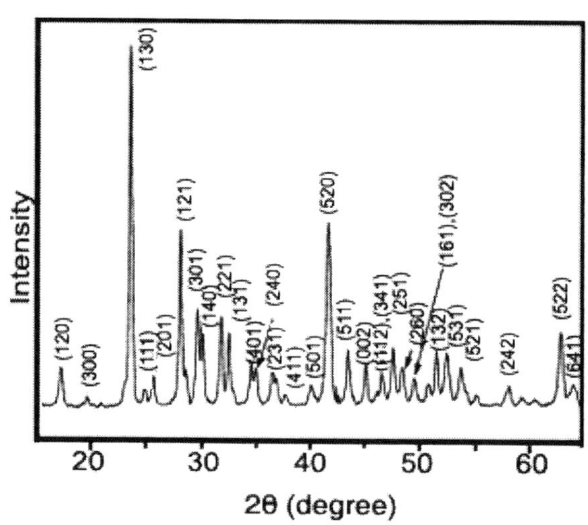

Figure 14. XRD pattern of the synthesized $(Bi(Bi_2S_3)_9I_3)_{0.667}$ needlelike crystals [Reprinted from Ref. 61, Copyright 2003, with permission from Elsevier].

For the formation of submicron Se and Te rod-like crystals, the feature of the present polyol route is the high boiling point of ethylene glycol, which crystallizes the as-obtained products at relatively low temperature and short reaction time. Both reaction time and reaction temperature have influences on the morphology of the obtained products. If the reaction was performed at low temperature, the yield of Se product decreased dramatically and they are of tube-like morphology instead of rod-like morphology (Figure 13a). High temperature and long reaction time are better for the crystallinity of the products but would result in the increase of diameters to the microscale (Figure 13).

BiSI and its related chalcogenide halide compounds belonging to the V–VI–VII group ternary compounds are ferroelectric, electro-optic, electric-mechanical and photo-electric semiconductors. Many studies in these compounds have focused on the synthesis and characterization of BiSI. However, few reports could be found on the synthesis and characterization of complexed $(Bi(Bi_2S_3)_9I_3)_{0.667}$ due to the difficulty in obtaining stoichiometric compounds since it was first reported by Miehe in 1971.[60] Here, $(Bi(Bi_2S_3)_9I_3)_{0.667}$ needlelike crystals were obtained by refluxing in ethylene glycol solution.[61] We used $Bi(NO_3)_3 \cdot 6H_2O$, $(NH_2)_2CS$ and NaI as the source materials. As shown in Figure 14, the XRD pattern can be indexed to hexagonal cell of $(Bi(Bi_2S_3)_9I_3)_{0.667}$, with lattice parameters a=15.66 Å, c=4.02 Å. The SEM image indicates that the needles have diameter of 400 nm and length up to 15 μm (Figure 15).

Figure 15. SEM image of the $(Bi(Bi_2S_3)_9I_3)_{0.667}$ needlelike crystals [Reprinted from Ref. 61, Copyright 2003, with permission from Elsevier].

We have reported the synthesis of cubic phase ZnSe with sphere morphology in the enthylenediamine-assisted polyol process when refluxing the source material in ethylene glycol solution. However, we found that when we added some polyethylene glycol 200 into the solution, $(ZnSe)_2 \cdot EN$ will be obtained as shown in Figure 16a.[62] If the $(ZnSe)_2 \cdot EN$ precursor is annealed at 500 °C in N_2 gas, XRD result shows the formation of hexagonal ZnSe product (Figure 16b).

SEM images of the as-prepared $(ZnSe)_2 \cdot EN$ precursor are shown in Figures 17a,b. The low-magnification image (Figure 17a) shows that large-scale disk-like crystals are obtained when PEG-200 was added to the present route. Most of the disk-like crystals are of uniform hexagonal morphology. The typical SEM image of the hexagonal disk-like crystal is shown in Figure 17b, which clearly indicates that the crystal is about 1.5 μm in length of side. And the thickness of these disk-like crystals is about 100 nm on average. Figure 17c and 17d are the SEM images of the ZnSe product after annealing the $(ZnSe)_2$·EN precursor. A low-magnified image (Figure 17c) shows that annealed at 500 °C, the sample is also with the hexagonal-shaped morphology. But from a high-magnified image (Figure 17d), one can see that this morphology is quite different with the simple aggregates of individual nanoparticles, for that each hexagonal-shaped crystals are composed with little nanocrystals connected with each other and we called it the quasi-network structure. And the pores in the quasi-network may be caused by the release of EN from the $(ZnSe)_2$·EN precursor under high temperature.

In the present polyol process, no $(ZnSe)_2$·EN precursor could be obtained without the addition of PEG-200. When we substituted PEG-200 with other polyols, from oligomeric polyols, such as DEG, TEG, to higher polyethylene glycols, such as PEG400, PEG 600, PEG20000, no $(ZnSe)_2$·EN precursor could be obtained. XRD pattern showed that the product in all these solvents is cubic phase ZnSe. It has been well known that the non-ionized surfactant PEG forms chainlike structure due to the assembly in the reaction system. Surfactant PEG with lower molecular weights gets more flexible and looser aggregate.[63]

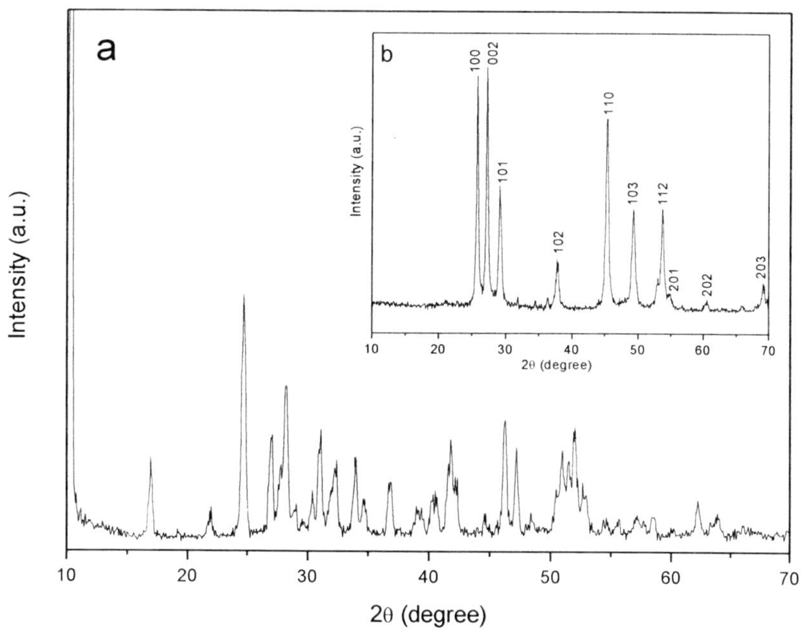

Figure 16. (a) XRD pattern of the synthesized $(ZnSe)_2$·EN precursor. (b) XRD pattern of the annealed sample, indicating the formation of hexagonal ZnSe product [Reprinted from Ref. 62, Copyright 2004, with permission from MRS].

Therefore, we can imagine that the loose and flexible aggregate of PEG-water (coming from EN and EG solution) divided reaction space into linear reaction fields when PEG and a small quantity of water were mixed together. The best polyol in the synthesis of (ZnSe)·2EN in the present process is PEG200 just as proved by numerous experiments.

Figure 17. (a,b) SEM image of the obtained $(ZnSe)_2$·EN precursor. (c,d) SEM image of the annealed samples [Reprinted from Ref. 62, Copyright 2004, with permission from MRS].

3.5. Preparation of Sulfides Compounds by the Rapid Microwave Polyol Process

Since 1986, microwave irradiation as a heating method has found a number of applications in chemistry.[64] The microwave synthesis, which is quite fast, simple and environmentally friendly, has been widely used in various fields including molecular sieve preparation, plasma chemistry, analytical chemistry catalysis, inorganic complexes synthesis and organic reactions, etc. More importantly, it is a rapidly developing area of research, which has attracted much attention as a new method for preparing nanoscale inorganic compounds.[65] If the microwave technique could be combined with the simple refluxing polyol process, transition metal chalcogenide nanocrystallites may be prepared at much lower cost. Manipulations and reactions were carried as follows: appropriate amount of metal salts and thiourea were used to obtain metal sulfides (Table 4).[66-68]

For example, IV–VI intermetallic compounds, SnS and SnS_2 are increasingly important because of their specially semiconducting properties. For example, flaky SnS with the orthorhombic phase is a narrow band gap semiconductor with an optical band gap of 1.08 eV, which is a kind of ideal material for photoelectric energy conversion. And SnS thin films have also exhibited conversion efficiency similar to the one found for silicon films (up to 25%) in photovoltaic devices.[69,70] On the other hand, layered SnS_2 with the layer CdI_2-type structure has a layer band gap of about 2–3 eV.Broad band gap leads to photoconductions, and makes it potential to be a candidate in solar cells and opto-electronic devices.[71-73]

Table 4. Results of the products from the reactions between metal salts and thiourea by the microwave polyol process

Reactants	Products	Phase	Morphology
$CdCl_2 2.5H_2O$+Tu	CdS	Hexagonal	nanoparticles
$Zn(NO_3)_2 \cdot 6H_2O$+Tu	ZnS	Hexagonal	nanoparticles
$CoCl_2 \cdot 6H_2O$+Tu	CoS	Hexagonal	nanoparticles
$PbCl_2$+Tu	PbS	Cubic	dendrite
$CuCl_2 \cdot 2H_2O$+Tu	CuS, $Cu_{2-x}S$	/	nanoparticles
$Bi(NO_3)_3 \cdot 5H_2O$ +Tu	Bi_2S_3	Orthorhombic	radiate Nanorods
$SbCl_3$+Tu	Sb_2S_3	Orthorhombic	nanorods
$AgNO_3$+Tu	Ag_2S	Monoclinic	nanoparticles
$SnCl_2 \cdot 2H_2O$+Tu	SnS	Orthorhombic	flakes
$SnCl_4$+Tu	SnS_2	Hexagonal	layer-rolled

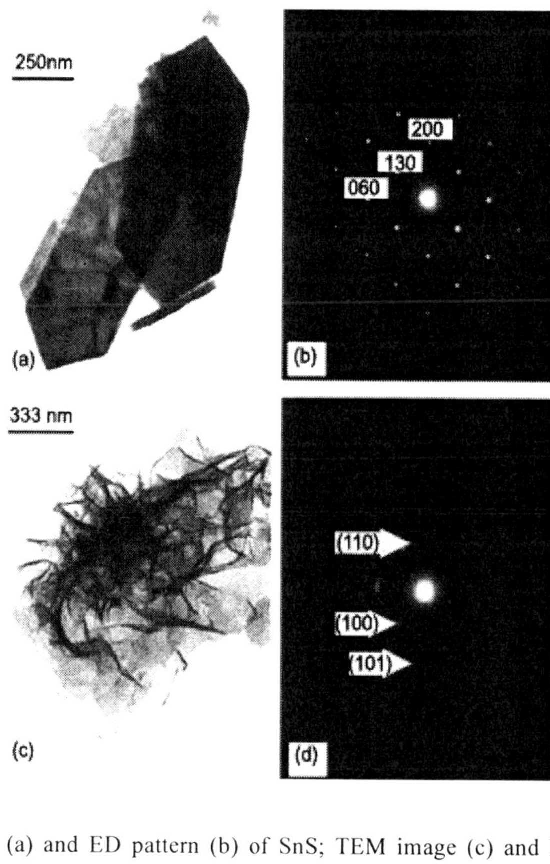

Figure 18. TEM image (a) and ED pattern (b) of SnS; TEM image (c) and ED pattern (d) of SnS_2 [Reprinted from Ref. 66, Copyright 2004, with permission from Elsevier].

Figure 18 shows the different morphologies of the as-prepared SnS and SnS_2 products. From Figure 18a, one can see that the orthorhombic phase SnS crystallines consist of nanoflakes with hexagonal shapes. The corresponding SAED pattern (Figure 18b) shows that the flaky SnS is single-crystal. Figure 18c shows the layer-rolled crystals morphology of hexagonal phase SnS_2, which is different from that of the obtained SnS. The corresponding

SAED pattern shown in Figure 18d shows the relatively good crystallinity of the SnS_2 layer-rolled crystals.

The proposed mechanism of microwave-assisted ethylene glycol synthesis of metal sulfides nanocrystallites may be summarized as below: firstly, the strong complexation between M^{n+} and thiourea lead to the formation of M-Thiourea complexes in the microwave synthesis, which prevent the production of a large number of free S^{2-} in the solution and will be favorable for the formation of the products. Secondly, the M-Thiourea complexes undergo thermal decomposition under the microwave irradiation to produce metal sulfides.

In this microwave synthesis, ethylene glycol (EG) and microwave irradiation play key roles for the preparation of metal sulfides. EG acting as both reaction media and dispersion media can efficiently adsorb and stabilize the surface of the particles and favor producing monodispersed metal sulfides crystals with good dispersivity. Meanwhile, EG with the high permanent dipole is an excellent susceptor of the microwave irradiation, which can take up the energy from the microwave field and get the polar reaction solution heated up to high temperature instantaneously.[74,75] All these help to the decomposition of M-Thiourea complexes and the formation of the products. Secondly, microwave irradiation as a heating method affords a uniform growth environment for the formation of nanosized metal sulfides. With microwave irradiation of reactants in polar solution, temperature and concentration gradients can be avoided, providing a uniform environment for the nucleation and the growth of the products. So it can be assumed that microwave irradiation not only provides the energy for the decomposition of the M-Thiourea complexes, but also greatly accelerates the formation of metal sulfides as a result of the fast and homogenous heating effects of microwaves.

3. CONCLUSION

Transition metal chalcogenides nanocrystallites are highly interesting materials from many points of views. They are important optoelectronic, luminescent, and semiconducting materials. In this chapter, we reviewed the synthesis and characterization of transition metal chalcogenides by the refluxing polyol process. The rapid refluxing polyol process is proved to be a general process for the synthesis of chalcogenide nanocrystallites with many kinds of morphologies, such as nanoparticles, nanorods, nanowires etc. Transition metal chalcogenides nanocrystallites can also be synthesized by combining the refluxing polyol process with the microwave techniques.

REFERENCES

[1] Iijima, S. *Nature* 1991, 354, 56.
[2] Golbstein, A. N.; Echer, C. M.; Alivisatos, A. P. *Science* 1992, 256, 1425.
[3] Alivisatos, A. P. *Science* 1996, 271, 933.
[4] Murray, C. M.; Norris, D. J.; Bawendi, M. G. *J. Am. Chem. Soc.* 1993, 115, 8706.
[5] Weller, H. *Angew. Chem.* 1996, 35, 1079.
[6] Li, W. Z.; Xie, S. S.; Qian, L. S.; Chang, B. H.; Zou, B. S.; Zhou, W. Y.; Zhao, R. A.; Wang, G. *Science* 1996, 274, 1701.

[7] Wang, Z. L. *Adv. Mater.* 1998, 10, 13.
[8] Rao, C. N. R.; Muller, A.; Cheetham, A. K. *the Chemistry of Nanomaterials: Synthesis, Properties and Applications*, Wiley-VCH, 2004.
[9] Zhang, J. Z.; Wang, Z. L. *Self-Assembled Nanostructures*, Kluwer Academic Publisher, 2002.
[10] Lauhon, L. J.; Gudiksen, M. S.; Lieber, C. M. *Phil. Trans. R. Soc. Lond. A.* 2004, 362, 1247.
[11] Xia, Y. N.; Yang, P. D.; Sun, Y. G.; Wu, Y. Y.; Mayer, B.; Gates, B.; Yin, Y. D.; Kim, F.; Yan, H. Q. *Adv. Mater.* 2003, 15, 353
[12] Law, M.; Goldberg, J.; Yang, P. D. *Annu. Rev. Mater. Res.* 2004, 34, 83.
[13] Jun, Y. W.; Lee, J. H.; Choi, J. S.; Cheon J. *J. Phys. Chem. B* 2005, 109, 14795.
[14] Shen, G. Z.; Tang, K. B.; Qian, Y. T.; Chen, D. *Int. J. Nanotechnology* 2004, 1, 366.
[15] S.M. Stuczynski, Y.U. Know, M.L. Steigerwald, J. *Organomet. Chem.* 1993, 449, 167.
[16] Feng, P. Y.; Bu, X. H.; Zheng, N. F. *Accounts Chem. Res.* 2005, 38, 293.
[17] Rajamathi, M.; Seshadri, R. *Current Opin. Solid State and Mater. Sci.* 2002, 6, 337.
[18] Greenwood, N. N.; Earnshaw, E. A. *Chemistry of the Elements*, Pergamon, Oxford, 1990.
[19] Nickless, G. *Inorganic Chemistry of Sulfur*, Elsevier, London, 1968, p. 670.
[20] Qi, L. M.; Colfen, H.; Antoniette, M. *Nano Lett.* 2001, 1, 61.
[21] Walton, R. I. *Chem. Soc. Rev.* 2002, 31, 230.
[22] Gorer, S. Penner, S. M. *J. Phys. Chem. B.* 1999, 103, 5750.
[23] Turco Lierri, V.; Rossi, M.; Arrigo, G. D.; Manno, D.; Micocci, G. *Appl. Phys. A.* 1999, 69, 369.
[24] Zhu, J. J.; Liu, S.; Palchik, O.; Koltypin, Y.; Gedanken, A. *J. Solid State Chem.* 2003, 153, 342.
[25] Shen, G. Z.; Cho, J. H.; Yoo, J. K.; Yi, G. C.; Lee, C. J. *J. Phys. Chem. B* 2005, 109, 9294.
[26] Shen, G. Z.; Lee, C. J. *Cryst. Growth and Design* 2005, 5, 1085.
[27] Jiang, Y.; Meng, X. M.; Liu, J.; Hong, Z. R.; Lee, C. S.; Lee, S. T. *Adv. Mater.* 2003, 15, 1195.
[28] Li, Q.; Wang, C. R. *Appl. Phys. Lett.* 2003, 83, 359.
[29] Parak, W. J.; Gorion, D.; Pellegrino, T.; Zanchet, D.; Micheel, C.; William, S. C.; Boudreau, R.; Le Gros, M. A.; Larabell, C. A.; Alivisatos, A. P. *Nanotechnology*, 2003, 14, R15.
[30] Aldana, J.; Lavelle, N.; Wang, Y. J.; Peng, X. G. *J. Am. Chem. Soc.* 2005, 127, 2496.
[31] Yu, S. H.; Yang, J.; Han, Z. H.; Zhou, Y.; Yang, R. Y.; Qian, Y. T.; Zhang, Y. H. *J. Mater. Chem.* 1999, 9, 1283.
[32] Li, B.; Xie, Y.; Huang, J. X.; Liu, Y.; Qian, Y. T. *Chem. Mater.* 2000, 12, 2614.
[33] F. Fievet, J.P. Lagier, M. Figlarz, *MRS Bull.* 1989, 29.
[34] L.K. Kurihara, G.M. Chow, P.E. Schoen, *Nanostruct. Mater.* 1995, 5, 607.
[35] P. Toneguzzo, G. Viau, O. Acher, F. Fievet-Vincent, F. Fievet, *Adv. Mater.* 1998, 10, 1032.
[36] S. Sun, C.B. Murray, D. Weller, L. Folks, A. Moser, *Science* 2000, 287, 1989.
[37] Shen, G. Z.; Chen, D.; Tang, K. B.; Li, F. Q.; Qian, Y. T. *Chem. Phys. Lett.* 2003, 370, 334.

[38] Shen, G. Z.; Chen, D.; Tang, K. B.; Liu, X. M.; Huang, L. Y.; Qian, Y. T. *J. Solid State Chem.* 2003, 173, 232.
[39] Shen, G. Z.; Chen, D.; Tang, K. B.; Huang, L. Y.; Qian, Y. T.; Zhou, G. E. *Inorg. Chem. Commun.* 2003, 6, 178.
[40] Shen, G. Z.; Chen, D.; Tang, K. B.; Qian, Y. T. *Nanotechnology* 2004, 15, 1530.
[41] Chen, D.; Shen, G. Z.; Tang, K. B.; Jiang, X.; Huang, L. Y.; Jin, Y.; Qian, Y. T. *Mater. Res. Bull.* 2003, 38, 509.
[42] Shen, G. Z.; Chen, D.; Tang, K. B.; Qian, Y. T. *J. Cryst. Growth* 2003, 252, 199.
[43] Shen, G. Z.; Chen, D.; Tang, K. B.; Fang, Z.; Sheng, J.; Qian, Y. T. *J. Cryst. Growth* 2003, 254, 75.
[44] Shen, G. Z.; Chen, D.; Tang, K. B.; Qian, Y. T. *J. Cryst. Growth* 2003, 257, 293.
[45] Snmanta, L.K.; Chatterjee. S. *Phys. Stat. Sol. B* 1994, 182, 85.
[46] Birkmire, R.W.; Eser. E. *Annu. Rev. Mater. Sci.* 1997, 27, 625.
[47] Thiel; F.A. *J. Electrochem. Soc.* 1982, 129, 1570.
[48] Karanjai, M.K.;Dasgupta; D. *Thin Solid Films* 1987, 155, 309.
[49] Yu, S.H.; Shu, L.; Yang, J.; Han, Z.H.; Qian, Y.T.; Zhang, Y. H. *J. Mater. Res.* 1999, 14, 4157.
[50] Shen, G. Z.; Chen, D.; Tang, K. B.; Jiang, X.; Qian, Y. T. *J. Cryst. Growth* 2003, 252, 350.
[51] Shen, G. Z.; Chen, D.; Jiang, X.; Tang, K. B.; Liu, Y. K.; Qian, Y. T. *Chem. Lett.* 2003, 32, 426.
[52] Shen, G. Z.; Chen, D.; Tang, K. B.; Qian, Y. T. *J. Cryst. Growth* 2003, 257, 276.
[53] Maier, H.; Daniel, D. R. *J. Electron. Mater.* 1971, 6, 693.
[54] Baxter, C. R.; Melennam, M. *J. Vac. Sci. Technol.* 1975, 12, 110.
[55] Valiukonis, G.; Guseinova, D. A.; Krivaite, G.; Sileika, A. *Phys. Status Solidi B* 1986, 139, 299.
[56] Zheng, X.W.; Xie, Y.; Zhu, L. Y.; Jiang, X. C.; Jia, Y. B.; Song, W. H.; Sun. Y. P. *Inorg. Chem.* 2002, 41, 455.
[57] Hulliger, F. *Structural Chemistry of Layered-Type Phase.* , Reidel, Dordrecht, 1976.
[58] Rajapure, K.; Lokhande, C.; Bhosele, C. *Thin Solid Films* 1997, 311 114.
[59] Shen, G. Z.; Chen, D.; Tang, K. B.; Qian, Y. T. *Mate. Res. Bull.* 2004, 39, 2077.
[60] Miehe, G.; Kupcik V. *Naturwissenschaften* 1971, 58, 219.
[61] Shen, G. Z.; Chen, D.; Tang, K. B.; Huang, L. Y.; Qian, Y. T. *J. Cryst. Growth* 2003, 249, 331.
[62] Shen, G. Z.; Chen, D.; Tang, K. B.; Qian, Y. T. *J. Mater. Res.* 2004, 19, 1367.
[63] Wang, W. Z.; Zhan, Y. J.; Wang, G. H. *Chem. Commun.* 2001, 727.
[64] Mingos; D.M.P. *Res. Chem. Intermediat.* 1994, 20, 85.
[65] Saskia, A.G. *Chem. Soc. Rev.* 1997, 26, 233.
[66] Chen, D.; Shen, G. Z.; Tang, K. B.; Lei, S. J.; Zheng, H. G.; Qian, Y. T. *J. Cryst. Growth* 2004, 260, 469.
[67] Chen, D.; Shen, G. Z.; Tang, K. B.; Jiang, X.; Huang, L. Y.; Jin, Y.; Qian, Y. T. *Inorg. Chem. Commun.* 2003, 6, 710.
[68] Chen, D.; Shen, G. Z.; Tang, K. B.; Sheng, J.; Fang, Z.; Liu, X. M.; Zheng, H. G.; Qian, Y. T. *Mater. Chem. Phys.* 2003, 82, 206.
[69] Nair, M. T. S.; Nair, P. K. *Semicond. Sci. Technol.* 1991, 6, 132.
[70] Prince, M. B. *J. Appl. Phys.* 1955, 26, 534.

[71] Lokhande, C. D. *J. Phys. D: Appl. Phys.* 1990, 23, 703.
[72] Domingo, G.; Itoga, R. S.; Cannewurf, C. R. *Phys. Rev.* 1966, 143, 536.
[73] Chen, D.; Shen, G. Z.; Tang, K. B.; Liu, Y. K.; Qian, Y. T. *Appl. Phys. A* 2003, 77, 747.
[74] Tu, W. X.; Liu, H. F. *Chem. Mater.* 2000, 12, 564.
[75] Rao, K. J.; Vaidhyanatham, B.; Ganguli, M.; Ramakrishnan, P. A. *Chem. Mater.* 1999, 11, 882.

In: Frontal Nanotechnology Research
Editor: Martin V. Berg, pp. 131-159

ISBN: 1-60021-212-3

Chapter 6

TiO_2 NANOSTRUCTURED MATERIALS: DESIGN, CHARACTERIZATION AND APPLICATIONS

Aicheng Chen**, Xinsheng Peng and Peter Holt-Hindle
Department of Chemistry, Lakehead University, Thunder Bay, Ontario, Canada

ABSTRACT

Nanostructured materials have attracted tremendous attention in both fundamental studies and applications due to their unique physical and chemical properties. Of the various types of nanostructured materials, TiO_2 nanomaterials are extensively studied for their broad range of applications. A number of methods have been developed for preparing TiO_2 nanoparticles, nanorods, whiskers, nanowires, nanotubes and mesoporous TiO_2 films. In this chapter, we describe two novel methods recently developed in our laboratory to grow TiO_2 nanostructured materials directly on titanium (Ti) substrate. Dense and well-aligned TiO_2 nanorod arrays form on a Ti substrate when using organic molecules, such as ethanol, acetaldehyde, acetone and dibutyltin dilaurate, as the oxygen source in the oxidation of Ti at high temperatures. In contrast, polycrystalline grains are formed when pure oxygen and formic acid are used as the oxygen source; elongated micro-crystals are produced when water vapor is used as the oxygen source; whereas the use of argon with a low concentration of oxygen produces random nanofibers growing from the ledges of the TiO_2 grains. Measurements of the water contact angle of these samples show that the wetting properties of the TiO_2 films are dictated by their structure, varying from a hydrophilic surface (polycrystalline grains) to hydrophobic (random nanofibers), and towards a highly-hydrophobic surface (well-aligned nanorod arrays). Secondly, we have synthesized large-scale $Na_2Ti_6O_{13}$ nanoplates, nanowires, and continuous nanowire network films directly on Ti substrate hydrothermally. The morphology of the formed $Na_2Ti_6O_{13}$ nanostructures is easily tuned by varying the experimental parameters of temperature, reaction duration and NaOH concentrations. Our study shows that the synthesized $Na_2Ti_6O_{13}$ nanostructures are easily converted into $H_2Ti_3O_7$ nanostructures, a desirable precursor for the fabrication of various TiO_2-based nanomaterials such as anatase, a mixture of anatase and rutile, and rutile TiO_2 nanomaterials, with shape-preservation through an ion-exchange process. The approaches described in this chapter provide simple and novel methods to synthesize various large-scale TiO_2-based nanostructured materials. This tunable growth of TiO_2 nanostructures is desirable for the promising applications of TiO_2

* E-mail address: achen@lakeheadu.ca; Fax: (807) 346-7775; Tel: (807) 343-8318. Corresponding author.

nanomaterials in the development of photo-catalysts, optical devices, sensors and self-cleaning coatings.

1. INTRODUCTION

Nanocrystalline materials have attracted tremendous attention in many disciplines [1-6]. Nanocrystals exhibit novel electronic, magnetic, optical, chemical, and mechanical properties, making them highly attractive for many important technological applications. Of the various types of nanocrystals, titanium oxide (TiO_2) and TiO_2-based materials are widely investigated for applications in photovoltaic cells, batteries, electrochemical sensors, optical emitters, photonic crystals, catalysis and photocatalysis, selective adsorption, ion exchange, ultraviolet blockers, smart surface coatings, and as functional filling materials in textiles, paints, paper and cosmetics [7-18]. TiO_2 exists mainly in three crystallographic forms: rutile (tetragonal, a = b = 4.584 Å, c = 2.952 Å), anatase (tetragonal, a = b = 3.782 Å, c = 9.502 Å), and brookite (rhombohedrical, a = 5.436 Å, b = 9.166 Å, c = 5.135 Å). However, only rutile and anatase play a role in the applications of TiO_2 and thus they have been extensively studied with surface science techniques. The surface science aspects of rutile and anatase TiO_2 have been comprehensively reviewed lately by Diebold [7]. The photoactivity of TiO_2 is one of its most attractive properties. Several review papers have discussed the technical and scientific aspects of TiO_2 photocatalysis [19-22]. Compared to anatase, rutile TiO_2 has attracted less attention in producing photocatalysts [23, 24] and dye-sensitized solar cells (DSCs) [25]. This is attributed to the fact that the rutile phase exhibits lower photo-electrochemical performance than the anatase phase, which is primarily due to the differences in their electronic structures. Most of the basic understanding of the electronic structure of TiO_2 surfaces is discussed by Hedrich and Cox [26]. On the other hand, rutile TiO_2 has some advantages over anatase, such as higher chemical stability, a higher refractive index and cheaper production cost [27]. Rutile TiO_2 is also preferred for optical applications. Rutile single crystals are used as polarizers in many optical devices such as isolators, modulators, switches and interferometers [28].

By far, the most actively pursued applied research on TiO_2 is its use for the photo-assisted degradation of organic molecules. TiO_2 photocatalysis provides a promising technology for the purification and treatment of both contaminated air and water [29]. TiO_2 catalysts have also been used for selective oxidation of olefins, alkyl-substituted benzenes, and alkanes at room temperature [30, 31]. TiO_2 photocatalysis has been applied in the treatment of air stripping of gas for the remediation of contaminated soil and solvents from contaminated ground water [32, 33]. When TiO_2 is exposed to UV light, electrons in the uppermost valence band jump to the conduction band, creating conduction band electrons and valence band holes. In most cases, the valence band holes and conduction band electrons simply recombine liberating heat or light, which is known as recombination. The efficiency of the photocatalytic process is measured as quantum yield, which is defined as the number of reaction events occurring per photon absorbed. In an ideal system, the quantum yield is proportional to the rate of the charge transfer processes and inversely proportional to the sum of the charge-transfer rate and the electron-hole recombination rate. Without recombination, the quantum yield could approach the ideal value of one for UV light, with a small enough wavelength to activate the catalyst [33]. Recombination is responsible for the low quantum yields. A positive charged TiO_2 coated electrode can suppress the rate of hole-electron recombination since the electrode can act as a sink for the photogenerated electrons. This concept has been used to enhance the efficiency of formic acid destruction under UV illumination, with a positive bias potential applied to the electrode [34]. Recent research suggests that this approach can achieve higher efficiency with TiO_2 photoanodes [35]. Although this application has achieved some degree of success, the true potential of the process has not been reached, probably because the resistance path for the electrons is relatively long and the surface area for reaction is low. An electron generated in one TiO_2 particle must hop or tunnel through boundaries from one TiO_2 conduction band to another until it finally enters an electrode which has a limited surface area. Depending on how far it has to travel to the electrode, an electron will still have an opportunity to interact with holes present in other TiO_2 particles and may recombine.

The partial oxidation of light alkanes and aromatics by molecular oxygen is a key process in the chemical industry [36, 37]. Unfortunately, partial oxidation reactions have always challenged industrial chemists because many oxidation reactions tend to proceed past the desired product to low- or no-value complete combustion products. Other selective oxidation reactions rely on the use of highly reactive chlorinated intermediates with the cost of producing chlorinated byproducts. Currently, UV-light-assisted oxidation shows promise for the oxidation of the cyclohexane and epidation of small olefins, whose selectivities differ from those obtained by other oxidation routes [38, 39]. This route offers a potential method for synthesis, provided that the expected product can be obtained within an acceptable quantum yield [40-42]. However, the reaction selectivity is still low due to the poor selectivity of the photocatalysts. An improvement in oxidation reaction activity and selectivity of a few percent could lead to a tremendous increase in efficiency, reduction of by-products, and improve the economics of production. Studying the relationship between reaction activity and the size of nanomaterials may lead to the improvement of their reaction selectivity.

The quantum yield for TiO_2-based photocatalysis is still very low. An initial approach to shifting the optical response of TiO_2 from UV to the visible spectral range by doping TiO_2 with transition-metal elements has been proposed [43-49]. However, metal doping has several drawbacks. The doped materials suffer from thermal instability and the metal centers act as electron traps, reducing the photocatalytic efficiency. Furthermore, the preparation of transition-metal-doped TiO_2 requires expensive ion-implantation facilities [50, 51]. Recent studies have shown that the desired narrow band gap of TiO_2 can be achieved by using anionic dopant species rather than metals ions [52-53]. Asahi et al. have demonstrated that TiO_2 films can be doped with nitrogen by sputtering methods and exhibit thereafter enhanced photoactivity in the visible spectral range [54-56]. The substitutional doping of nitrogen is very effective as its p states contribute to the band gap narrowing by mixing with O 2p states. Considerable effort has been made to dope TiO_2 thin films and powders with nitrogen by annealing TiO_2 at elevated temperatures under NH_3 flow. Nevertheless, this doping process only results in a small amount (2%) of nitrogen incorporation in micrometer-sized TiO_2 systems [54].

While notable advances have been made in these areas, three broad challenges still remain: (i) to increase the quantum yields, (ii) to improve the reaction selectively to produce desired products when used for chemical synthesis, and (iii) to shift the irradiation light from UV to visible light as UV light accounts for only a small portion (8%) of solar energy compared to visible light (45%). To address these three challenges, a number of methods have been developed for preparing TiO_2 thin films, for instance, electron beam evaporation [57], low temperature oxidation of metal Ti thin film in a mixture of water and oxygen gas [58], chemical vapor deposition [59, 60], physical vapor deposition [61], dip coating [62], spin coating [63], and atmosphere pressure chemical vapor deposition [64]. The thin films are usually composed of TiO_2 nanoparticles. Recently, one-dimensional TiO_2-based nanostructures have begun to attract a great deal of attention because of their promising applications in many areas, such as highly efficient photocatalysis [65] and photovoltaic cells [66, 67]. Two main approaches have been reported for preparing titania nanorods and nanowires, one using a template synthesis [68-71] and the other employing a hydrothermal reaction from TiO_2 [72-78]. Tian et al. applied a hydrothermal method to synthesize TiO_2-based nanotubes with TiO_2 nanoparticles as seeds [79]. Akbar and co-workers developed a process for 'carving' oriented arrays of single crystal TiO_2 nanofibers from bulk TiO_2 crystals [80]. Nanostructured TiO_2 is also produced using the sol-gel method, where titanium alkoxide or halide ($TiCl_4$, TiF_4) is hydrolyzed in the presence of a template, such as nanospheric nanorods or anodic porous alumina [18, 68, 81-86]. The growing widespread utilization of these advanced nanomaterials demands novel methods for the fabrication of large-scale TiO_2-based nanostructures. The new technology should involve a minimal number of steps and be able to manufacture large amounts through continuous production. It is well known that the use of aqueous suspensions limits practical applications because of the difficulties in separating TiO_2 fine particles and recycling the photo-catalyst. Many techniques have been proposed for the immobilization of TiO_2 on solid substrates to overcome this problem [87, 88]. Generally speaking all the proposed methods involve a tedious procedure: first prepare TiO_2 nanostructures from solution, then re-disperse them into certain agents and finally precipitate, paste or spin coat them onto a substrate.

In this chapter, we summarize our recent work on the development of templateless and seedless methods to directly prepare large arrays of oriented TiO_2 nanomaterials on Ti substrates. We have synthesized well-aligned rutile TiO_2 nanorod arrays on Ti substrates by a simple gas-solid reaction process using small organic molecules as the oxygen source. Our experimental results show that tunable and oriented rutile nanostructures are formed by oxidizing metallic titanium at high temperatures using different small organic molecules as the oxygen source, and that the wetting properties strongly depend on the surface structure. The growth mechanism of the TiO_2 nanostructures is discussed. Secondly, we have synthesized $Na_2Ti_6O_{13}$ nanostructures using metal Ti as the Ti source in NaOH solution under hydrothermal conditions. This study demonstrates that the morphology of the products can be easily tuned by controlling the reaction parameters of temperature, reaction duration, and the NaOH concentration. Through the ion-exchanged process, these $Na_2Ti_6O_{13}$ nanostructures are converted into $H_2Ti_3O_7$ nanoplates, nanowire arrays and nanowire networks, while preserving the morphology of the nanostructures. The formed $H_2Ti_3O_7$ nanostructures are further annealed and reconstructed into anatase and rutile TiO_2 at 450 and 750 °C, respectively. The corresponding optical properties and photocatalytic activity of the synthesized TiO_2 nanostructures is presented. In addition, electrochemical methods are promising in fabricating one-dimensional TiO_2 nanotubes. The electrochemical anodization of titanium and the applications of TiO_2 nanotubes in Green Chemistry and electrochemical biosensoring is also briefly addressed.

2. EXPERIMENTAL SECTION

2.1. Thermal Oxidation Method

The thermal oxidation process used in the study is described in our previous reports [89-91]. Briefly, the oxide films were grown on Grade 1 titanium substrate. Small titanium plates of 30 x 15 x 0.8 mm^3 were first degreased using acetone, then washed with distilled water, etched in 30 wt% HCl at 80 °C for 10 minutes to remove the oxide layer on the surface, then completely washed with distilled water, and finally dried in a vacuum oven at 40 °C. The oxidation was carried out in a horizontal tube furnace equipped with a ceramic tube (diameter 35 mm, length 700mm). Prior to heating, the pretreated Ti plates were placed into the ceramic tube, and the system was purged using ultrapure argon (99.999%) at a flow rate of 200 sccm for 4 hours. The furnace was heated to a predetermined temperature at a rate of 30 °C/minute. Then a mixture of the ultrapure argon (200 sccm) and the selected oxygen source was introduced into the system for a pre-determined period of time. Finally the system was cooled to room temperature under a purge of the ultrapure argon. Oxygen, water vapour, formic acid, ethanol, acetaldehyde, acetone and dibutyltin dilaurate (DBTDL) were used as the oxygen sources. DBTDL, $(C_4H_9)_2Sn(OOC_{11}H_{23})_2$ is a commercially available tin organic compound that easily decomposes at high temperatures.

2.2. Hydrothermal Method

The hydrothermal method was used to prepare $Na_2Ti_6O_{13}$ nanostructures directly from Ti in NaOH solution. Commercial Ti (Grade 1) plates sized 3 cm (width) x 3 cm (length) x 0.8 mm (thickness) were used as the Ti source and the substrate. Sodium hydroxide solutions were prepared by dissolving NaOH in nanopure water (18.2 MΩ.cm). To investigate the influence of the NaOH concentration, six samples were prepared in the following steps: (i) after being washed, the Ti plates were put into Teflon autoclaves filled with 40 mL of x M sodium hydroxide aqueous solution, where x is 0.0, 0.01, 1.25, 2.5, 7.5 and 10.0; (ii) heated up to 150 °C at a rate of 10 °C per minute; (iii) held at 150 °C for 15 hours. All the heating rates in this study are the same. To investigate the effect of temperature, four samples were prepared under the same NaOH concentration (2.5 M) and the same length of heating time but at different temperatures: 125, 150, 175 and 200 °C. To determine the effect of varying time, five samples were synthesized under the same NaOH concentration (2.5 M) and at the same temperature (150 °C) but were kept for 0.5, 1.5, 15, 30 and 48 hours. After the autoclave cooled naturally to room temperature, all the samples were taken out and immersed into Nanopure water for three days, then washed several times in pure water and air dried. We then put the synthesized $Na_2Ti_6O_{13}$ samples into 1.0 M HNO_3 solution for three days. The resulting $H_2Ti_3O_7$ nanostructures, from an ion-exchanged process [92-94], were washed several times with Nanopure water and dried at room temperature. Some synthesized hydrogen titanate samples were transferred into a furnace and treated at 450, 600 and 750 °C for 5 hours in air and converted into anatase, a mixture of anatase and rutile, and rutile TiO_2 nanostrutures, respectively.

2.3. Characterization

The synthesized nanostructures were characterized by X-ray diffraction (XRD) (Philips PW 1050-37 10 Diffractometer with Cu Ka radiation), scanning electron microscopy (SEM) (JEOL JSM 5900LV), transmission electron microscopy (TEM) and high-resolution TEM (HRTEM) (JEOL 2010F TEM/STEM) using a 2048 x 2048 CCD camera equipped with a GIF imaging filter. The HRTEM was equipped with an energy dispersive X-ray spectrometer (EDS, Oxford Links ISIS). For SEM observation, the samples were cut into 10 mm x 10 mm squares and pasted onto an Al holder using carbon conducting paste; then a conducting gold film was sputtered on the surface. Specimens for the TEM and HRTEM investigation were scratched from the substrate and ultra-sonicated in ethanol, then a droplet of the suspension was placed on a holey copper grid covered with porous carbon film. A dye, Eosin Y, was employed as a probe to study the optical and photocatalytic properties of the $H_2Ti_3O_7$ nanowires and TiO_2-based nanostructures. The synthesized $H_2Ti_3O_7$ nanowires, anatase and rutile TiO_2 nanostructures were immersed in an ethanol solution containing Eosin Y (9 mM) for 30 minutes. The nanomaterials loaded with dye were irradiated using an UV-Xe lamp (365 nm and 5 mW/cm^2). After irradiation for selected periods of time, diffuse reflectance spectra were recorded to examine the changes of the absorption of Eosin Y using Cary5e UV-VIS.

2.4. Electrochemical Measurements

TiO_2 nanotubes were fabricated by anodizing Ti sheets in a dilute HF solution, and then we co-immobilized horseradish peroxidase (HRP) and thionine on the TiO_2 nanotube arrays by immersing the Ti/TiO_2 electrodes in a mixture of HRP and thionine solution for biosensoring [18]. Electrochemical measurements were performed with a Solartron 1287 potentiostat and a Solartron 1252B frequency response analyzer. A three-electrode system was used for all electrochemical experiments, employing one of the above electrodes, platinum wire and an Ag/AgCl electrode (3 M KCl) as the working, auxiliary and reference electrodes, respectively. All experiments were performed at room temperature (22±2 °C) with a 0.1 M phosphate buffer (PB) solution as the background electrolyte. All experimental solutions were deoxygenated by bubbling ultra-pure argon for 15 minutes and maintained under an argon atmosphere during the course of the experiments. Amperometric experiments were performed in a stirred cell with the successive addition of H_2O_2 into 8.0 ml supporting solution while the electrode potential was set at –450 mV.

3. Results and Discussion

3.1. TiO_2 Nanostructures Prepared by Thermal Oxidation of Ti

3.1.1. The Effect of Oxygen Source

Figure 1 presents the general SEM images of the samples synthesized at 850 °C for 1.5 hours using different oxygen sources. Figures 1a and 1b show that polycrystalline grains were formed when pure oxygen gas and formic acid were used as the oxygen sources, respectively.

When water vapor was used as the oxygen source, as shown in Figure 1c, large and long micro-crystals with diameters of about one micrometer and length of about five micrometers

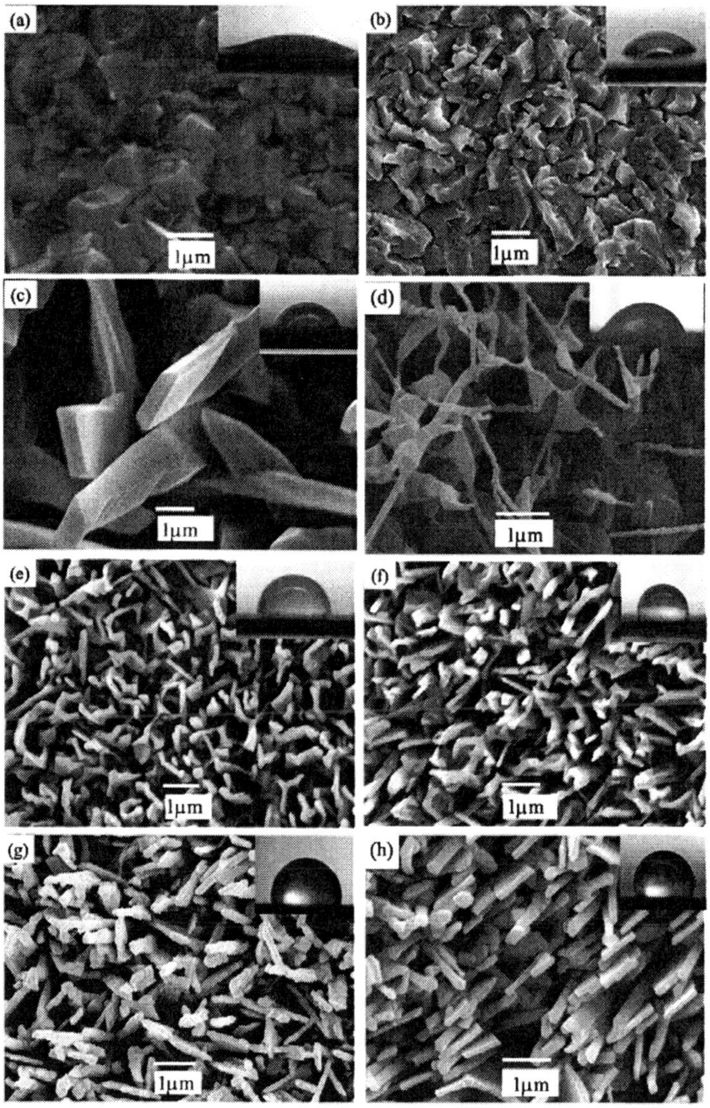

Figure 1. SEM images of the TiO2 samples prepared by oxidizing Ti using (a) O2, (b) HCOOH, (c) H2O, (d) a mixture gas of Ar (200 sccm) and O2 (1 sccm), (e) CH3CH2OH, (f) CH3CHO, (g) CH3COCH3, and (h) dibutyltin dilaurate as oxygen sources at 850 °C for 1.5 hours. The insets show the corresponding contact angle of water droplets on their surfaces. (Figure 1a, b, c, e, f and g reprinted with permission from [91]).

were formed. Figure 1d shows that long and random nanofibers were formed when diluted oxygen (a mixture gas of Ar and O_2) was used as the oxygen source. In contrast to Figure 1a and 1b (polycrystalline films), 1c (elongated micro-crystalline fibers) and 1d (long nanofibers), oriented and aligned TiO_2 nanorod arrays were yielded as seen in Figure 1e, 1f, 1g and 1h, when ethanol, acetaldehyde, acetone and DBTDL were used as the oxygen sources, respectively. As shown in Figure 1e, the nanorods formed using ethanol as the oxygen source are plate-like with approximately 150 nm thickness, 500 nm width and 1 micrometer length. Figure 1f demonstrates that the nanorods formed in acetaldehyde are similar to those formed in ethanol, although slightly larger. Figure 1g shows that the nanorods formed in acetone have two typical morphologies: plate-like and column-like. As shown in Figure 1h, more and larger plate-like nanorod arrays with thickness of approximately 150 nm, width of several hundred nanometers and length of a few micrometers were formed when DBTDL was used as the oxygen source. All these SEM images demonstrate that the morphology of the formed titanium oxide nanomaterials is tunable by choosing different oxygen sources.

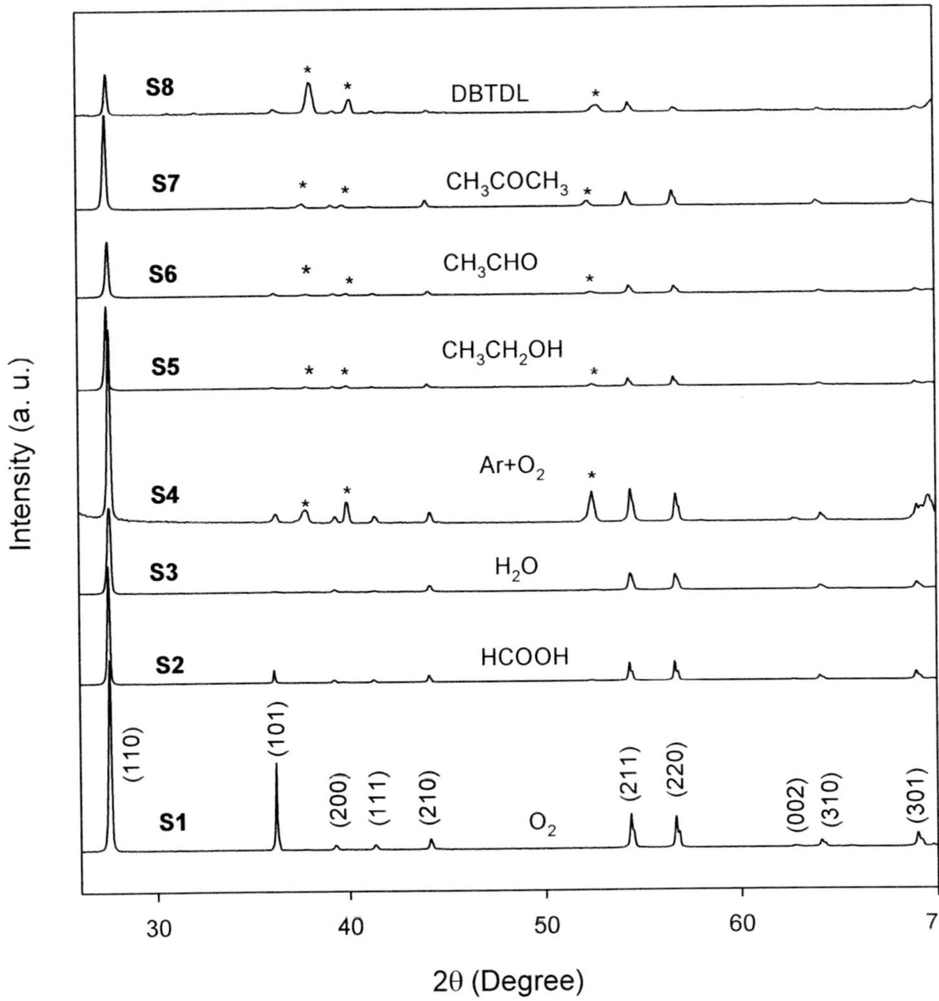

Figure 2. XRD patterns of the samples synthesized by oxidizing Ti using (S1) O_2, (S2) HCOOH, (S3) H_2O, (S4) a mixture gas of Ar (200 sccm) and O2 (1 sccm), (S5) CH_3CH_2OH, (S6) CH_3CHO, (S7) CH_3COCH_3, and (S8) dibutyltin dilaurate as oxygen sources at 850 °C for 1.5 h.

Figure 2 presents the corresponding XRD patterns of those samples shown in Figure 1. The diffraction peaks, (110), (101), (200), (111), (210), (211) and (220) can be indexed to those of the tetragonal rutile TiO_2 phase, with lattice constants a = 4.593 Å, c = 2.959 Å and a space group of P42/mnm (136), (JCPDS file No. 21-1276). The peaks marked with a star are derived from the Ti substrate. All of the XRD spectra show that the formed titanium oxide films have the same rutile phase. The changes of the intensity ratio from the (110) to (101) peaks shown in Figure 2 are consistent with the aforementioned SEM observations presented in Figure 1. The TiO_2 growth is performed from the random polycrystalline films S1, S2 and S4, to the oriented nanofibers S3, and to the well-aligned nanorod arrays S5, S6, S7, and S8. In addition, the peaks indexed to the Ti substrate appear in the XRD spectra of the samples

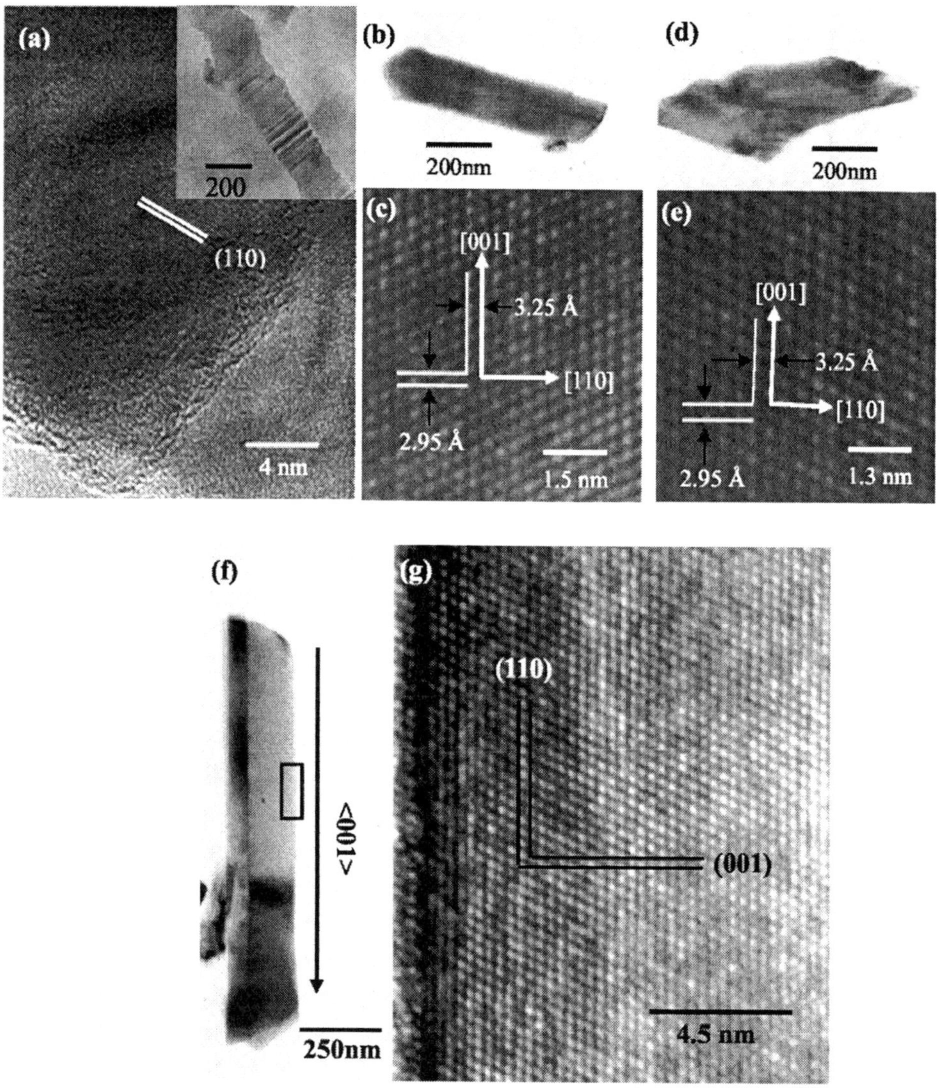

Figure 3. TEM and HRTEM images: (a) the chain-like nanofibers of S4 fabricated using diluted oxygen gas as the oxygen source; (b) and (c) a sample nanorod; (d) and (e) a nanoplate of S7 synthesized using acetone as the oxygen source; (f) and (g) an individual rod of S8 formed using dibutyltin dilaurate as the oxygen source at 850 °C for 1.5 hours. The inset to Figure 1a is a chain-like nanofiber of S4 fabricated using diluted oxygen gas as the oxygen source. (Figure 3b, 3c, 3d and 3e reprinted with permission from [90]).

S4-S8; but they are invisible in the spectra of the samples S1-S3. This indicates that the thickness of the TiO_2 films formed on the Ti substrate is greater when oxygen gas, formic acid and water vapor were used as the oxygen source versus those formed using diluted oxygen, ethanol, acetaldehyde, acetone and DBTDL as the oxygen source.

The detailed crystal structure, morphology and composition of S4, S7 and S8 were characterized using TEM and HRTEM. Figure 3a presents HRTEM images of the chain-like nanofibers (as shown in the inset to Figure 3a) of S4 fabricated using the diluted oxygen gas as the oxygen source. This nanofiber is crystalline with an inter-planar spacing of 0.325 nm, in agreement with the *d* value of the (110) planes of the rutile TiO_2 crystal. Figure 3b and 3d present low magnification TEM images of S7 prepared using acetone as the oxygen source. The corresponding HRTEM images are shown in Figure 3c and 3e. The planes with the lattice space of 0.325 nm and 0.295 nm (shown in Figure 3c and 3e) are assigned to the (110) and (001) planes of rutile TiO_2 crystal, respectively. These results show that the TiO_2 nanorod fabricated with acetone as the oxygen source is single crystalline with growth along the <001> direction. Figure 3f presents a TEM image of an individual rod of S8 (using DBTDL as the oxygen source) with a rectangle cross-section similar to the SEM results (Figure 1h). Figure 3g, the corresponding HRTEM image recorded from the rectangle zone, reveals that the rod is single crystalline with inter-planar spacings of 0.325 nm (110) plane and 0.295 nm (00 1) plane of the rutile TiO_2. These results indicate that these nanorods also grow along the <001> direction.

3.1.2. Temperature Effect

We further studied the temperature effect on the formation of the TiO_2 nanostructures. The oxidation behaviour of Ti at different temperatures using acetone as the oxygen source was investigated. Figure 4 shows the representative SEM images of the samples synthesized at (a) 650 °C, (b) 700 °C, (c) 750 °C, (d) 800 °C and (e) 850 °C for 1.5 hours using acetone as the oxygen source. Some small nanodots were formed at 650 °C as seen in Figure 4a. Increasing the temperature results in a significant increase in the dimension and the coverage of the formed nanostructures. Sparse and short aggregates composed of several nanoplates were formed at 700 °C (Figure 4b); dense and long aggregates were produced at 750 °C (Figure 4c). Further increasing the temperature to 800 °C, continuous, large and plate-like nanorod arrays were formed on the Ti substrate (Figure 4d). At 850 °C, continuous, long, column-like and plate-like nanorods were formed as seen in Figure 4e. These SEM images show that the reaction temperature significantly affects the morphology of the formed titanium oxide nanostructures. Figure 4f presents the corresponding XRD spectra of the five samples (Figures 4a to 4e). All the peaks (110), (101), (200), (111), (210), (211), (220) are indexed to those of the tetragonal rutile TiO_2 phase (JCPDS file No. 2 1-1276). Thus, all these formed titanium oxide films have the same phase. The peaks marked with a star are derived from the Ti substrate. The significant increase of the intensity of the TiO_2 peaks and the obvious decrease of the intensity of the Ti peaks from 650 °C to 850 °C indicates that the yield amount/thickness of the titanium oxide films on the Ti substrate increases with the increase of the reaction temperature. These XRD results presented in Figure 4f are consistent with the SEM observations shown in Figures 4a to 4e.

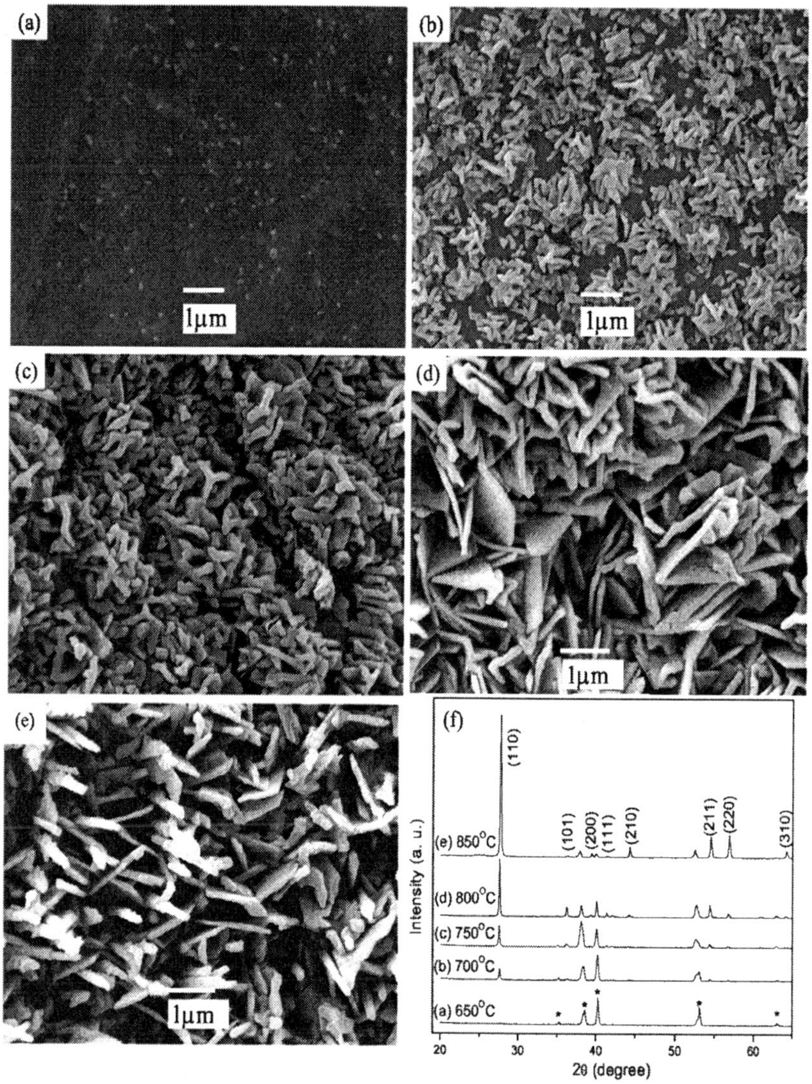

Figure 4. SEM images of Ti plates oxidized in acetone at (a) 650, (b) 700, (c) 750, (d) 800, (e) 850 °C for 1.5 hours. (f) The corresponding XRD results of the samples as shown in a to e.

Figure 5 presents the Raman spectra of the samples synthesized at (a) 650 °C, (b) 700 °C, (c) 750 °C, (d) 800 °C and (e) 850 °C for 1.5 hours using acetone as the oxygen source. The bands at 446.5 and 607.9 cm^{-1} are indexed to the Raman active modes E_g and A_{1g} of rutile TiO_2, respectively [95]. There is no observable TiO_2 Raman peak shown on the spectrum of Figure 5a, where the temperature is 650 °C (the lowest one among the five investigated temperatures) and only sparse TiO_2 nanodots were formed as seen in Figure 4a. By increasing the temperature from 700 °C (Figure 5b) to 850 °C (Figure 5e), the intensity of the Raman bands centered at 446.5 and 607.9 cm^{-1} dramatically increases, indicating that the yielded TiO_2 significantly increases with the increasing of the reaction temperature. This is consistent with the XRD studies shown in Figure 4f.

3.1.3. Growth Mechanism

Ti can easily react with oxygen to form a TiO_2 film. But for the growth of titanium oxide, the question is where the oxidation occurs, at the Ti and TiO_2 interface, on the TiO_2 surface or inside the TiO_2 film. If oxidation occurs at the Ti and TiO_2 interface, then the oxygen species nccds to move to the interface through the formed TiO_2 film.

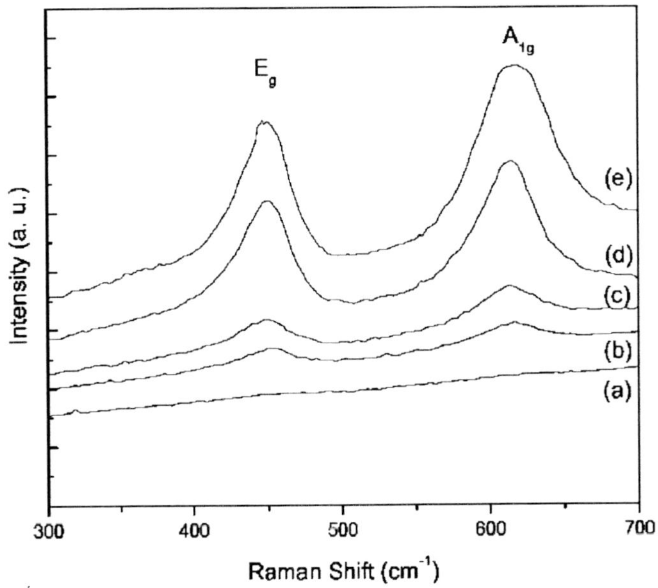

Figure 5. Raman Spectra of the samples prepared using acetone as the oxygen source at (a) 650, (b) 700, (c) 750, (d) 800, (e) 850°C for 1.5 hours. (Reprinted with permission from [91]).

On the other hand, if the oxidation occurs on the formed TiO_2 surface, it requires the Ti species to move to the oxide surface. The third alternative is that both the Ti and oxygen species are involved in the diffusion processes and encounter each other inside the oxide film. It is expected that the mechanism which dominates the growth process of TiO_2 depends on the applied temperature, the concentration of oxygen and the thickness of the TiO_2 film, and would affect the structure of the TiO_2 film. The oxidation of titanium in pure oxygen and air has been widely studied. The structure of the oxide layer (pin-shaped, lamellar and acicular crystals [96], pyramid crystals [97], nanoparticle thin films [98]) generally depends on the duration and temperature of the oxidation process. Several possible mechanisms for the oxidation of titanium on metallic surfaces have been proposed [96-98]. Badescu and Momirlan report that there is a predominant diffusion of oxygen through the grain boundaries of the oxide layer and the oxidation site of metal-oxide at temperatures in the range of 800-1000 °C; while at temperatures above 1000 °C, a considerable increase in the diffusion of Ti cations through the oxide layer is observed [96]. On the other hand, Czerwinski and Szpunar examined the surface topography of oxide layers on titanium using atomic force microscopy and X-ray diffraction and show that TiO_2 growth at 900 °C is controlled by the surface diffusion of Ti cations to the ledges located on the sidewalls of pyramidal grains [97]. Our experimental results presented show that the morphology of the synthesized TiO_2 nanostructures strongly depends on the oxygen source and reaction temperature, providing new insight into the growth mechanisms of TiO_2 nanostructures. Since the melting point of Ti and rutile TiO_2 is very high, at 1660 and 1840 °C, respectively, the vapour pressure of Ti and rutile TiO_2 at our experimental temperatures should be very low. Thus, the growth of rutile TiO_2 nanorods is unlikely via either vapor-liquid-solid (VLS) [99] or vapour-solid (VS) mechanisms [100]. The VLS mechanism is also rejected due to the fact that no spherical particles were found at the tips of the nanorods; furthermore, the VS mechanism is also ruled out because all of the nanorods we observed grew directly from the Ti substrate and we did not find any traces of rutile TiO_2 anywhere else in the furnace tube. Titanium can easily react with oxygen to form a TiO_2 film. For the different oxygen sources, the initial formation of TiO_2 film is expected to be very similar (oxygen species reacting with Ti to form TiO_2). Then titanium oxide may grow at the oxide/gas interface, inside of the oxide film, and/or at the oxide/metal interface. It is known that grain boundaries are the most favorable paths for Ti and oxygen species diffusion [101]. It is expected that the structure of the formed titanium oxide depends on: (i) the diffusion rate of active oxygen species from the oxide/gas interface to the oxide/metal interface through the grain boundaries; (ii) the diffusion rate of the Ti species from the oxide/metal interface to the oxide/gas interface through the grain boundaries; (iii) the diffusion rate of the active oxygen species and Ti species on a specific crystal plane; and (iv) the amount of available active oxygen species.

Under the same temperature, the diffusion rate of the Ti species can be assumed to be the same. The oxygen source thus plays a key role in TiO_2 growth. Formic acid can be easily decomposed. When oxygen or formic acid is used as the oxygen source, a large amount of active oxygen species is available. The diffusion of the active oxygen species dominates the growth process and the oxidation occurs mainly at the oxide/metal interface, resulting in a thick polycrystalline TiO_2 film as shown in Figure 1a and 1b. In contrast, when ethanol, acetaldehyde, acetone or DBTDL is used as the oxygen source, the produced active oxygen species is expected to be much less; the diffusion of the Ti species to the oxide/gas interface via the network of the grain boundaries of the thin oxide layer predominates. Then the diffused Ti species react with the selectively adsorbed oxygen species on different planes of rutile TiO_2 [102]. This is supported by the formation of aligned nanorod arrays along the <001> direction on the substrate. The available active oxygen species is expected to be less at the bottom than that at the top of the nanorods; thus, smaller and longer rods are formed. The observation that TiO_2 growth is predominated by Ti diffusion is further supported by the effect of reaction time on the formed TiO_2 nanostructures [89]. When acetone was used as the oxygen source at 850 °C, short and oriented dots emerged with the reaction time of 10 minutes; dense and short nanorod arrays were formed with the reaction time of 30 minutes and long nanorod arrays were produced when the reaction time was increased to 60 minutes. Moreover, the dramatic difference of the formed TiO_2 nanostructures at different temperatures indicates that the diffusion rate of the Ti species is the determining factor for the growth of TiO_2 nanostructures. As shown in Figure 4, the different morphologies are due to the different diffusion rates of Ti atoms from the substrate to the oxide surfaces. At the low temperature of 650 °C very few nanodots are formed, because both the Ti diffusion rate and oxidation rate are very low. At 700 and 750 °C the surface oxidation mainly produces rutile nanoplates since the increased diffusion rate results in a faster growth rate. When the temperature is increased to 800 °C, the increased diffusion rate of Ti facilitates the TiO_2 nanoplate growth such that the aligned nanoplates uniformly cover the whole substrate surface. When the temperature is further increased to 850 °C, the TiO_2 nanorod arrays are formed; moreover, the Ti diffusion rate is high enough to move Ti species from the Ti substrate to the nanorod tips and to sustain the nanorod growth. On the other hand, when water and diluted oxygen gas is used as the oxygen source, TiO_2 polycrystalline and micro/nano-fibres are formed as shown Figure 1 c and 1 d, respectively; indicating that both active oxygen species diffusion and Ti species diffusion are heavily involved in the course of the TiO_2 growth. The formation of the fiber-like structures is due to the difference in diffusion rate of Ti atoms along the different crystal directions of TiO_2 [7].

3.1.4. Wetting Properties

The surface wettability conversion of TiO_2 has been widely investigated based on the change of the water contact angle before and after UV irradiation using single crystal [103], polycrystalline films [102], and nanocrystal films [57]. However, very little work has been done on how the surface micro-structures of TiO_2 films affect their wetting properties. The insets to Figure 1 show that the wetting abilities of these oxide films strongly depend on their morphology. The TiO_2 surfaces formed using oxygen gas, formic acid, water and diluted oxygen as the oxygen source are hydrophilic; and their water contact angles increase from 10° (Figure 1 a), to 30° (Figure 1b), to 40° (Figure 1 c), and to 60° (Figure 1 d). They are all smaller than the water contact angle of a smooth rutile TiO_2 surface, 70° [103]. In contrast, the TiO_2 films formed using acetaldehyde, ethanol, acetone and DBTDL as the oxygen source are hydrophobic as shown in the insets to Figure 1e, 1f, 1g and 1h, respectively. The water contact angles of the TiO_2 films produced using ethanol and acetone as the oxygen source are very similar (~120°) (Figure 1 e and Figure 1 g), which is slightly smaller than the TiO_2 film formed using DBTDL (~125°) as the oxygen source (Figure 1h), but slightly larger than the TiO_2 film formed using acetaldehyde as the oxygen source (~108°) (Figure 1f). It is known that roughness has a significant influence on the wetting behavior of a given surface [104-106]. The dramatic difference of the wetting properties of the formed TiO_2 films using different oxygen sources can be explained by two different models: Wenzel's Equation [107] and Cassie and Baxter's Law [108]. In Wenzel's Equation, the apparent water contact angle θ^* is given by:

$$\text{Cos}\theta^* = r\cos\theta \qquad (1)$$

Where r is the roughness factor (the ratio of total surface area to the projected area in the horizontal plane), and θ is the contact angle measured on a flat surface. Wenzel's Equation assumes that liquid fills up the rough surface forming a fully wetted contact and that there is no air entrained in the pits of the rough surface below the liquid. As seen in Figure 1(a), the formed TiO_2 polycrystalline grains are big and dense, and the height of the hollows presented on the surface is very low. Thus, the air trapping ability of the formed TiO_2 surface, using pure oxygen gas as the oxygen source, is very low. The very small water contact angle (~10°) is consistent with Wenzel's Equation. When this θ^* is substituted into Wenzel's Equation with the θ value 70° for a smooth TiO_2 surface, the calculated r is 2.88, which is in agreement with the surface roughness r as estimated by our SEM observations. However, Wenzel's Equation is invalid for the samples shown in Figures 1e – 1h, where the surfaces are hydrophobic with the apparent angle θ^* over 100°. The high hydrophobicity of these surfaces originates from the air trapped in the interspace of the nanorod arrays similar to that reported for Gold bead arrays [105], Carbon nanotube arrays [109] and SnO_2 nanoflower arrays [110]. This is consistent with Cassie and Baxter's Law [108]:

$$\cos \theta^* = f(\cos\theta + 1) - 1 \quad (2)$$

Where f is the ratio of the liquid-solid contact area to the projected surface area. Cassie and Baxter's Law describes the apparent contact angle θ^* at a heterogeneous surface as being composed of two different materials. The surfaces shown in Figure 1e – 1h are composed of two phases: solid (TiO_2) and gas phase (trapped air). When the θ^* is substituted into Cassie and Baxter's Law with $\theta = 70°$ for a smooth TiO_2 surface, the calculated f is 0.51 for the surface shown in Figure 1e and 0.37 for the surfaces shown in Figures 1f and 1g, and 0.27 for the surface shown in Figure 1h. This is consistent with their SEM images. The slightly higher θ^* for the surfaces shown in Figures 1b, 1c and 1d compared to the surface shown in Figure 1a results from partial air trapping and is similar to the transition from Wenzel's model to Cassie and Baxter's air trapping model [109]. All these results show that the change of the surface structure from polycrystalline grains to well-aligned nanorod arrays switches the dominant wetting mode from Wenzel's to Cassie and Baxter's and results in the significant change of the wetting behavior from hydrophility to hydrophobicity.

3.2. TiO_2 Nanostructures Synthesized by Hydrothermal Method

3.2.1. Preparation of Na2Ti6O13 Nanowire Arrays

Figure 6 shows the SEM images of the sample synthesized in a 2.5 M NaOH solution under hydrothermal conditions at 150 °C for 15 hours at (a) a low magnification and (b) a high magnification. Nanowire arrays are uniformly formed in a large-scale on the titanium substrate. These nanowires have diameters of tens of nanometers, and length of about 2 micrometers. The crystal phase of these nanowire arrays is characterized using XRD. A representative XRD pattern is shown in Figure 7. All the diffraction peaks can be indexed to $Na_2Ti_6O_{13}$ with monoclinic lattice (JCPDS 37- 951, a = 15.12, b = 3.738 and c = 9.16,

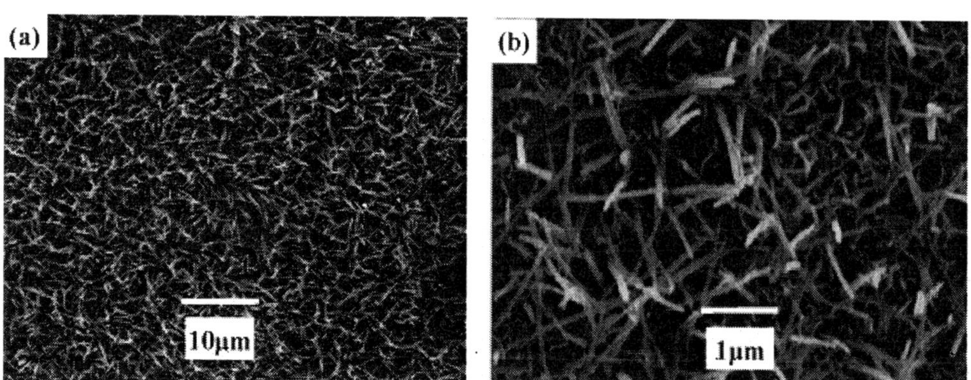

Figure 6. (a) low and (b) high magnification SEM images of the Ti surface after hydrothermal treatment in 2.5 M NaOH at 150 °C for 15 hours.

β = 99.3°), demonstrating that the as-synthesized nanowires have the structure of $Na_2Ti_6O_{13}$. The general reaction is expressed as Equation 3:

$$6\,Ti + 2\,NaOH + 11\,H_2O \rightarrow Na_2Ti_6O_{13} + 12\,H_2 \quad (3)$$

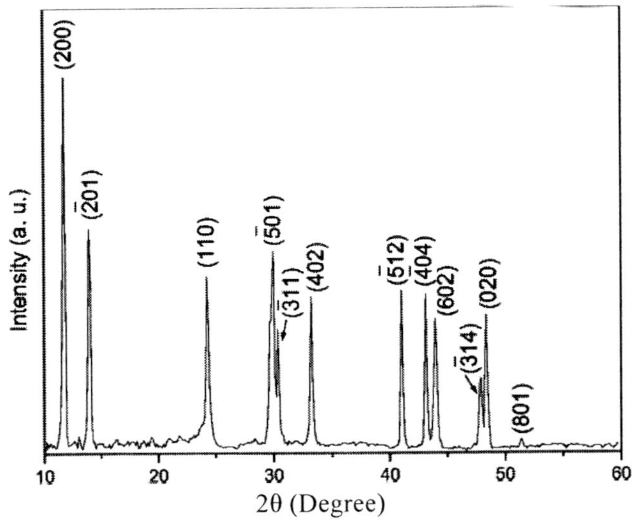

Figure 7. A representative XRD pattern of the as-synthesized $Na_2Ti_6O_{13}$ nanowire arrays prepared in a 2.5 M NaOH solution at 150 °C for 15 hours.

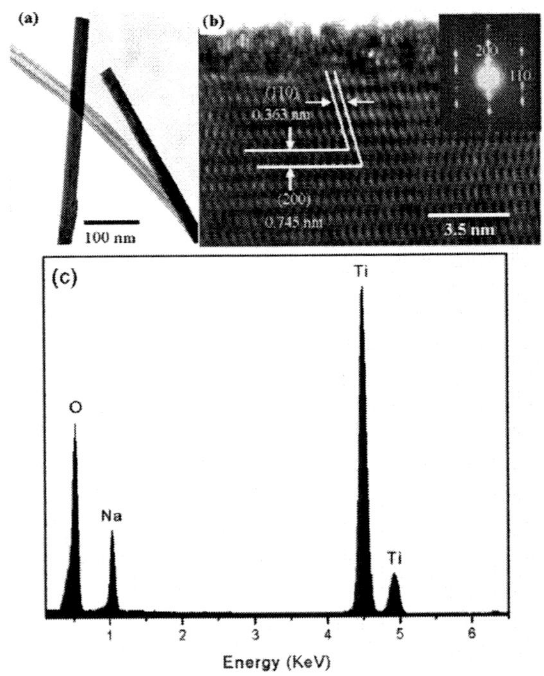

Figure 8. (a) TEM and (b) HRTEM and SAED (the inset) of synthesized $Na_2Ti_6O_{13}$ nanowires synthesized in 2.5 M NaOH at 150 °C for 15 hours. (c) The corresponding EDS of one of these $Na_2Ti_6O_{13}$ nanowires.

The crystal structure and the composition of these nanowires were further characterized using TEM, HRTEM, SAED and EDS. Figure 8a shows the TEM image of wire-like nanostructures with diameters of about 30 nm. Figure 8b, the HRTEM image recorded perpendicularly to the nanowire shows that the nanowires have a layered lattice structure with inter-planar spacing of 0.363 and 0.745 nm, in excellent agreement with the *d* values of the (110) and (200) planes of the $Na_2Ti_6O_{13}$ crystal. The corresponding SAED pattern presented in the inset of Figure 8b shows the monoclinic *hk0* spots of $Na_2Ti_6O_{13}$. All these results confirm that the formed $Na_2Ti_6O_{13}$ nanowires are single crystals and grew along the <110> direction. Figure 8c presents the corresponding EDS spectrum, showing that the nanowire is composed of 9.61% Na, 29.12% Ti and 61.2 % O. This is very close to the chemical ratios of the elements in $Na_2Ti_6O_{13}$ and further supports our XRD results described above.

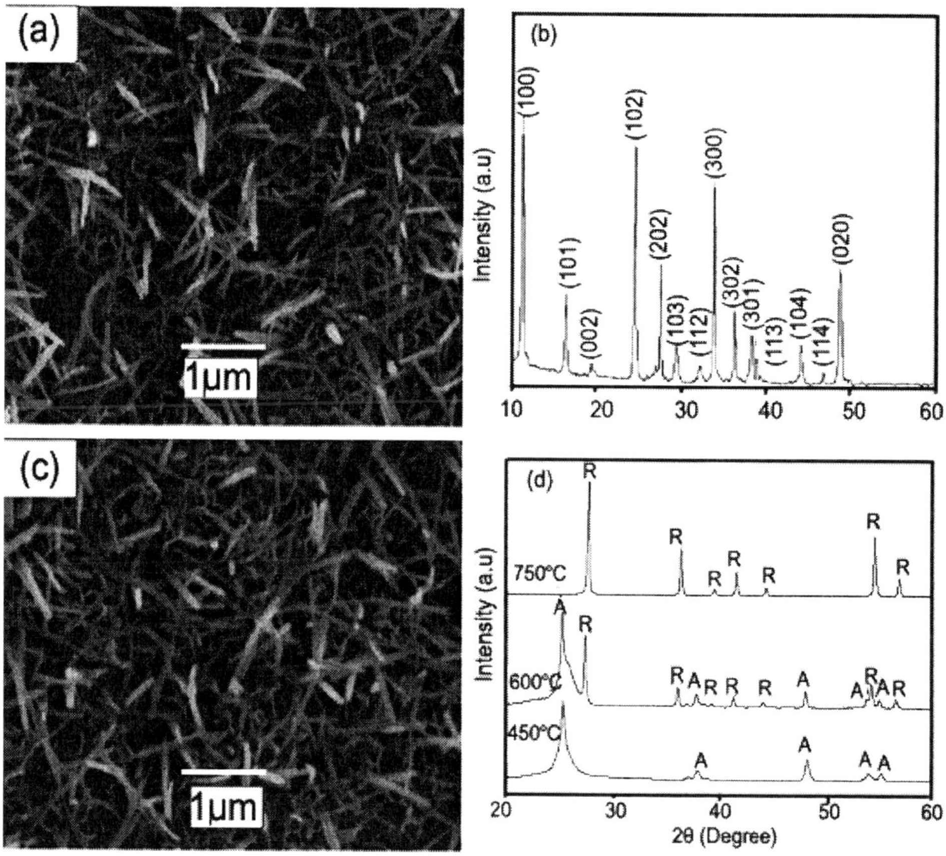

Figure 9. (a) SEM image and (b) XRD pattern of the $H_2Ti_3O_7$ nanowire arrays synthesized from $Na_2Ti_6O_{13}$ nanowires arrays through ion exchanging in 1 M HNO3 solution for 3 days. (c) SEM images of the sample synthesized by annealing $H_2Ti_3O_7$ nanowire arrays at 750 °C for 5 hours. (d) the XRD patterns recorded from the three samples synthesized by annealing $H_2Ti_3O_7$ nanowire arrays at 450 °C, 600 °C and 750 °C.

To protonate the $Na_2Ti_6O_{13}$ nanostuctured materials, we simply immersed the synthesized monoclinic $Na_2Ti_6O_{13}$ samples in 1 M HNO_3 solution for three days. Figure 9a shows a typical SEM image of the $Na_2Ti_6O_{13}$ sample after the acid treatment. The morphology of the nanowire arrays is maintained without any obvious change before and after the acid treatment (as compared to Figure 6b). The XRD pattern of this sample is shown in Figure 9b. All the peaks can be indexed to the monoclinic $H_2Ti_3O_7$ phase (JCPDS 3 6-654) with lattice constants a = 8.08, b = 3.725, c = 9.17 and β = 101.27. EDS analysis shows that there is no Na element present in the sample after the acid treatment and that the composition of the synthesized nanowires is composed of 30.25% Ti and 69.75% O, which is close to the ratio of the Ti and O in $H_2Ti_3O_7$. The absence of a Na peak in the EDS spectra further demonstrates that the $Na_2Ti_6O_{13}$ is completely converted into $H_2Ti_3O_7$ during the ion-exchange process. These results indicate that the $Na_2Ti_6O_{13}$ nanowires are successfully converted into $H_2Ti_3O_7$ nanowires with morphology-preservation through an ion-change process, where the Na_+ ions in $Na_2Ti_6O_{13}$ are replaced by H^+ or H_3O^+. The general reaction is expressed as Equation 4:

$$Na_2Ti_6O_{13} + 2\ HNO_3 + H_2O \rightarrow 2\ H_2Ti_3O_7 + 2\ NaNO_3 \quad (4)$$

3.2.2 Synthesis of Anatase and Rutile Nanowire Arrays

Figure 9c shows a typical SEM image of the synthesized $H_2Ti_3O_7$ nanowires after being annealed for 5 hours at 750 °C. It is interesting to note that the morphology of the sample does not change; it is still very similar to the $Na_2Ti_6O_{13}$ and $H_2Ti_3O_7$ nanowires. The XRD patterns recorded from the three samples synthesized by annealing $H_2Ti_3O_7$ nanowire arrays at 450 °C, 600 °C and 750 °C are presented in Figure 9d. At 450 °C, all the peaks shown in the XRD spectrum can be indexed to anatase (JCPDS 2 1-1272), demonstrating that the $H_2Ti_3O_7$ nanowires are completely converted into anatase nanowires. At 750 °C, all the XRD peaks can be indexed to rutile (JCPDS 4-673), indicating that the synthesized $H_2Ti_3O_7$ nanowire arrays are completely converted into rutile nanowire arrays. In addition, the XRD spectrum shows that a mixture of anatase and rutile nanowires are formed from the $H_2Ti_3O_7$ nanowires after being annealed at 600 °C. The general reaction is expressed as Equation 5:

$$H_2Ti_3O_7 \rightarrow 3\ TiO_2 + H_2O \quad (5)$$

We did further characterization on the formed nanowires. Figure 10 presents the typical TEM, HRTEM and SAED results of the samples (a)-(b) as-synthesized $H_2Ti_3O_7$, after being further annealed at 450 °C (c and d) and 750 °C (e and f) for 5 hours. The size of the $H_2Ti_3O_7$ nanowires as seen in Figure 10a is similar to that of the $Na_2Ti_6O_{13}$ nanowires shown in Figure 8a, with diameters of about 30 nm. In contrast, HRTEM analysis (Figure 10b) shows that the microstructure of the $H_2Ti_3O_7$ nanowires seen in Figure 10a is different from that of the $Na_2Ti_6O_{13}$ nanowires observed in Figure 8a. The lattice structure is parallel to the long axis of the $H_2Ti_3O_7$ nanowire. The distances of 2.25 and 1.88 between the adjacent lattice fringes can be assigned to the interplaner distances of monoclinic $H_2Ti_3O_7$ (004) and (020), respectively. The inset of Figure 10b shows the corresponding SAED pattern recorded from the same $H_2Ti_3O_7$ nanowire, indicating that the $H_2Ti_3O_7$ nanowire is a single-crystalline. As seen in Figure 10c, the diameter of the anatase nanowires is similar to that of $H_2Ti_3O_7$ nanowires in the range of several tens of nanometers. The HRTEM image shown in Figure 10d and the SAED pattern (the inset of Figure 10d) confirm that these nanowires are single-crystalline anatase with the growth along the <101> direction. Figure 10e shows that the formed rutile nanomaterials have an analogous nanowire structure to that of the anatase samples. The lattice image in Figure 10f recorded from one of the nanowires in Figure 10e has lattice fringes with inter planner distances of 3.25 and 2.51 Å, which is in excellent accordance with the *d* values of the rutile (110) and (101) planes. This is further supported by the SAED pattern as seen in the inset of Figure 10f. In addition, EDS analysis shows that the sample annealed at 450 °C is almost identical to that annealed at 750 °C. A quantitative analysis shows that all of the annealed nanowires are composed of Ti and O, with a ratio very close to 1:2. All of these results indicate that, through this facile annealing treatment, different phase titanium oxide based nanomaterials are synthesized and that the $Na_2Ti_6O_{13}$ nanostructures directly grown on a Ti surface under hydrothermal conditions can serve as a precursor to synthesize various desirable TiO_2 nanomaterials for photocatalytic applications. We further studied the effect of NaOH concentration, temperature and time on the morphology of these products synthesized under hydrothermal conditions.

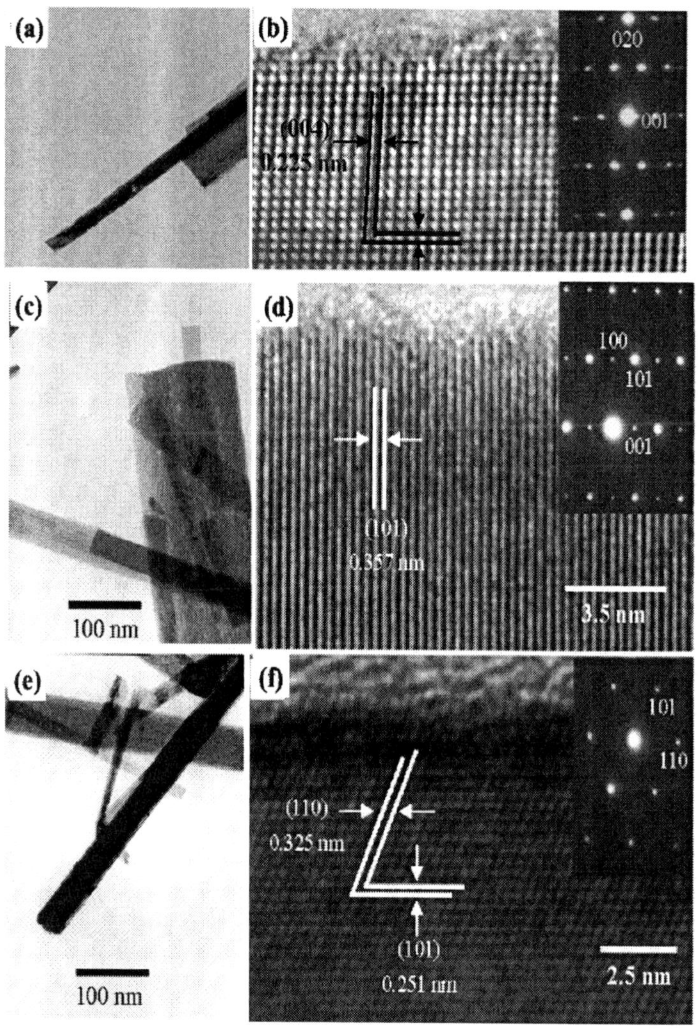

Figure 10. (a) TEM and (b) HRTEM and SAED (the inset) of $H_2Ti_3O_7$ nanowires; (c) TEM and (d) HRTEM and SAED (the inset) of the sample synthesized by annealing $H_2Ti_3O_7$ nanowires arrays at 450 °C for 5 hours; (e) TEM and (f) HRTEM and SAED (the inset) of the sample synthesized by annealing the $H_2Ti_3O_7$ nanowire arrays at 750 °C for 5 hours.

3.2.3. NaOH Concentration Effect

Figure 11 shows the SEM images of the surfaces of the Ti substrates after hydrothermal treatment at 150 °C for 15 hours in (a) 0.0, (b) 0.01, (c) 1.25, (d) 2.5, (e) 7.5, and (f) 10.0 M NaOH solutions. The Ti surface does not change as shown in Figure 11a in the absence of NaOH. At a low concentration of NaOH 0.01 M, particles and nanodots were formed on the Ti substrate as seen in Figure 11b. Increasing the NaOH concentration to 1.25 M, well-aligned dense and short nanorod arrays were formed (Figure 11c). Compared with the nanorod arrays formed in 1.25 M NaOH, longer and thinner nanowire arrays were produced in 2.5 M NaOH as shown in Figure 11d. Further increasing the NaOH concentration to 7.5 M, continuous nanowire network films were formed on the Ti surface as observed in Figure 11e. When a 10.0 M NaOH solution was used, the films formed on the Ti surface were composed of long and large nanowires as shown in Figure 11f. All of these results show that the morphology of the synthesized nanodots, nanorod arrays, nanowire arrays and continuous nanowire network films, are strongly affected by the concentration of NaOH solution.

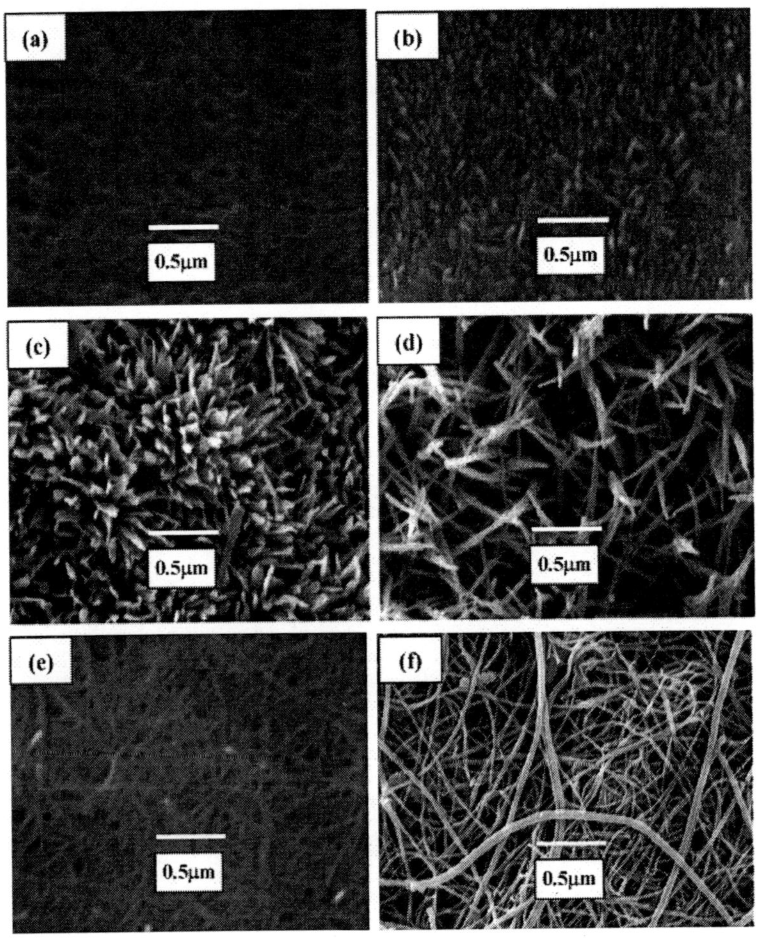

Figure 11. SEM images of the surface structures of the samples synthesized hydrothermally from Ti in (a) 0.0, (b) 0.01, (c) 1.25, (d) 2.5, (e) 7.5 and (f) 10.0 M NaOH solution at 150 °C for 15 hours.

3.2.4. Effect of Temperature and Reaction Time

Figure 12 shows the SEM images of the Ti substrates after the hydrothermal treatment in a 2.5 M NaOH solution at temperatures varying from 125 to 200 °C for 3 hours. At 125 °C, dense nanograss was formed as shown in Figure 12a and 12b. Increasing the temperature to 175 °C (Figures 12c and 12d), some flower-like structures were formed on top of the nanowire films. When the temperature was further increased to 200 °C, Figures 12e and 12f show that many nanospheres composed of irradiated short nanorods were grafted onto the surface of the continuous nanowire network films. These results indicate that the reaction temperature dramatically influences the structure of the hydrothermal products. We further investigated the influence of the reaction time in a 2.5 M NaOH solution at 150 °C. A mixture of nanoplates and a few nanowires were formed on the surface of the Ti substrate when the reaction time was 0.5 hours. Increasing the time to 1.5 hours, many nanowires accompanied with a few small nanoplates covered the Ti substrate. Large-scale nanowires were formed when the reaction time extended to 15 hours. When the time was further increased to 30 hours and 48 hours, continuous nanowire network films were formed. These results show that the reaction time is another critical parameter that affects the morphology of the formed nanomaterials.

3.3. Applications of Tio2 Nanomaterials

The photocatalytic and photoelectrochemical properties of semiconducting oxide materials provide a strong incentive for ongoing scientific research into Green Chemistry applications [17, 22, 111, 112]. TiO_2 belongs to one of the most extensively used and investigated systems and has the advantages of being inexpensive, chemically stable, and non-toxic. The photoactivity of TiO_2 is one of its technologically most attractive properties. Intense research was initiated by Fujishima and Honda's discovery that water can be photocatalytically split into hydrogen and oxygen [8]. The band gap of TiO_2 is relatively wide and its absorption properties are not well matched to the frequency spectrum of natural sun light. The functionalization of the TiO_2 surface with dye molecules has made TiO_2-based solar cells much more efficient for the conversion of solar energy into electric energy. Another special interest of the employment of TiO_2 is photo-degradation of organic pollutants. Nanostructured materials possess a large surface area and are promising in photocatalytic applications. Figure 13a shows the diffuse reflectance spectra of the nanowires, curve 1: anatase TiO_2 resulting from annealing $H_2Ti_3O_7$ nanowires at 450 °C, curve 2: a mixture of anatase and rutile TiO_2 synthesized by annealing $H_2Ti_3O_7$ nanowires at 600 °C, and curve 3: rutile TiO_2 fabricated by annealing $H_2Ti_3O_7$ nanowires at 750 °C. The reflection spectra show that the band gap of these nanowires changes in the following order:

$$\text{anatase} > \text{mixture of anatase and rutile} \approx \text{rutile}$$

These results show that different materials with different optical properties can be formed through this simple annealing treatment. The decomposition of Eosin Y, an organic dye, was employed as a probe to study the photocatalytical activity of these nanomaterials. To immobilize Eosin Y on the nanowire arrays, all of the samples were immersed in an ethanol solution containing Eosin Y (9 mM) for 30 minutes. Then, the samples loaded with dye were irradiated by an UV-Xe lamp (365 nm and 0.2 mW/cm^2). After irradiation with UV light for selected periods of time, diffuse reflectance spectra were recorded to examine the changes in

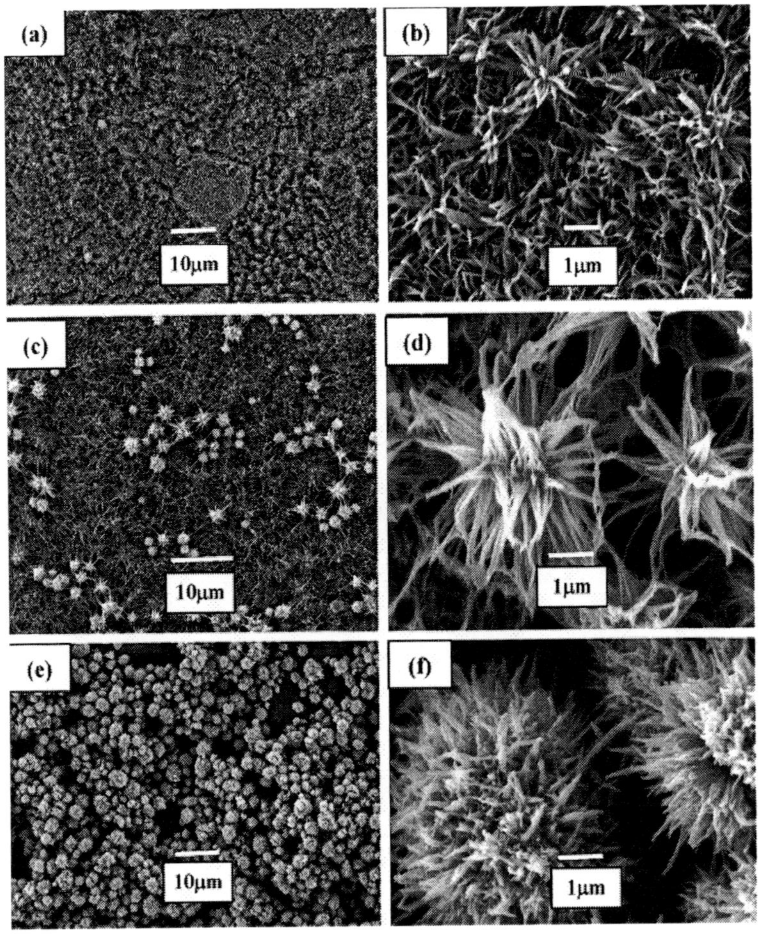

Figure 12. SEM images of the samples synthesized hydrothermally in 2.5 M NaOH solution at (a) – (b) 125 °C; (c) - (d) 175 °C; and (e) - (f) 200 °C for 3 hours.

absorption of Eosin Y. Figures 13b, 13c and 13d show the performances of the anatase nanowires, mixture of anatase and rutile nanowires, and rutile TiO_2 nanowiress for the photocatalytic decomposition of Eosin Y, respectively, after being irradiated for selected periods of time. Before the UV irradiation, as shown in curves 0 h, all the three samples show strong absorption in the wavelength ranging from 420 to 580 nm due to the immobilized Eosin Y. Upon increasing the irradiation time, the absorbance decreases monotonically, indicating that the Eosin Y adsorbed on the nanowires gradually decomposed; however, with different degrading rates. Comparison of all the diffuse reflectance spectra shows that the anatase TiO_2 nanomaterials possess the highest activity for Eosin Y photo-degradation (Figure 13b). After being annealed at higher temperatures, the photocatalytic activity of the nanostructured materials decreases from anatase (Figure 13b), to a mixture of anatase and rutile (Figure 13c), and to rutile TiO_2 nanowires (Figure 13d). This is in good accordance with current literature reports [112-114].

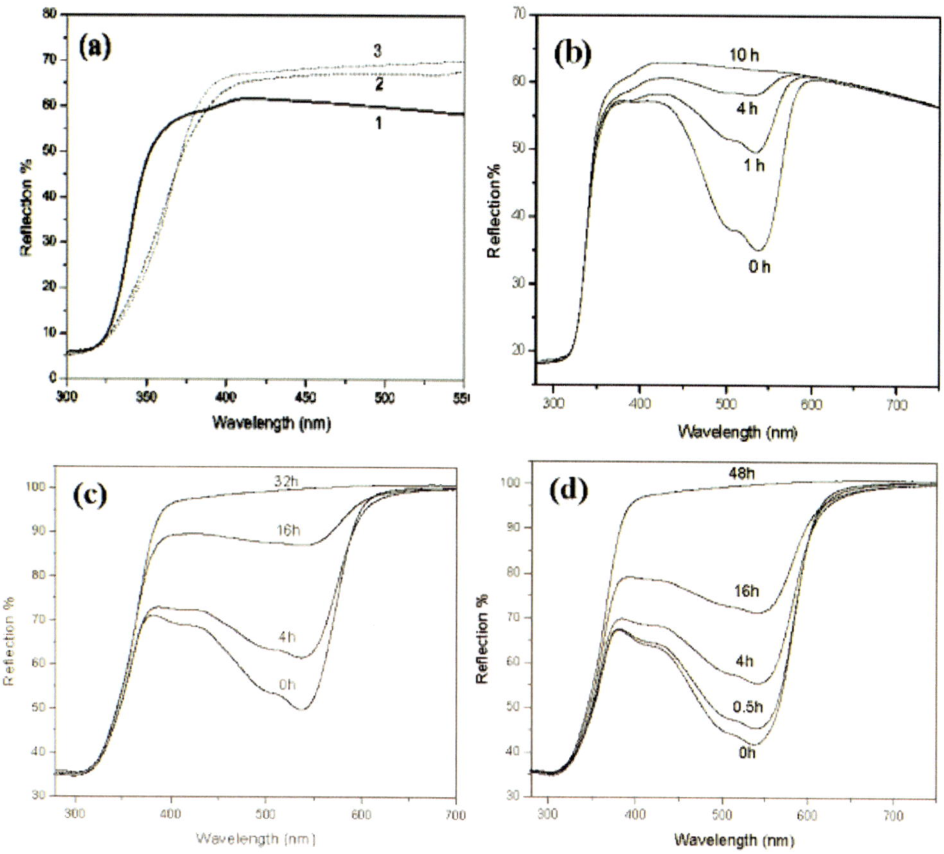

Figure 13. (a) Diffuse reflectance spectra of the samples: (1) anatase synthesized by annealing $H_2Ti_3O_7$ nanowires at 450 °C , (2) mixture of anatase and rutile fabricated by annealing $H_2Ti_3O_7$ at 600 °C, and (3) rutile synthesized by annealing $H_2Ti_3O_7$ at 750 °C; Diffuse reflectance spectra of the Eosin Y loaded samples: (b) anatase, (c) composite of anatase and rutile, and (d) rutile nanowires after irradiation with UV light (365 nm) for selected periods of time.

In addition, electrochemical anodization of titanium sheet has been widely investigated in synthesizing TiO_2 nanotubes [18, 115-121]. Solutions containing floride ions are used as the electrolytes since the dissolution of TiO_2 as soluble $[TiF_6]^{2-}$ is required for pore formation. For instance, Figure 14a shows the cross-sectional and top view of the ordered porous TiO_2 layers fabricated by Schmuki and co-workers using potentiostatic anodic etching of polycrystalline titanium [118]. The Ti sample was anodized at 20 V for 2 hours in 1M $(NH_4)_2SO_4$ + 0.5 wt% NH_4F after a potential sweep from open-circuit potential to 20 V at a scan rate of 0.1 V/s. The thickness of the porous layer is in the order of 2.5 μm with an average pore size of 100 nm. The morphology of the formed porous TiO_2 layer depends on the applied voltage and the electrolyte used. Nanotube arrays were synthesized by Bard et al. [121] by anodization of a Ti substrate in 1M $(NH_4)H_2PO_4$ + 0.5 wt% NH_4F at 15 V with the porous layer thickness in the order of 3 μm. The pore diameter was about 70 nm, and the wall thickness was 20 nm. In contrast, anodizing Ti substrate at 20 V in 1M $(NH_4)H_2PO_4$ + 0.5 wt% NH_4F only produced a porous TiO_2 film. Bard and co-workers further compared the photo-activity of the TiO_2 nanotube arrays, the porous TiO_2 film and one commercial P-25

TiO_2 photoanode under Xe lamp irradiation (~100 mW/cm^2) as shown in Figure 14b. The TiO_2 nanotube arrays produced much higher photocurrent than the nanoporous TiO_2 film and the 15-µm-thick P-25 TiO_2 film. This suggests that TiO_2 nanotube arrays can harvest solar light more effectively than the standard P-25 TiO_2 photoanode.

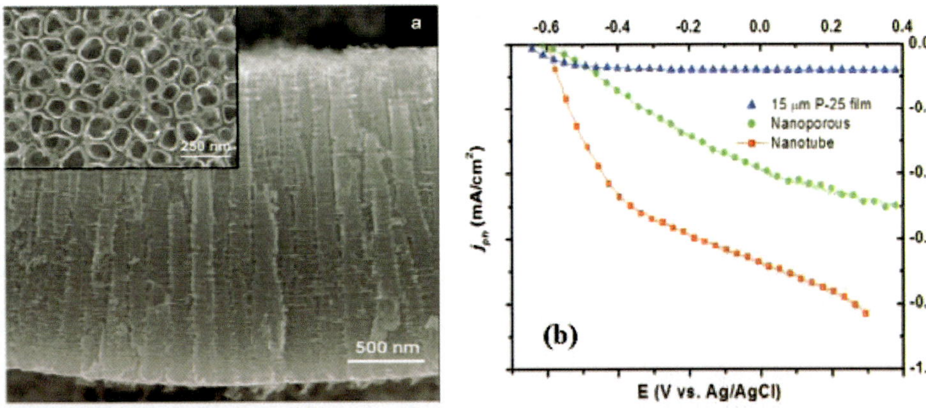

Figure 14. (a) Cross-sectional and top view of a 2.5µm thick self-organized porous TiO_2 layer (Ti samples were anodized at 20 V for 2 hours in 1M $(NH_4)_2SO_4$ + 0.5 wt% NH4F) (b) Variation of photocurrent density vs measured potential for three different TiO_2 photoanodes in 1 M KOH electrolyte. (Figure 14a reprinted with permission from [118]; Figure 14b reprinted with permission from [121]).

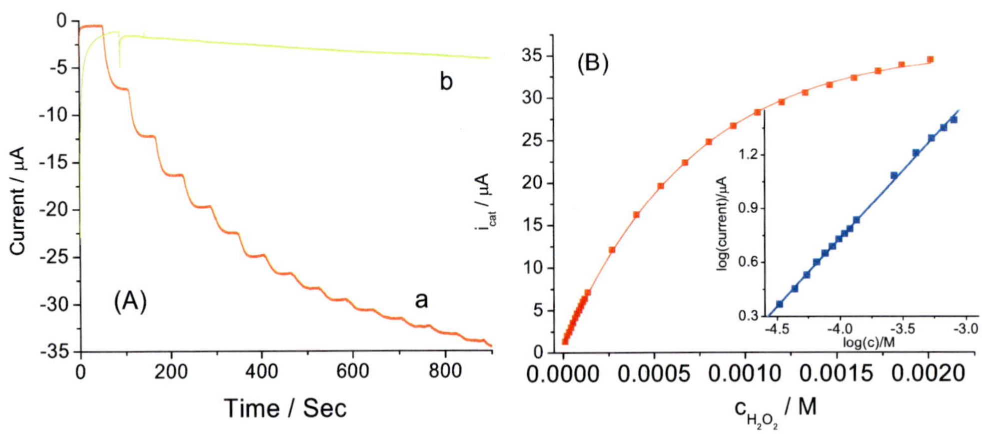

Figure 15. (A) Amperometric response of (a) Th/HRP/TiO_2 and (b) Th/TiO_2 at -0.45V upon successive additions of 0.134 mM H_2O_2 into 0.1 M pH 6.8 PB. (B) Plot of the reduction current versus the H_2O_2 concentration. (Inset) Linear calibration curve. (Reprinted with permission from [18]).

Due to non-toxicity, nanostructured TiO_2 materials show high biocompatibility and good retention of biological activity for protein binding [122-123]. Our recent studies have shown that TiO_2 nanotube arrays provide excellent matrices for the co-immobilization of thionine and horseradish peroxidase (HRP) [18]. Figure 15a presents the chronoamperometric response of the Th/HRP/TiO_2 (a) and Th/TiO_2 (b) electrode with successive additions of

0.134 mM H_2O_2 into 0.1 M pH 6.8 PB at the potential of −0.45 V. The sensor shows an increasing amperometric response to H_2O_2 from 1.1×10^{-5} to 2×10^{-3} M as shown in Figure 15b. The linear response range of the sensor to H_2O_2 concentration is from 1.1×10^{-5} to 1.1×10^{-3} M as seen in the inset of Figure 15b. The linear regression equation is log I (μA) = 3.679 + 0.736 log c (M), with a correlation coefficient of 0.9987. The detection limit is estimated as 1.2×10^{-6} M at 3σ based on the slope. These results show that the immobilized HRP on the TiO_2 nanotube arrays not only retains its bioactivity but also has high affinity for H_2O_2 reduction, resulting in the possibility of developing novel H_2O_2 sensors with very good stability.

4. CONCLUSION

In summary, simple and facile methods have been developed to directly synthesize TiO_2 nanostructures on titanium substrates by oxidizing Ti foil using small organic molecules as the oxygen source. The morphologies of the nanostructures can be controlled by varying the oxygen source, reaction temperature and time. Polycrystalline grains are formed when pure oxygen and formic acid are used as the oxygen source; elongated micro-crystals are produced when water vapor is used as the oxygen source; oriented and aligned TiO_2 nanorod arrays are synthesized when ethanol, acetaldehyde, acetone and DBTDL are used as the oxygen source. The diffusion of Ti atoms to the oxide/gas interface *via* the network of the grain boundaries of the thin oxide layer is the determining factor for the formation of well-aligned TiO_2 nanorod arrays. The nanorods all grow along the <001> direction through the surface diffusion of Ti atoms or ions to the tips of the nanostructures. The wetting properties of the formed TiO_2 nanostructured surfaces are dictated by their structure, varying from a hydrophilic surface to a strongly hydrophobic surface as the surface structure changes from polycrystalline grains to well-aligned nanorod arrays. The controlled formation of the TiO_2 nanorod arrays promises applications in the area of sensors and catalysts. It is anticipated that this method could be used to synthesize ordered nanorod arrrays of other oxides.

In addition, large-scale $Na_2Ti_6O_{13}$ nanoplates, nanowires, and continuous nanowire network films have been directly formed on Ti substrate hydrothermally for the first time. The morphology of the formed $Na_2Ti_6O_{13}$ nanostructures is easily tunable by varying the experimental parameters of temperature, reaction duration and the NaOH concentrations. This study has demonstrated that the synthesized $Na_2Ti_6O_{13}$ nanostructures are easily converted into $H_2Ti_3O_7$ nanostructures, a desirable precursor for fabrication of various TiO_2-based nanomaterials, with shape-preservation through an ion-exchange process. Anatase, a mixture of anatase and rutile, and rutile TiO_2 nanowires were formed when the $H_2Ti_3O_7$ nanowires were annealed at 450 °C, 600 °C and 750 °C, respectively. The formed TiO_2 anatase nanowires show much higher photocatalytic activity for degradation of Eosin Y, compared to the rutile TiO_2 nanowires. In this chapter, we have demonstrated that hydrothermal treatment of Ti substrate in NaOH solutions provides a simple and easy method to synthesize various large-scale TiO_2-based nanostructured materials, ready for practical applications such as the photocatalytic purification of wastewater. We have also illustrated that electrochemical anodization of titanium is an effective method in fabricating TiO_2 nanotubes. The formed TiO_2 nanotube arrays are very promisng in Green Chemistry applications and biosensoring.

ACKOWLEDGEMENTS

This work was supported by a Discovery Grant from the Natural Sciences and Engineering Research Council of Canada (NSERC). A. Chen acknowledges the Ontario Premier's Research Excellence Award.

REFERENCES

[1] Schimid, G. *Nanoparticle: from Theory to Application;* Wiley-VCH: Weinheim, Germany, 2004.

[2] Zhang, J. Z.; Wang, Z. L.; Liu, J.; Chen, S.; Liu, G.-Y. *Self-Assembled Nanostructures;* Kluwer Academic/Plenum Publishers: New York City, NY, 2003.

[3] Nirmal, M.; Brus, L. E. *Acc. Chem. Res.* 1999, 32, 407-414.

[4] Dumestre, F.; Chaudret, B.; Amiens, C.; Renaud, P.; Fejes, P. *Science* 2004, 303, 821-823.

[5] Sun, S.; Murray, C. B.; Weller, D.; Folks, L.; Moser, A. *Science* 2000, 287, 1989-1992.

[6] Peng, X. S.; Meng, G. W.; Wang, X. F.; Wang, Y. W.; Zhang, J.; Liu, X.; Zhang, L. D. *Chem. Mater.* 2002, 14, 4490 – 4493.

[7] Diebold, U. *Surface Science Reports* 2003, 48, 53-229.

[8] Fujishima, A.; Honda, K. *Nature* 1972, 238, 37-38.

[9] Schiavello, M. *Photoelectrochemistry, Photocatalysis, and Photoreactors: Fundamentals and Developments;* NATO ASI Series; D. Reidel Publishing Company: Dordrecht, The Netherlands, 1985.

[10] Ollis, D. S.; Al-Ekabi, H. *Photocatalytic Purification and Treatment of Water and Air;* Elsevier: Amsterdam, The Netherlands, 1993.

[11] Khan, S. U. M.; Akikusa, J. *J. Phys. Chem. B,* 1999, 103, 7184-7189.

[12] Licht, S.; Wang, B.; Mukerji, S.; Soga, T.; Umeno, M.; Tributsch, H. *J. Phys. Chem. B,* 2000, 104, 8920-8924.

[13] Wilcoxon, J. P. *Photocatalysis Using Semiconductor Nanoclusters,* In *Advanced Catalytic Materials;* Lednor, P. W.; Nagaki, D. A.; Thompson, L. T.; Materials Research Society Symposium Proceedings: Boston, MA, 1999.

[14] Serpone, N.; Pelizzetti, E. *Photocatalysis: Fundamentals and Applications;* Wiley: New York City, NY, 1989.

[15] Kozhukharov, V.; Vitanov, P.; Stefchev, P.; Kabasanova, E.; Kabasanov, K.; Machkova, M.; Blaskov, V.; Simeonov, D.; Tzaneva, G. *J. Environ. Prot. Eco.* 2001, 2, 107-115.

[16] Schiavello, M.; Dordrecht, H. *Photoelectrochemistry, Photocatalysis, and Photoreactors: Fundamentals and Developments;* Kluwer Academic: Boston, MA, 1985.

[17] Linsebigler, A. L.; Lu, G.; Yates, J. T. *Chem. Rev.* 1995, 95, 735-738.

[18] Liu, S.; Chen, A. *Langmuir* 2005, 21, 8409-8413.

[19] Mills, A.; Davies, H. R.; Worsley, D. *Chem. Soc. Rev.* 1993, 22, 417.

[20] Legrini, O.; Oliveros, E.; Braun, A. M. *Chem. Rev.* 1993, 93, 671.

[21] Heller, A. *Acc. Chem. Res.* 1995, 28, 503.

[22] Hoffman, M.; Martin, S.; Choi, W.; Bahnemann, D. *Chem. Rev.* 1995, 95, 69-96.

[23] Bonhote, P.; Gogniat, E.; Gräztel, M.; Ashrit, P. V. *Thin Solid Films,* 1999, 350, 269-275.

[24] Jia, J. G.; Ohno, T.; Matsumura, M. *Chem. Lett.* 2000, 908.

[25] Sun, J.; Gao, L.; Zhang, Q. *J. Am. Ceram. Soc.* 2003, 86, 1677-1681.

[26] Henrich, V. E.; Cox, P. A. *The Surface Science of Metal Oxides*; Cambridge University Press: Cambridge, 1996.

[27] Kim, K. J.; Benksten, K. D.; van de Lagemaat, J.; Frank, A. J. *J. Chem. Mater.* 2002, 14, 1042-1047.

[28] Sato, T.; Baba, K.; Hirozawa, T.; Kawakami, S. *IEEE J. Quantum Electron.* 1993, 29, 175-181.

[29] Hagfeldt, A.; Gratzel, M. *Chem. Rev.* 1995, 95, 49-68.

[30] Sahle-Demessie, E.; Gonzalez, M.; Wang, Z. M.; Biswas, P. *Ind. Eng. Chem. Res.* 1999, 38, 3276-3284.

[31] Mills, A.; Hunte, S. L. *J. Photochem. Photobiol.* 1997, *A*108, 1-35.

[32] Hoffmann, M. R.; Martin, S. T.; Choi, W.; Bahnemann, D. W. *Chem. Rev.* 1995, 95, 69-96.

[33] Linsebigler, A. L.; Lu, G.; Yates, J. T. Jr. *Chem. Rev.* 1995, 95, 735-758.

[34] Kim, D.; Anderson, M. *Environ. Sci. Technol.* 1994, 28, 479-483.

[35] Vinodgopal, K.; Kamat, P. *CHEMTECH* 1996, 26, 18-22.

[36] Wada, K.; Yoshida, K.; Watanabe, Y. *J. Chem. Soc., Faraday Trans.* 1995, 91, 1647-1654.

[37] Dartt, C. B.; Davis, M. E. *Ind. Eng. Chem. Res.* 1994, 33, 2887-2899.

[38] Maldotii, A.; Bartocci, C.; Amadelli, R.; Polo, E.; Battioni, P.; Mansuy, D. *J. Chem. Soc., Chem. Commun.* 1991, 20, 1487-1489.

[39] Groves, J. T.; Quinn, R. *J. Am. Chem. Soc.* 1985, 107, 5790-5794.

[40] Chen, Y. S.; Crittenden, J. C.; Perram, D. L.; Hand, D. W. *U.S. Provisional Application No.* 60/121,500 1998.

[41] Fox, M. A. *Acc. Chem. Res.* 1983, 16, 314-321.

[42] Ollis, D. F.; Hsiao, C. Y.; Budiman, L.; Lee, C. L. *J. Catal.* 1984, 88, 89-96.

[43] Shah, S. I.; Li, W.; Huang, C. P.; Jung, O.; Ni, C. *Proc. Natl. Acad Sci. U.S.A.* 2002, 99, 6482-6486.

[44] Xu, A.; Zhu, J.; Gao, Y.; Liu, H. *Chem. Res. Chin. Univ.* 2001, 17, 281.

[45] Wang, C.; Bahnemann, D. W.; Dohrmann, J. K. *Chem. Commun.* 2000, 16, 1539-1541.

[46] Wang, Y.; Hao, Y.; Cheng, H.; Ma, H.; Xu, B.; Li, W.; Cai, S. *J. Mater. Sci.* 1991, 34, 2773.

[47] Coloma, F.; Marquez, F.; Rochester, C. H.; Anderson, J. A. *Phys. Chem. Chem. Phys.* 2000, 2, 5320-5324.

[48] Altynnikov, A. A.; Zenkovets, G. A.; Anufrienko, V. F. *React. Kinet. Catal. Lett.* 1999, 67, 273-279.

[49] Umebayashi, T.; Yamaki, T.; Itoh, H.; Asai, K. *J. Phys. Chem. Solids* 2002, 63, 1909-1920.

[50] Wang, Y.; Cheng, H.; Hao, Y.; Ma, J.; Li, W.; Cai, S. *Thin Solid Films* 1999, 349, 120-125.

[51] Yamashita, H.; Honda, M.; Harada, M.; Ichihashi, Y.; Anpo, M.; Hirao, T.; Itoh, N.; Iwamoto, N. *J. Phys. Chem. B* 1998, 102, 10707-10711.

[52] Yu, J. C.; Yu, J. G.; Ho, W. K.; Jiang, Z. T.; Zhang, L. Z. *Chem. Mater.* 2002, 14, 3808-3816.
[53] Khan, S. U. M.; Al-Shahry, M.; Ingler, W. B. Jr. *Science* 2002, 297, 2243-2245.
[54] Asahi, R.; Morikawa, T.; Ohwaki, T.; Aoki, K.; Taga, Y. *Science* 2001, 293, 269-271.
[55] Iric, H.; Wanatabe, Y.; Hashimoto, K. *J. Phys. Chem. B* 2003, 107, 5483-5486.
[56] Burda, C.; Lou, Y. B.; Chen, X. B.; Samia, A. C. S.; Stout, T.; Gole, J. L. *Nano Lett.* 2003, 3, 1049-1051.
[57] Yang, T. S.; Shiu, C. B.; Wong, M. S. *Surf. Sci.* 2004, 548, 75-82.
[58] Lee, K. S.; Park, I. S. *Scripta Materialia* 2003, 48, 659-663.
[59] de Lucas, M. C.; Fabreguette, F.; Collin, S.; Bourgeois, S. *Int. J. Inorg. Mater.* 2000, 2, 255-259.
[60] Byun, D.; Jin, Y.; Kim, B.; Lee, J. L.; Park, D. *J. Hazard Mater.* 2000, 73, 199-206.
[61] Damiriv, D.; Bally, A. R.; Ballif, C.; Homes, P.; Schmid, P. E.; Sanjines, R.; Levy, F.; Parvulescu, V. I. *Appl. Catal.* 2000, 25, 83-92.
[62] Wang, X. P.; Yu, Y.; Hu, X. F.; Gao, L. *Thin Solid Films* 2000, 371, 148-152.
[63] Kato, K.; Tsuzuki, A.; Taoda, H.; Torrii, Y.; Kato, T.; Batsugn,Y. *J. Mater. Sci.* 1994, 29, 5911-5915.
[64] Hoyer, P. *Langmuir* 1996, 12, 1411-1413.
[65] Burnside, S. D.; Shklower, V.; Barbe, C.; Comte, P.; Arendse, F.; Brookes, K.; Grätzel, M. *Chem. Mater.* 1998, 10, 2419-2425.
[66] Yoo, S.; Akbar, S. A.; Sandhage, K. H. *Adv. Mater.* 2004, 16, 260-264.
[67] Liu, S. M.; Gan, L. M.; Lu, L. H.; Zhang, W. D.; Zheng, H. C. *Chem. Mater.* 2002, 14, 1391-1397.
[68] Uchida, S.; Chiba, R.; Tomiha, M.; Masaki, N.; Shirai, M. *Electrochemistry* 2002, 70, 418-420.
[69] Imai, H.; Takei, Y.; Shimizu, K.; Matsuda, M.; Hirashima, H. *J. Mater. Chem.* 1999, 9, 2971-2972.
[70] Chae, W. S.; Lee, S. W.; Kim, Y. R. *Chem. Mater.* 2005, 17, 3072.
[71] Liu, S. M.; Gan, L. M.; Liu, L. H.; Zhang, W. D.; Zeng, H. C. *Chem. Mater.* 2002, 14, 1391-1397.
[72] Kasuga, T.; Hiramatsu, M.; Hoson, A.; Sekino, T.; Niihara, K. *Langmuir* 1998, 14, 3 160-3163.
[73] Kasuga, T.; Hiramatsu, M.; Hoson, A.; Sekino, T.; Niihara, K. *Adv. Mater.* 1999, 11, 1307-1311.
[74] Zhang, Y. X.; Li, G. H.; Jin, Y.; Zhang, X. Y.; Zhang, J.; Zhang, L. D. *Chem. Phys. Lett.* 2004, 365, 300-304.
[75] Lei, Y.; Zhang, L. D.; Meng, G. W.; Li, G. H.; Zhang, X. Y.; Liang, C. H.; Chen, W.; Wang, S. X. *Appl. Phys. Lett.* 2001, 78, 1125-1127.
[76] Zhang, S. L.; Zhou, J. F.; Zhang, Z. J.; Du, Z. L.; Vorontsov, A. V.; Jin, Z. S. *Chin. Sci. Bull.* 2000, 45, 1533-1536.
[77] Seo, D. S.; Lee, J. K.; Kim, H. *J. Cryst. Growth* 2001, 229, 428-432.
[78] Zhang, Q. H.; Gao, L. A.; Sun, J.; Zheng, S. *Chem. Lett.* 2002, 2, 226-227.
[79] Tian, Z. R. R.; Voigt, J. A.; Liu, J.; McKenzie, B.; Xu, H. F. *J. Am. Chem. Soc.* 2003, 125, 12384-12385.
[80] Adachi, M.; Murata, Y.; Yoshikawa, S. *Chem. Lett.* 2000, 8, 942-943.

[81] Adachi, M.; Okada, I.; Ngamsinlapasathian, S.; Murata, Y.; Yoshikawa, S. *Electrochemistry* 2002, 70, 449-452.

[82] Hippe, C.; Wark, M.; Lork, E.; Schulz-Ekloff, G. *Microporous Mesoporous Mater.* 1999, 31, 235-239.

[83] Khitrov, G. *MRS Bull.* 2000, 25, 8, 3-4.

[84] Zhang, M.; Bando, Y.; Wada, K. *J. Mater. Sci. Lett.* 2001, 20, 167-170.

[85] Zhang, M.; Bando, Y.; Wada, K. *J. Mater. Res.* 2001, 16, 1408-1412.

[86] Li, D.; Xia, Y. N. *Nano Lett.* 2003, 3, 555-560.

[87] Corcoran, D. J. D.; Tunstall, D. P.; Irvine, J. T. S. *Solid State Ionics* 2000, 136-137.

[88] Pozo, R. L.; Baltanás, M. A.; Cassano, A. E. *Catal. Today* 1997, 39, 219-231.

[89] Peng, X.; Chen, A. *J. Mater. Chem.* 2004, 14, 2542-2548.

[90] Peng, X.; Chen, A. *Appl. Phys. A* 2005, 80, 473-476.

[91] Peng, X.; Wang, J.; Thomas, D. F.; Chen, A. *Nanotechnology* 2005, 16, 2389-2395.

[92] Fiest, T. P.; Davies, P. K. *J. Solid. State Chem.* 1992, 101, 275-295.

[93] Suzuki, Y.; Yoshikawa, S. *J. Mater. Res.* 2004, 19, 982-985.

[94] Armstrong, A. R.; Armstrong, G.; Canales, J.; Bruce. P. G. *Angew. Chem., Int. Ed.* 2004, 43, 2286-2288.

[95] Parker, J. C.; Segel, R. W. *Appl. Phys. Lett.* 1990, 57, 943-945.

[96] Badescu, V.; Mormirlan, M. *J. Crystal Growth* 1996, 169, 309-316.

[97] Czerwinski, F.; Szpunar, J. A. *Micron* 1998, 29, 201-206.

[98] Gracia, F.; Holgado, J. P.; Contreras, L.; Girardeau, T.; González-Elipe, A. R. *Thin Solid Films* 2003, 429, 84-90.

[99] Peng, X. S.; Zhang, L. D.; Meng, G. W.; Wang, X. F.; Wang, Y. W.; Wang, C. Z.; Wu, G. S. *J. Phys. Chem. B* 2002, 106, 11163-11167.

[100] Peng, X. S.; Meng, G. W.; Zhang, J.; Wang, X. F.; Wang, Y. W.; Wang, C. Z.; Zhang, L. D. *J. Mater. Chem.* 2002, 12, 1602-1605.

[101] Smeltzer, W. W. *Mater. Sci. Forum* 1988, 29, 151.

[102] Sakai, N.; Fujishima, A.; Watanabe, T.; Hashimoto, K. *J. Phys. Chem. B* 2003, 107, 1028-1035.

[103] Wang, R.; Sakai, N.; Fujishima, A.; Watanabe, T.; Hashimoto, K. *J. Phys. Chem. B* 1999, 103, 2188-2194.

[104] Zhang, X.; Shi, F.; Yu, X.; Liu, H.; Fu, Y.; Wang, Z. Q.; Jiang, L.; Li, X. Y. *J. Am. Chem. Soc.* 2004, 126, 3064-3065.

[105] Shiu, J. Y.; Kuo, C. W.; Chen, P. L.; Mou, C. Y. *Chem. Mater.* 2004, 16, 561-564.

[106] Yoshimisu, Z.; Nakajima, A.; Watanabe, T.; Hashimoto, K. *Langmuir* 2002, 18, 5818-5822.

[107] Wenzel, R. N. *Ind. Eng. Chem.* 1936, 28, 988-994.

[108] Cassie, A. B. D.; Baxter, S. *Trans. Faraday Soc.* 1944, 40, 546-551.

[109] Otten, A.; Herminghaus, S. *Langmuir* 2004, 20, 2405-2408.

[110] Chen, A.; Peng, X.; Koczkur, K.; Miller, B. *Chem. Commun.* 2004, 17, 1964-1965.

[111] Peng, X.; Chen, A. *Adv. Funct. Mater.* Adfm200500464 (*in press*).

[112] Serpore, N.; Pilezzeti, E. *Photocatalysis: Fundamentals and Applications;* Wiley-Interscience: New York City, NY, 1989.

[113] Hosono, E.; Fujihara, S.; Kakiuchi, K.; Imai, H. *J. Am. Chem. Soc.* 2004, 126, 7790-7791.

[114] Cozzoli, P. D.; Comparelli, R.; Fanizza, E.; Locia-Curri, M.; Agostiano, A.; Laub, D. *J. Am. Chem. Soc.* 2004, 126, 3868-3879.

[115] Diwarld, O.; Thompson, T. L.; Zubkov, T.; Goralski, E. G.; Walck, S. D.; Yates, J. T. Jr. *J. Phys. Chem. B* 2004, 108, 6004-6008.

[116] Gong, D.; Grimes, C. A.; Varghese, O. K; Hu, W. C.; Singh, R. S.; Chen, Z.; Dickey, E. C. *J. Mater. Res.* 2001, 16, 3331-3334.

[117] Wang, W. Z.; Varghese, O. K.; Paulose, M.; Grimes, C. A.; Wang, Q. L.; Dickey, E. C. *J. Mater. Res.* 2004, 19, 417-422.

[118] Balaur, E.; Macak, J. M.; Taveira, L.; Schmuki, P. *Electrochem. Commun.* 2005, 7, 1066-1017.

[119] Macak, J. M.; Tsuchiya, H.; Ghicov, A.; Schmuki, P. *Electrochem. Commun.* 2005, 7, 1133-1137.

[120] Beranek, R.; Tsuchiya, H.; Sugishima, T.; Macak, J. M.; Taveira, L.; Fujimoto, S.; Kisch, H.; Schmuki, P. *Appl. Phys. Lett.* 2005, 87, 243114.

[121] Park, J. H.; Sungwook, K.; Bard, A. J. *Nano Lett.* 2006, 6, 24-28.

[122] Topoglidis, E.; Cass, A. E. G.; Gilardi, G.; Sadeghi, S.; Beaumont, N.; Durrant, J. R. *Anal. Chem.* 1998, 70, 5111-5113.

[123] Zhang, J. K.; Cass, A. E. G. *Anal. Biochem.* 2001, 292, 307-310.

In: Frontal Nanotechnology Research
Editor: Martin V. Berg, pp. 161-187
ISBN: 1-60021-212-3

Chapter 7

MOLECULAR FOUNDATIONS FOR ONSET WEAR. A REVIEW

Mario D'Acunto[1]

Department of Chemical Engineering and Materials Science, University of Pisa, via Diotisalvi 2, 56100 Pisa, Italy

and

Department of Physics, Univeristy of Pisa, largo B. Pontecorvo 3, 56127 Pisa, Italy

ABSTRACT

One definition of wear in a tribological sense is that it is the progressive loss of mass from the operating surface of a body as a result of relative motion at the surface. Generally, tribology may be defined as the study of friction wear and lubrication, but in comparison with friction forces or lubrication methods, very little work has been done on the extremely important area of wear and surface damage. For systems consisting of common materials (metals, polymers, ceramics), there are at least four main mechanisms by which wear and surface damage can occur between solids in relative motion: abrasive- and/or adhesive-, chemical- or corrosive-, fatigue-wear mechanisms. The occurrence of such different wear mechanisms involves the fundamental mechanics of molecular and supra-molecular scale. As a consequence, the understanding of surface properties on nanoscale level should be generated to have also a satisfactory knowledge of responses of materials and machine components on macrometer scale. Investigation of the fundamental characteristics of wear at the nanoscale is complicated by some factors and forces that act on such scale level have not yet been addressed in the tribology of macro-systems. Since these forces are sensitive to the environment and surface condition of the specimens, it is quite difficult to determine them accurately. Further, quantification of wear is not straightforward since the amount of wear is often too small to be detected by surface-sensitive instruments. Nanowear measurements have been object of rapid development of precise measuring tools following to the introduction of scanning probe microscopies, such as atomic force microscopy (AFM). The advent of AFM opened new opportunity to study wear mechanisms on nanoscale, making possible to simulate one surface sliding on other one and controlling many physical parameters during the contact dynamics. Further, AFM made possible to a have a direct connection between experimental data and theoretical models.

[1] E-mail address: m.dacunto@ing.unipi.it

The chapter is devoted to give a panoramic lecture of some experimental results and models on the onset wear occurrence on nanoscale.

1. INTRODUCTION

The atomistic mechanisms and dynamics of the interactions of two materials during relative motion, need to be understood in order to develop fundamental understanding of adhesion, friction, wear, indentation and lubrication processes. At most solid-solid interfaces of technologies relevance, contact occurs at many asperities. Consequently the importance of investigating single asperity contacts in studies of the fundamental mechanical and tribological properties of surfaces and interfaces on nanoscale has long been recognized. The recent emergence and proliferation of scanning probe microscopies (SPMs) and computational techniques for simulating tip-surface interactions and interfacial properties, has allowed systematic investigations of interfacial problems with high resolution as well as ways and means for modifying and manipulating nanoscale structures. These advances have led to the appearance of the new field of nanotribology, which pertains to experimental and theoretical investigations of interfacial processes occurring during adhesion, friction, wear, scratching, nanoindentation and thin-film lubrication at sliding surfaces on scales ranging from the atomic- and molecular- to the microscale [Bhushan et al., 1995; Bhushan, 2003]. The nanotribological studies are needed to develop fundamental understanding of interfacial phenomena on atomic, molecular and supramolecular scale and to study interfacial phenomena in nanostructures used in the area commonly defined nanotechnology. Friction and wear of lightly loaded nanocomposites are highly dependent on the surface interactions involving few atomic layers. Consequently, these structures are generally lubricated with molecularly thin films.

A definition of wear mechanisms can be as follows: the succession of events whereby atoms, products of chemical conversion, fragments, debris, are induced to leave the system and identified in a manner that embody or suggest solutions [Ludema, 2001]. These solutions may include choice of materials, choice of lubricants, choice of contact condition, choice of the manner of operation of the mechanical system. This definition includes also: a) the conditions imposed upon surfaces, such as contact stress, shear stress and number of cycles of these stress applications; b) the changing composition of substances in the sliding interfaces due to both sliding and these will likely influence the friction and shearing stresses imposed in sliding contact; c) the processes whereby particles (atoms, molecules) are separated from a surface; d) the circulation and path of those particles before they are expelled as debris. There are more than 24 common terms to describe wear mechanisms [Ludema 1996]. Generally, such wear mechanisms are listed in six categories of terms. Among the most important categories, we limit to discuss the most important, such as adhesion based wear mechanisms, fatigue derived wear mechanisms and abrasion.

Adhesive wear is the most difficult term to define. It may denote a particular type of material loss due to high local friction (which is often attributed to adhesion) and is a tempting term to use because high local friction produces tearing and fragmentation whereas lubricants diminish tearing. Some time, lubricated wear is taken to be the opposite of adhesive wear. An often overlooked aspect of adhesive wear is that if adhesion is effective then the particles torn loose by adhesion will re-adhere and not be expelled as debris.

Wear mechanisms deriving from cyclic stressing imply fatigue of materials such as fretting. Fretting is the oscillatory sliding motion of small slip amplitude between two nominally stationary contacting surfaces. It occurs when two quasi-static surfaces in contact are actually subjected to deformation and/or vibrations. Damage caused by fretting is one of the severe problems in engineering since it contributes substantially to shorten life of mechanical components. Analogous to fretting is the delamination mechanisms, which describes a type of wear debris developed by low cycle fatigue when surfaces are rubbed repeatedly by a small shaped slider.

Abrasive-wear mechanisms denote a wide class of wearing mechanisms occurring when a hard solid, particle, or asperity rubs against a softer surface. Examples of abrasive wear include the smoothing of wood by sandpaper, the action of emery paper in polishing metals, grinding processes, and harmful effects of hard particles (e.g. sand, dirt, carbide or silica inclusions, oxides and work-hardened wear debris) in lubricated systems. Abrasive wear is a form of microcutting, involving a hard particle or shape which indents, grooves, and then cuts material out of the surface.

On macroscale, wear has been the subject of much study and many general mechanisms are reasonably well understood [Hutchings, 1992]. On nanoscale, this is yet not so. The mechanisms are largely unknown and not readily determined. Nevertheless, the occurrence of wear processes at nanoscale has been demonstrated [Kim et al., 1991; Bhushan and Koinkar, 1994; Hu et al., 1995; Carpick and Salmeron, 1997; Schöfer and Santer, 1998; Gåhlin and Jacobson, 1998].

Let us consider the possible mechanisms of wear to be essentially two types: abrasive- and adhesive-wear mechanism. On macroscale, if the contribution to wear is essentially of adhesive nature, it is possible to quantify the volume of wear for plastic contacts as

$$V = \frac{k_{adh} L x}{H} \tag{1}$$

where k_{adh} is a non-dimensional wear coefficient dependent on the materials in contact, L is the load applied, H is the hardness coefficient and x is the sliding distance. Equation (1) is commonly referred as Archard's equation [Archard, 1953]. It suggests that the wear rate is independent of the apparent area, directly proportional to the applied load and the wear rate is constant with sliding distance (or time). These properties seem to be analogous to Amontons' laws of friction. Furthermore, equation (1) suggests that the probability of decohesion of a certain volume of material and formation of a wear particle increases with real area of contact, A_r (=L/H) and the sliding distance. An equation of the form similar to Archard's equation for adhesive wear is found to cover a wide range of abrasive situations

$$V = \frac{k_{abr} L x}{H} \tag{2}$$

In analogy to macroscopic scale case, the basic mechanisms for the occurrence of adhesive- or abrasive-wear mechanisms quantifying the wear volume rate correspondent to material loss during the scanning probe tip have been performed in micro/nanoscale [Chung and Kim, 2003].

Several techniques are used for fundamental studies of tribological features. Proximal probes- including scanning tunnelling microscopy (STM), AFM and the surface forces apparatus (SFA) that operate in ultrahigh vacuum (STM, AFM) and in controlled environments (AFM, SFA) – constitute the basic imaging and force measuring tools. These techniques have made possible the study of single-asperity contacts in a controlled fashion, eliminating complications encountered in more traditional methods, such as pin-on-disk, indenters, profilometers, etc., in which the contact is made between rough surfaces involving a large and uncontrolled number of asperities, figure 1. The chapter is structured as follows: in section 2, prominent wear mechanisms on nanoscale are discussed; in section 3, activation process describing onset wear on the nanoscale will be presented.

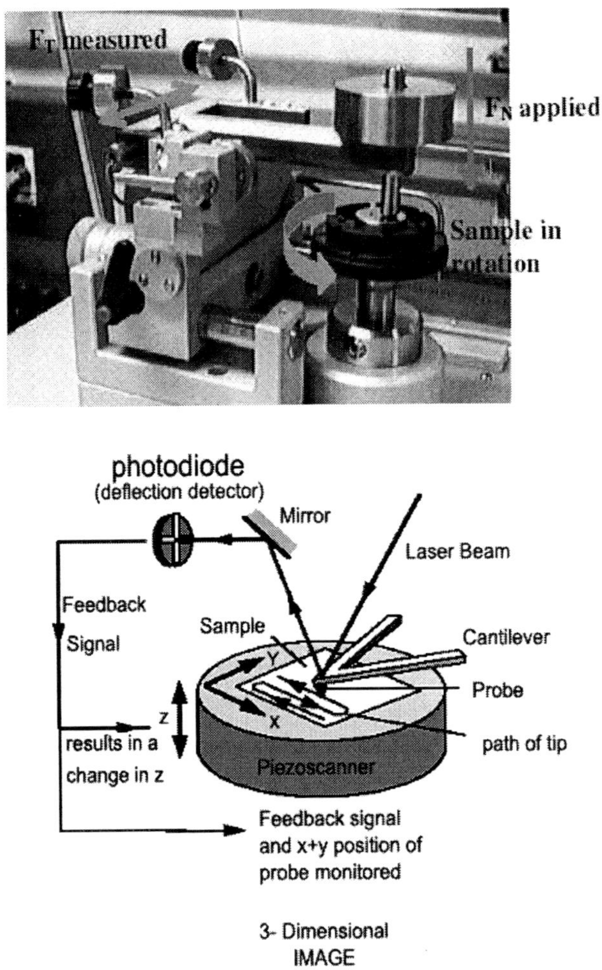

Figure 1. (Upper image) typical wear machine where a flat or a sphere shaped indenter is loaded on to the test sample with a precisely known weight. A pin is mounted on a stiff lever, designed as a frictionless force transducer. As the sample is either rotating or sliding, resulting frictional forces acting between the pin and the sample are measured by very small deflections of the lever. Wear coefficients for both the sample and material are calculated from the volume of the material lost during a specific friction run. (Reprinted from CSM instruments, Switzerland, with permission). Below, typical experimental set up for AFM configuration.

2. MOLECULAR AND SUPRAMOLECULAR FEATURES FOR WEAR

One main SPM involved method for the observation of the nanowear processes is given by the production of defects in sliding contacts making use of the probe tip. Generally, the probe tip is used to scan repeatedly the same area of the sample. AFM has been used to study such processes on mica both in controlled atmosphere [Kopta and Salmeron, 2000] and in aqueous environment [Helt and Batteas, 2005]. Muscovite mica, a layered alumino-silicate in the form of $KAl_2(Si_3AlO_{10})(OH)_2$ has been frequently employed as a model surface for the investigation of the atomic scale relationship between friction [Erlandsson et al., 1988; Carpick et al., 1996a], adhesion [Carpick et al., 1996b; Xu and Salmeron, 1998; Barthel and Roux, 2000] and wear [Miyake, 1994; Hu et al., 1995] by the AFM and SFA [Israelashvili, 1992] measurements due to the ability to readily generate large domains of atomically smooth surface. The layered sheet structure of mica has a ~1nm periodicity normal to the *(001)* crystal plane, which is defined by the repeating boundary layers of K^+, figure 2. The K^+ ions electrostatically bind the layers togheter and act to neutralize the net negative charge associated with the partial substitution of Si with Al (on average one out of every four) in the basal plane. The K^+ interlayer demarcates the preferred cleavage planes and defines the atomic boundaries along which shearing of the layers most readily occurs. Under aqueous conditions, the outmost K^+ layer is exchanged into solution and therefore has yet to be imaged due to its ready displacement. With the K^+, ions solvated, the outermost mica surface exposes a hexagonally arrayed pattern of oxygen atoms with a 0.52nm periodicity.

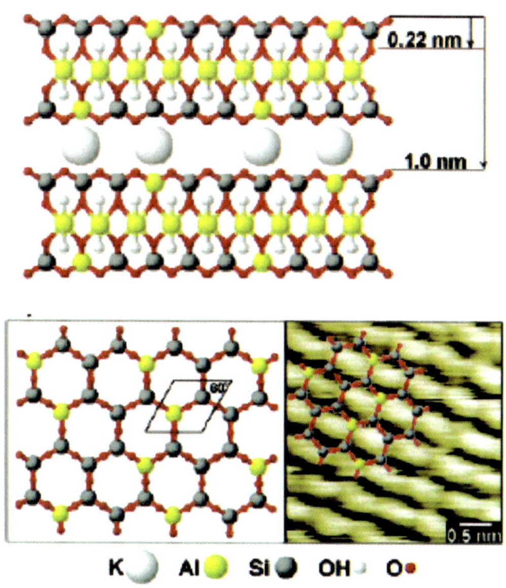

Figure 2. Cross-section depicting micas layered structure along the *c* axis with interlayer distances referenced to the SiO_3/AlO_3 tetrahedral basal plane (upper image). A graphic representation and an unfiltered AFM topographic image (3.4nm×3.4nm) of the mica (001) crystal plane showing the 0.52nm lattice periodicity (lower image). (Reprinted from [Helt and Batteas, 2005], with permission).

AFM studies of mica surfaces have shown that under sufficiently high loads the surface becomes damaged and ~0.2 or ~1nm deep holes are formed, depending on the imaging conditions. Wear scars have been produced by repeatedly scanning over the same region at a fixed load. Point defects are created as a result of the lowering of the activation energy for bond breaking by the applied stresses [Dickinson et al., 1997]. In the tip sample contact area, the number of defects created is given by [Kopta and Salmeron, 2000]

$$N_{def}(L) = t_{res} n_0 A(L) \nu e^{\{-[\varepsilon_0 - \delta\varepsilon(L)]/k_B T\}} \quad (3)$$

where t_{res} is the residence time of the tip, n_0 ($\sim 10^{20}m^{-2}$) the surface density of atoms, and $A(L)$ the contact area as a function of total load L, ν is the attempt frequency ($\sim 10^{13}s^{-1}$) to overcome the energy barrier ε_0 to break a *Si-O* bond, $\delta\varepsilon(L)$ is the decrease in activation energy due to the applied stress, which can be estimated by multiplying the force acting on an atom by the resulting bond stretch. The average force per atom is given by the pressure $P(L)$ in the contact area divided by n_0. The relative deformation (bond stretch) $\Delta l/l$, should be on the same order as the Hertzian indentation depth h, divided by the characteristic length of the affected area, which is on the order of the contact radius $r(L)$: $\Delta l/l \approx h/r(L)$, with $h \approx r^2(L)/2R$ for a tip radius $R >> h$. Thus the decrease in activation energy due to the applied stress $\delta\varepsilon(L)$ changes as $L^{2/3}$. Wear damage will occur when the defect density, $N_{def}/n_0A(L)$, reaches a critical value η_{crit}, and gives rise to nucleation of a hole. Thus the number of consecutive scans necessary to cause a visible damage (0.2nm holes) on the muscovite mica surface is given by the equality of critical value with the product of number scans with the defect density. Such threshold value for number scans can be expressed by the following cumulative curve

$$N_{scans} = AL^{-2/3} e^{-BL^{2/3}} \quad (4)$$

where A and B are two constants.

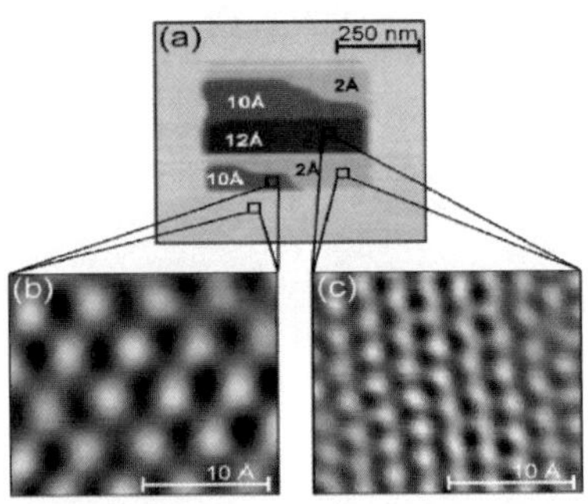

Figure 3. AFM images of a hole scratched into muscovite mica with an AFM tip. The images were recorded making use of a silicon cantilever with a nominal force constant of 2.1N/m and a nominal tip radius of 50nm was used. The hole was created after 8 consecutive scans with a load of 230nN, over an area of 500×500nm^2 at a scan speed of 60ms/line. (a) Image size is 0.75×0.75μm2. The hole obtained with these parameters shows 2, 10 and 12Å deep regions (denoted by gray levels). Marked with a square are regions where the atomic lattice resolution images, shown Fourier filtered in (b) and (c), were obtained. The periodicities in these images are 5. Å (b) and 3.0 Å (c). The latter is rotated by~30°. (Reprinted from [Kopta and Salmeron, 2000], with permission).

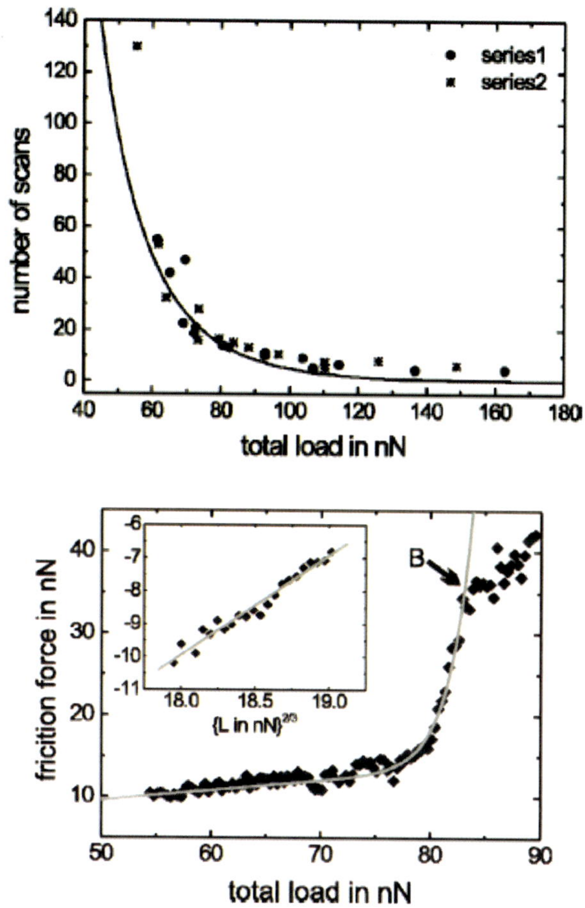

Figure 4. In the upper image, it is shown the number of consecutive scans necessary to cause visible damage to the mica surface (2Å holes). Data from two series of experiments measured with the same tip (force constant 0.4N/m) on two consecutive days are shown. In the lower image, friction vs load curve obtained during the creation of 2Å and 10Å deep holes. Rapid increase in friction is observed above 75nN load, corresponding to a large increase in the production of defects. When the density of defects reaches a critical value, the friction levels off and a wear scar detectable by AFM is produced. The inset shows the region, where the friction increases rapidly, in a semilog representation: $\ln\{[F(L) - cL^{2/3}]/L^{2/3}\}$ is plotted vs $L^{2/3}$ to allow for a simple linear fit of the coefficients B, eq.(4). (Adapted from [Kopta and Salmeron, 2000], with permission).

Helt and Batteas reported the wear of mica surfaces under aqueous environments of varying pH [Helt and Batteas, 2005]. Using the AFM, they have evaluated the simultaneous changes in surface topography, friction and adhesion at the mica surface under controlled pH conditions. By varying pH and applied loads, the incidence and mechanisms of the observed wear were found to be influenced dramatically. In fact, with the appropriate control of applied load, number of scans, and pH, they have been able, for the first time, to directly observe defect production, consisting in the breaking of terminal surface bonds, as detected in the form of a surface restructuring prior to the formation of 2Å pits or scars, generally, considered gross wear. A further result shown by Helt and Batteas is that the degree of wear can be completely controlled by modifications of imaging conditions and pH. Under low pH conditions (pH ≤3), the wear threshold pressure relating to the formation of 2Å pits, is equivalent to that of a simply humid environment and complete inhibition of wear is observed for tests conducted at the threshold pressure for higher pH studies.

At the highest pH conditions (pH≈8), the mica surface undergoes rapid wear. Even under low loads, wear scars ~2Å in depth form readily. The threshold pressure for wear at pH=8 was found to be nearly 1.2GPa. With additional scans (>10), wear scars of 10Å or more are rapidly formed, figure 5. Interestingly, the depth of the wear scars formed under these pH conditions is found to scale roughly linearly with the number of scans for wear scars of up to 60-70Å in depth.

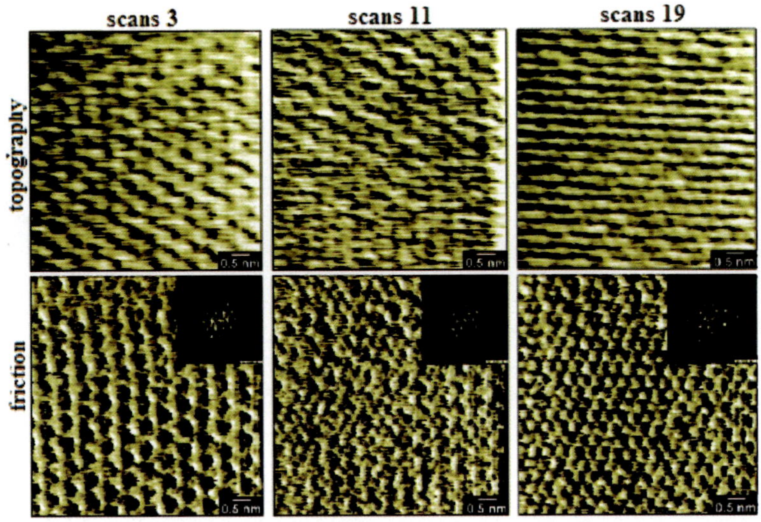

Figure 5. Topography and frictional images monitoring the tribochemical surface reconstruction of the native mica lattices from a 5.2Å (scans 3) to an ~3 Å (scans 19) periodicity under the local stress of an AFM tip. Images represent the 3rd, 11th and 19th scans during continuous scanning at a rate of 33Hz over a 7×7 nm^2 region at pH=5 and an applied load of 9nN (mean pressure of 1.25GPa). The 2D FFT insert of the friction micrographs illustrates the lattices transition where the intermediate state (11th scan) possesses a mixed 3Å and 5.2Å periodicity. (Reprinted from [Helt and Batteas, 2005], with permission).

Dickinson and co-worker performed a series of fundamental experiments on nanowear on single-crystal calcite in satured CaCO3 solutions [Park et al., 1996a; Park et al., 1996b; Dickinson et al., 1997], to quantify the shear stress required to detach single crystal salt particles from a soda lime glass substrate under controlled atmospheres of known humidity [Dickinson et al, 1999], and corrosive wear on a single-crystal brushite ($CaHPO_4 \cdot 2H_2O$) in aqueous solution [Scudiero et al., 1999]. Generally, in such experiments the tribological stimulus was provided by the scanning probe tip itself. In the case of single-crystal calcite, the dissolution and growth of calcite crystals in aqueous solution have been extensively studied due to their importance in mineral formation, global CO2 exchange, and interactions with environmental contaminants. Localized corrosive wear can readily be induced by the SFM tip on calcite surfaces imaged in aqueous solutions. Localised dissolution along pre-existing steps can also be produced by simply indenting the surface near the edge of an etch pit. Wear produced by drawning the probe tip back and forth in a repetitive, linear fashion across a monolayer step can be unambiguously interpreted in terms of volume activated, double kink nucleation in the strain field of the SFM tip. Experimental details can be found in [Dickinson et al, 1997]; here, we like to discuss the activation process.

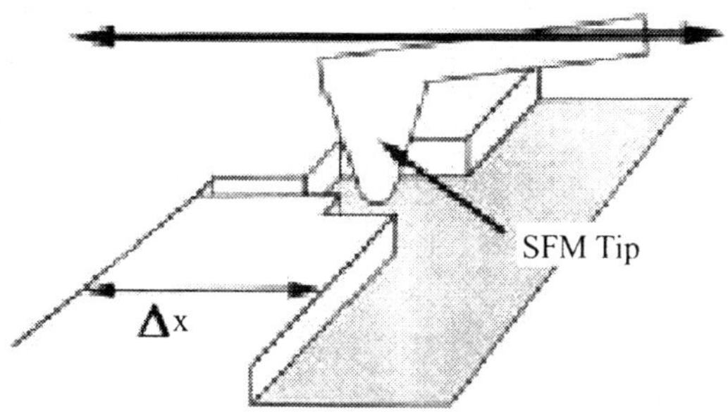

Figure 6. Schematical sketch of wear geometry of the experiment design from Dickinson and co-workers. The SFM tip is drawn back and forth in a linear fashion across the edge of pre-existing etch pit. (Reprinted from [Dickinson et al., 1997], with permission).

Wear tracks can be easily imaged in a geometry as presented in figure 6. The wear track growth rate ($\Delta x/\Delta t$) is a strong function of contact force. In figure 8, it is presented a plot of the wear track growth rate for wear tracks force.

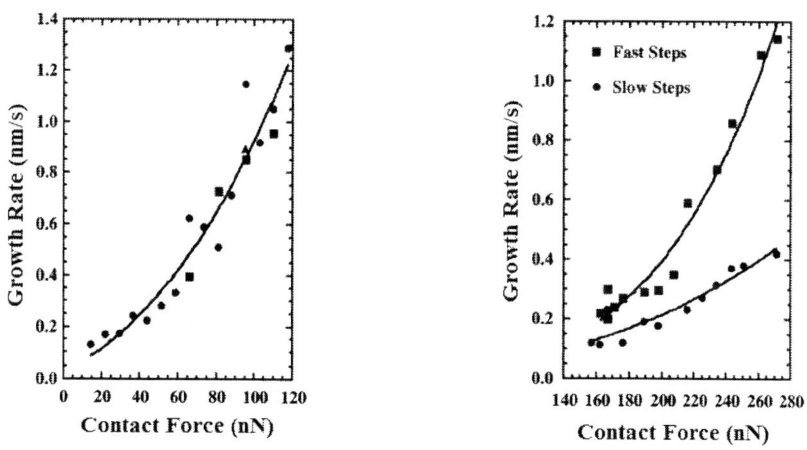

Figure 7. (a) Wear track growth rate versus contact force during flow of solution and repeated linear scanning. The data points represented by different symbols were made on different calcite samples with different cantilevers. The continue line is a least-squares fit of eq. (5) to the data. The measurements employed 500×500nm2 scans acquired at a tip velocity of 12μm/s. (b) Wear track growth rate for both fast and slow steps versus contact force under high degree of saturation. As in the case (a), the continue lines represent least-squares fits of eq. (5) to the data. (Reprinted from [Dickinson et al., 1997], with permission).

The magnitude of the stresses surrounding the SFM tip can be estimated assuming isotropic, elastic behaviour. Under such assumptions, the maximum tensile stress involves the radial components, σ_r, along the circle where the SFM tip contacts the substrate. Double-kink nucleation (the removal of ions from initially complete or intact steps) should be enhanced over a modest range of distances from the tip. The radial stress, σ_r, can be calculated by the Hertzian relation [Johnson, 1985]:

$$\sigma_r = \frac{(1-2v)}{\pi}\left(\frac{2PE^2}{9R^2(1-v^2)^2}\right)^{1/3} \tag{5}$$

where v is the Poisson's ratio and E is the reduced Young's modulus (approximately 58GPa), R is the tip radius and $P=F_N/real\ area\ of\ contact$ is the normal pressure. With an applied normal force of 100nN yields a maximum radial stress of about 500MPa. The dependence of the growth rate of the wear track V, on contact force is readily modelled with an Arrhenius-type expression

$$V = V_0 \exp\left[-\left(\frac{E_{act} - \upsilon^* \sigma_r}{k_B T}\right)\right] \tag{6}$$

where V_0 is the appropriate pre-exponential, E_{act} is the zero stress activation energy for double-kink nucleation and υ^* is an activation volume. The best-fit curve of eq. (6) to the data of fig. 7a (unsatured solution), corresponds to $V_0\exp(-E_{act}/k_BT)\approx 5.8$pm/s and $\upsilon^* \approx 4.4\times10^{-29}m^3$. This activation volume is slightly larger than the average volume per ion in the $CaCO_3$ lattice ($3.3\times10^{-29}m^3$), making it reasonable to suppose that this activation volume corresponds to the displacement of one or perhaps two ions from a step site. A similar analysis of the wear track growth rates at slow steps under satured conditions (fig. 7b) yields $\upsilon^* \approx 4.4\times10^{-29}m^3$. This agreement is consistent with the expectation that the degree of saturation controls the lifetime of nucleated kinks and not the nucleation rate.

The description of length of the wear track in terms of the rate of mechanically enhanced double-kink nucleation was found particularly efficient to explain the corrosive wear of single-crystal brushite in aqueous solution. In the case of atomic-scale corrosion, an understanding of fundamental effects is particularly necessary, requiring accurate details of interactions of the water molecules with stressed surfaces. The measurements of E_{act}, υ^*, and the factors for the key dissolution processes (such as pit nucleation, double-kink nucleation) provide an experimental foundation for a complete predictive model for corrosive wear of crystalline materials.

Chemical factors have been shown to be relevant for the particles removal, as in the case of single-crystal NaCl salt [Dickinson et al., 1999]. The nominal shear stress required to remove such particles from a soda lime glass substrate with an SFM probe tip is a strong function of particle size and relative humidity. At 3% of relative humidity (RH), particles larger than about 500nm on an edge could not be removed by the probe tip at any accessible contact force. Increasing the RH to 30% dramatically reduces the stress required for particle removal and promotes the removal of much larger particles. Further increases in RH yield smaller decreases in the failure stress. The humidity dependence of the shear stress can be described in terms of the expected change in interfacial energy due to water absorption. The stress required for particle removal scales roughly with the inverse square root of the particle area. Fracture mechanics can help to suggest that this is associated with an initial flaw size that is roughly proportional to their area. An improved understanding for particle removal could be an important factor in a number of technologies including integrated circuit manufacture and optical component manufacture. The experiments of Dickinson and co-workers showed in special way how the combination of chemical and mechanical effects plays a key role for tribological features on molecular scale.

Recently, a series of nanowear tests where the SPM in UHV has been used to realize and detect atomic-scale abrasion on KBr(001) have been prformed [Gnecco et al, 2002; Socoliuc et al., 2003]. They showed how the debris extracted from the KBr(001) surface by a silicon tip is reorganized on the atomic scale, figure 8. The dependence of wear vs time was investigated scratching across a single line for a long time at fixed load and velocity (5120 scans at F_N=11.2nN and scanning velocity of 2.5μm/s). The mean lateral force $<F_L>$ increased continuously from an initial value as a function of the number of scratches N. The behavior of the lateral force is a clear demonstration of the abrasive character of the wear process. In fact, the ions seem to be removed and released by pairs or by small cluster of ion pairs, and no dramatic crack events were observed in thousands of scans, and the regular arrangement of the deposited material resembles an epitaxial growth of thin ionic films, where the ions are piled up pair by pair. The ion removal can also be recognized in the irregularity of the friction loops, where the number of displaced ion pairs is comparable to the number of slip events, as the ions were detached when the tip jumps from one stick position to the next one. Generally, the stick-slip appears more irregular compared to atomic stick-slip observed without wear.

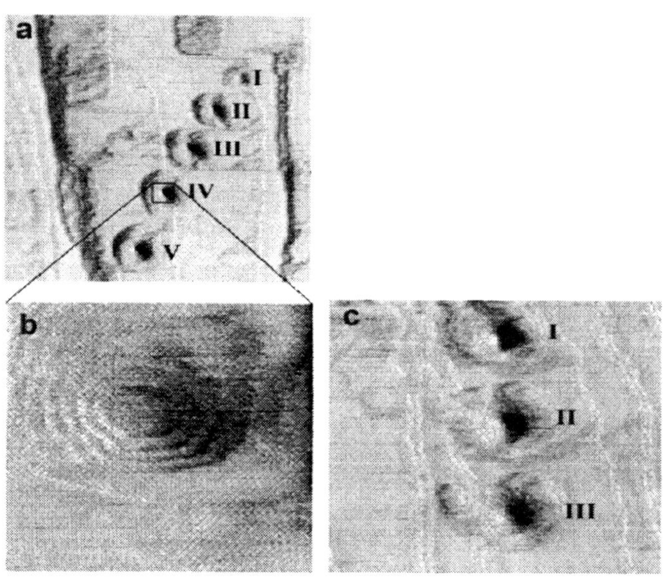

Figure 8. (a) Lateral force images of pits and mounds produced by 256 scratches on 5×5nm2 areas with FN =5.7, 10.0, 14.3, 18.6 and 22.8nN (up to down) and v=25nm/s. Frame size: 150nm. (b) Lateral force image of the fourth pit. Frame size: 20nm. (c) Lateral force images of pits and mounds produced by 256 scratches on 5×5nm2 areas with v=25, 50 and 100nm/s (upper lo lower) and FN=14.1nN. Frame size: 74nm.

The AFM scratch process was also used to produce nanoscale regular patterns on KBr(001) in UHV. In such tests, wear of single atomic layers has been observed when the AFM tip is repeatedly scanned across a line. The initially flat surface is rearranged in a quasiperiodic pattern of mounds and pits, figure 9. The distance between the pits was observed to be about 40nm when normal forces of a few nNs are applied, and it slowly increases with the load. If a square area is scanned, a pattern of regular ripples was formed. Ripple formation due to perturbations acting on a surface is commonly observed on macroscopic scale. Well-known examples are given by wind-blown sand dunes observed in the desert and in the submarine sands [Bagnold, 1941]. In the case of the continuous scratching of AFM probe tip, such features can be interpreted within an erosion-diffusion process induced by a periodic increase of the strain produced by the scanning tip.

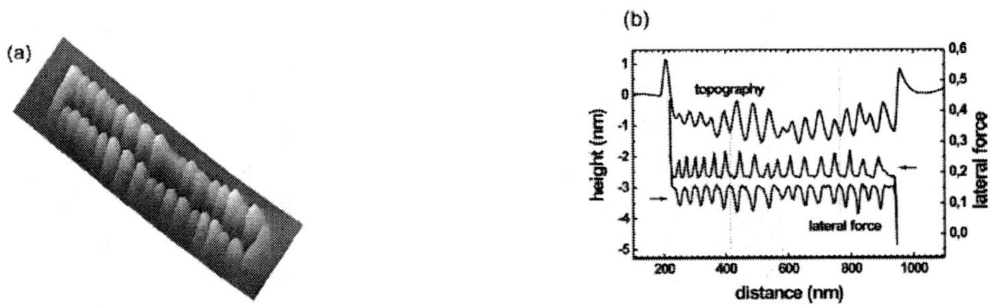

Figure 9. (a) Topography image of a groove formed on KBr(001) after 512 scratches along the (100) direction with a normal load of FN=26.6nN and a scanning velocity of 2.67μm/s. the groove is surrounded by a pattern of 34 mounds aligned along the two sides, and by two mounds piled up at its ends. The height difference between the top of the mounds and the bottom of the pits in the groove is about 2.5nm. The distance between two adjacent pits or mounds is about 40nm, as well as the distance between the two rods or mounds. (b) Cross section of the topography along the ground of the groove and lateral force acquired while scratching the groove. The lower and upper parts of the lateral force loop were obtained while scanning forwards and backwards, as indicated by the arrows. The dotted lines indicate positions in the pits where the tip is decelerated as derived from an increase of the lateral force in either direction. (Reprinted from [Socoliuc et al, 2003], with permission).

Figure 10. On left, topography map of a region scratched with the following loads (up to down): 1.7, 5.3, 8.8, 12.4, 15.9, 19.5, 23, 26.6, 30.1nN; frame size: 1.2μm. On right, development of lateral force with time while scratching a single line of 800nm length forwards and backwards. In horizontal direction the lateral force during forward scan is plotted, where darker colour means stronger lateral force; the vertical direction is the time axis. (Reprinted from [Socoliuc et al., 2003], with permission).

One common recognized explanation of the results showed in the various experiments mentioned in the section converge on the attribution of onset wear as an activated process, where the stripping off atomic or molecular debris is due essentially to the scanning probe tip induced stresses. The next section is devoted to examine thoroughly the activated process when applied to the onset wear problem.

3. Onset Wear as an Activated Process

In the preview section, there have been shown some experimental results linking the onset wear to the activation rate for the breaking of bonds and consequent stripping off of atomic debris. The onset wear so described seems to be very close to activation for chemical reactions. Ever since Kramers seminal paper [Kramers, 1940], the fluctuational escape over a potential barrier has been a paradigm for a thermal activation process. Kramers understood chemical reactions, as the climbing of an energy barrier between two energy wells in a large configuration space, where the first well represents the state of the reactant molecules and the second well those of the product molecules. Activated escape underlies diffusion in crystals, protein folding, nucleation rate in phase transitions and provides a paradigm for activated chemical reactions. In a wear experiment, we can be mainly interested to quantify the amount of mass stripped off from a surface in relative motion with another generally harder surface. As a consequence of the wearing debris is formed, one possibility is to describe the change of spatial location for the defects produced during the damaging as described in figure 11 and figure 12.

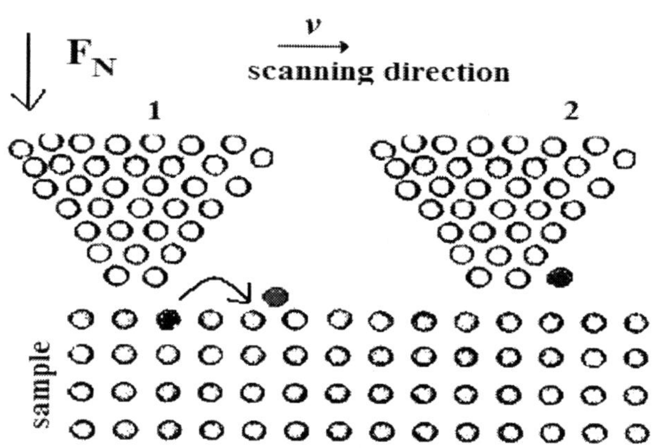

Figure 11. Schematic sketch on the production of debris (defects) due to abrasion wear-mechanism or adhesion wear-mechanism. The scanning tip in contact with the sample produces a damage on the regular surface moving the black atom, relocating in a new position (red atom), or adsorbed by the scanning probe tip (blue atom). The AFM-induced activated possible detachment of the black atom should be described by a conditional probability $P_1(x,t_1/x_0,t_0)$, where at time t the atomic debris is in the x position after the tip stress contact is made starting from x_0 position before tip contact, and the possibility of relocation on the native surface or adhesion to the stressing tip after the contact should be given by an other conditional probability, $P_2(y,t_2/x,t_1)$. The whole conditional probability is the convolution between these two probability functions.

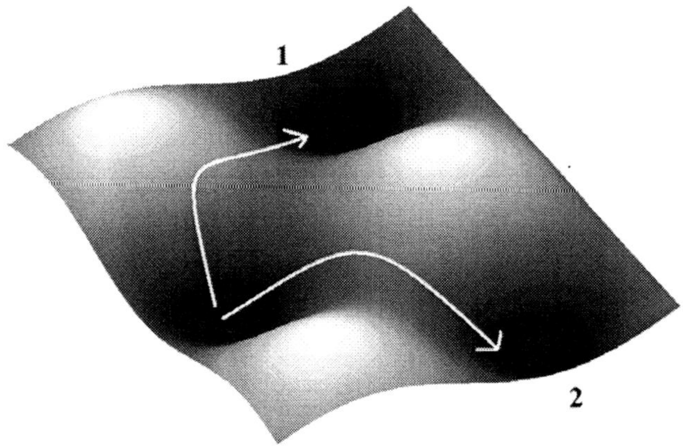

Figure 12. Schematic sketch to describe the well potential transition applied to describe the possible location of a debris stripped off from the native surface during sliding motion between two interacting surfaces, as in figure 11. For molecular scale contact, the debris has molecular dimensions, and the different final location of debris are connected to the specific wear-mechanisms. As a consequence, the final wells 1 and 2 represent different surfaces (in analogy to figure 11, re-positioning on sample surface or tip surface) and wear mechanisms.

In activated process, meanly we have a system of interest at thermal equilibrium connected to a driving force responsible for the activation of process from the system of interest. The mechanism responsible can be readily understood for adiabatically slow driving, where the driving frequency is small compared to the relaxation rate in the absence of fluctuations and the system of interest remains in quasi-equilibrium. For systems in thermal equilibrium, the fluctuation probabilities are given by the activation law, $W \propto \exp(-E/k_BT)$, where E is the activation energy of escape, that is tuned by driven force, and even where the small fluctuation of amplitude $|\delta E|$ is small compared to E, it may still substantially exceed k_BT, in which case W will be changed very strongly. In our picture, W represents the probability that a portion of mass (debris) can be detached by one surface when it is in relative motion respect to other one surface. The formation of wear debris is a random process and stochastic picture should be the most adequate tool to approach quantitative results. A stochastic process is the correspondence between a random experiment and a group of functions, as a consequence, the system can be predicted in terms of probability as for example the velocity of a Brownian particle [van Kampen, 1983; Risken, 1984].

Effectively, the connection between Brownian motion and debris can be made only a level of product of wear process, when a debris is a formed detached from the native surface and its mechanical features are subject to environmental fluctuations. Two systems in contact, such as the AFM tip and the scanned sample share a common surface. Each exerts an equal and opposite force on the other determined by the pressure it exerts on every elements of the surface of separation as required by the physics of an open systems [Bader, 2000]. Like general approach to open systems, we can consider a sample of interest in contact with external or environmental forces. In the tip-sample contact, the external forces are represented by the tip-induced stresses. Some general assumption can be made in order to simplify the tip-sample systems looking at the different time scale for the motion of the AFM tip and the atomic or molecular components of the sample investigated. The AFM has an effective cantilever mass around 10^{-11} Kg, many orders of magnitude larger than the atomic mass unit ($\sim 10^{-27}$ Kg). The resonant frequency of the cantilever is around 10^5 Hz, while the atomic vibration frequency of the order of 10^{13} Hz. Practically, in the contact area we have two kinds of dynamics, the slow dynamics of the cantilever and the fast vibrations of atoms. At the time scale of picoseconds, we encounter thermal vibrations of atoms, while the cantilever is stationary. On contrary, at the time scale of microsecond, we observe the scanning vibrations of the cantilever in contact with a cluster of atoms. If the relaxation time of this atomic cluster is much smaller than the cantilever time scale, then we will always observe an equilibrium cluster of atoms that follows the slow motion of the cantilever. Therefore, the energy modes between the cantilever and the atomic substrate can be decoupled. In Molecular Dynamics simulations, integration of dynamic equations of motion for the cantilever and the atomic substrate can be performed separately. As the consequence, we can limit to study the possible escape process for the sample atoms considering the effect of the AFM tip in terms of induced stresses lowering the energy activation for the surface atoms and considering the tip rigid.

For example, let us consider a general situation where the fundamental wear-mechanisms are abrasive and adhesive wear mechanisms. The quantification of wear due to adhesive or abrasive forces has been calculated in terms of rate of transition between two subsequent well potential. The problem can be reduced to a bi-dimensional question. Along the z-axis, the vertical forces are defined by a van der Waals term leading to the adhesion mechanism. Shear forces along the x-axis, i.e., the direction of motion of the tip leading to the abrasive mechanism for the atomic debris. The occurrence of one or both of two mechanisms depends on the intensity of the forces along the two dimensions, respectively. Such situation can be resumed by the sketch in figure 12, where is shown the simultaneous double-well for the two wear processes. In both the cases, the wear volume rate could be calculated within the microscopic diffusive processes can be expressed by the formula [Bassani and D'Acunto, 2000; D'Acunto 2003; D'Acunto 2004]

$$\Psi = n_{tot} w \Theta \tag{7}$$

where n_{tot} is the total number of possible debris lying on the primitive sample surface, w is rate of transition associated to the activated process, and Θ is the mean debris volume (pair ionic atomic volume). In experiments performed on ionic crystal, the mean debris volume can be identified in ionic pairs, so the wear volume rate can be reduced in the calculation of wear rate for activated de-bonding under applied tip stresses. Equation (7) should be considered an equivalent of macroscopic wear volume equation (1) and (2) on nanoscale. In a simplified vision, wear contributions due to adhesive- and abrasive-wear mechanisms can be considered separated and summarised as $\Psi=\Psi_{abr}+\Psi_{adh}=(n_{abr}+n_{adh})\Theta$, where Ψ_{abr} and Ψ_{adh} are the abrasive- and adhesive-wear volumes, respectively, and n_{abr} n_{adh} are the number of atoms involved for the two distinct wear mechanisms. In reality, the overlap of the two wear processes can be easily supposed. The experimental results performed on ionic compound in ultrahigh vacuum (UHV) by Gnecco et al., (2002), show clearly that $n_{abr} >> n_{adh}$ to be valid, i.e., the abrasive-wear mechanism is stronger compared to the activation of adhesive-wear mechanism. The advantage of modelling the interface between two surfaces by a double-well potential is that the wear volume can be quantified essentially in terms of rate of transition between the well minima. Moreover, the rate and (as a consequence) the wear volume depend on the potential barrier height, and the potential barrier height would be connected to experimental and materials quantities such as hardness, elasticity, load applied, and time (or sliding velocity). Obviously, the wear volume will be a function of the energy barrier. Essentially, vertical forces lead to adhesion-wear mechanism (transition from surface sample to tip), and shear forces lead to abrasion-wear mechanism (transition from surface sample to surface sample). Independently by the wear mechanisms activated, the rate of transition in equation (7) is given anyway by the Kramers picture

$$w_{abr,adh} \approx e^{-\Delta E_{x,z}/k_B T} \tag{8}$$

where the index is referred to x-dimension for abrasion and z-dimension for adhesion. To find the force of interaction between an SFM tip and a flat sample, the interatomic potential has to be summed over the whole number of atoms involved in the contact between sample and tip. The accurate derivation of such rate of transition for a double well potential model were derived in [Bassani and D'Acunto, 2000; D'Acunto 2003; D'Acunto, 2004].

The probabilistic approach to the production of wear debris can be performed within the classical double-well potential with time fluctuating barrier. The barrier is fluctuating because during the contact between the tip and the sample we have two different time scales. Here, we limit to comment the wear rate as obtained as function of the frictional state in the initial well potential. Each crossing particle is defined by a frictional term, proportional to $\gamma\dot{x}$, where $\dot{x}$ is the velocity of the particle living in the initial well and γ is the damping factor. The physical nature of such frictional term can be explained in terms of irreversible transferring of energy to the environment, other particles or phonon states. Nevertheless, there is another irreversible process, which costs energy leading to a non-conservative lateral force during the contact between the probe tip and the sample. This supplement of friction is manifest during the transition between well potential. The total amount of lost energy for the sample can be quantified like

$$F = n_A w \gamma \dot{x} / L \tag{9}$$

where L is the spatial distance between the minima of the well potential, n_A described the atom cluster on the native surface, (initial well potential) and $n_A w$ is the debris formed during the contact with the tip. The frictional state of particles in the initial well modifies the wear rate. In this sense, we can define a correlation between molecular friction and adhesive wear. In condition of high internal viscosity, i.e., when the cluster velocity relaxes much more rapidly than the position, constant wear rate as

$$w(t) = \frac{\Omega^2}{2\pi\gamma} \exp\left(-\frac{\Delta U(t)}{k_B T} \right) \tag{10}$$

where Ω is instantaneous effective oscillation frequency and ΔU is the energy barrier between the two wells. As can be expected, the wear rate depends by an Arrhenius factor and it is interesting to note that the rate does not depend on the shape of the potential barrier, but only on the difference of the top barrier. The total number of particles passing the barrier in one second is equal to $w \cdot n_A$. This is the case of high viscosity. It is also possible to treat in a similar fashion the limiting case of low viscosity with result obtaining the expression [Risken, 1984]:

$$w(t) = \gamma \frac{\Delta U(t)}{k_B T} \exp\left(-\frac{\Delta U(t)}{k_B T} \right) \tag{11}$$

where now, the wear rate is directly proportional to the friction term γ. Equations (10) and (11) can be slightly modified if we consider additive fluctuating terms to potential barriers due to noise of other nature respect to thermal noise due to the complexity of dynamic interaction between tip and sample in a sliding process, where anomalous diffusion processes take place [D'Acunto, 2003]. The presence of such fluctuations in the potential barrier is important since it may be hypothesised that in certain catalytic reactions, the catalyst-environment interaction could give rise to effects which can be represented with an effective potential which contains a supply noise coupled with position variable [Stein, et al., 1990; San and Wortis, 2004].

As mentioned above, the combined effects of the sliding tip and interatomic forces can be summarized in terms of shearing forces and vertical forces. The first ones could be considered responsible for the detachment of the ionic atoms from its crystal site and subsequent dragging until a new equilibrium crystal site is reached. The new position is achieved when the equilibrium is approached, this new position is a step on the flat atomic surface, figure 11. The ionic crystal shows that the process involves pair by pair ionic atoms. Vertical forces are due to the interaction of the tip probe approaching the surface sample. This interaction is led by a van der Waals force and stimulates the ionic atoms to follow the tip adhering to it, figure 11. In terms of effective wear rendering, the adhesion causes a vacancy along the crystal surface. Following the equation (10), the abrasive- and adhesive-wear mechanisms can be quantified considering the rate of transition as given by Kramers picture for the ionic atoms in a two-dimensional well potential. The correspondent height of energy barriers determines the incidence of wear both for abrasion or adhesive case. In the case of shear forces, the energy barrier can be deduced by the surface potential U_0, and a reasonable energy barrier along the x-direction $\Delta E_x \approx U_0$.

At the minimum value of the barrier, the vertical force is given essentially by the energy binding of the attractive term of the Van der Waals force. The apex tip is considered a cylinder of radius $R \cong 10$nm, the potential barrier can be expressed like $\Delta E_z \approx AR^{1/2} r_0 / z^{3/2}$, where z is the distance of separation between the surface and the tip and A is the Hamaker constant [Israelachvili, 1992]. The distance of separation, z, is directly proportional to the applied load. After the minimum value of the barrier is reached, correspondent to the minimum position z_{min}, the first term of a Lennard-Jones potential is not negligible. Moreover, the temperature should increase the rate lowering the well potential. Although the effect of temperature could engrave in a different way on the shear forces compared to the vertical forces. Nevertheless, the role of thermal fluctuations should be investigated more appropriately. Let us consider a volume of 120nm×120nm×1.8nm, the density of atoms involved in the wear processes is given by the summation on the first three layers of the quantity $n_L \exp(-(i-1))$, where n_L is the number of atoms for layer, and i=1,2,3, denotes the layers. The exponential correction is introduced because the atoms lying on the inner layer are less involved in the possible wear mechanism compared to the first layer one. The lattice constant is considered to be a=0.6nm, any layer contains 200 atoms, and the total number of atoms could be involved in the wear process for the volume considered is nearly 6×10^4. The calculation for an applied load of F_N=0.65nN is performed at room temperature, T=298°K, the thermal energy is $k_BT=0.041\times10^{-19}$J, the lattice potential barrier is $U_0=0.432\times10^{-19}$J, and the Hamaker constant $A=10^{-19}$J. Inserting these values in the expression of the barriers ΔE_x and ΔE_z we obtain that $\Delta E_x/k_BT\cong19$, while $\Delta E_z/k_BT$ is nearly 15 times $\Delta E_x/k_BT$. Because the functional dependence of the rates by the ratio between the energy barrier and the thermal energy is $x\exp(-x)$, the second term correspondent to vertical transitions toward the tip is practically negligible. The rate of atoms passing a potential barrier and undergoing a wear mechanism is given dividing the wear volume by the atomic volume

$$V/\Theta = nw_{abr} \approx n\left(\frac{2\pi/T_0T_1}{\beta}\exp\left(-\frac{U_0}{k_BT}\right)\right) \tag{12}$$

For a load of F_N=0.65nN and a cycle of 1000 scans, equation (12) gives nearly 2400 atoms involved in the abrasion-wear mechanism, i.e., the 3.9% of the total number of atoms, while the number of atoms undergoing an adhesive-type process is practically irrelevant. Unfortunately, it is difficult to perform a direct comparison with experimental results obtained by Gnecco et al., 2002. In fact, their experimental wear tests were made at higher applied loads, as $F_N\approx5\div25$nN. Nevertheless, it is remarkable that this model describes in a direct manner that the incidence of abrasive-type mechanism for wear is the most prominent mechanism for wear on nanoscale for low applied loads.

At low loads, it can be interesting to know the effects of the tip on the atoms, which are not able to pass the energy barrier. These atoms do not form the material loss but can be subject to a lattice modification, like dislocations, and consequent formation of gradient stress.

3.1. Self-Assembled Monolayers as a Frame to Study Wear in Viscoelastic Materials

The use of viscoelastic materials such as polymeric materials constantly increases in the field of nanotechnology. These materials are softer than metallic and inorganic ones, and, because of that, they are easier to wear and deform. Therefore, the wear mechanisms occurring for viscoelastic materials are rather complex, and, generally, present more complications respect to metals or ceramics materials. As a consequence, such mechanisms should be analyzed in a simplified model. Self-Assembled Monolayers (SAMs) are very useful for the systematic modification of the physical, chemical and structural properties of a surface by varying chain length, tail group and composition. Many of these properties can be studied making use of the AFM, and the interaction between AFM probe tip and SAMs can be considered also an excellent reference to study fundamental properties of dissipation phenomena and onset wear for viscoelastic materials on the nanoscale.

In this subsection, we deal out a numerical study showing that the fundamental mechanism for the onset wear is a process of nucleation of domains starting by initial defects. A SAM surface repeatedly sheared by an AFM probe tip with enough applied loads shows the formation of progressive damages enucleating in domains. The AFM induced surface damages involve primarily the formation of radicals from the carbon chain backbones, but also the deformations of the chains resulting in changes of period lattice have to be taken in consideration. The nucleation of the wear domains starts generally by initial surface defects where the energy cohesion between chains is lower. Moreover, the presence of surface defects is consistent with changes in lateral force increasing the probability of the activation for the removal of carbon debris from the chain backbone. The quantification of progressive worn area can be performed making use of the Kolmogorov-Johnson-Mehl-Avrami (KJMA) theory for phase transition kinetic processes [Sear, 2004; Jun et al., 2005]. The advantage of the knowledge of general conditions for onset wear on SAM surfaces can help to know fundamental mechanisms for the tribological properties of viscoelastic materials, in solid lubrication applications and biopolymer mechanics.

When the AFM probe tip slides on a deformable viscoelastic surface, relaxation of inner polymer structure leads to inevitable contact damping, which makes surface forces non-conservative. The class of SAMs provides a useful tool for controlling both physical and chemical properties of a solid viscoelastic surface [Ulmann, 1991]. Since the pioneering paper of Nuzzo and Allara [Nuzzo and Allara, 1983], the fabrication of nanoscale structures using SAMs has attracted attention due to their scientific importance and potential applications [Haussling et al., 1991; Ulman, 1996; Schlesser et al., 1996; Tidwell et al., 1997; Lavrich et al., 1998]. SAM and Langmuir-Blodgett (LB) techniques are the most common methods for forming monomolecular films on solid substrates for use as model lubricants [Bliznyuk et al., 1998; Barrena et al., 1999; Bhushan and Liu, 2001]. In the case of LB films, the molecular interaction with the substrate is through weak van der Waals forces, and repeated shearing results in rapid wear of the molecules. On contrary, SAMs are ordered molecular assemblies formed by the chemical adsorption of an active surfactant on a solid surface and are therefore better candidates as lubricants. SAMs can be formed by immersion of an appropriate substrate into a solution of an active surfactant in an organic solvent. SAMs consist of three building groups: a head group that binds strongly to a substrate, a tail group that constitutes the outer surface of the film and a spacer chain (backbone chain) that connects the head and tail groups, figures 13-14. It is now well established that complete alkanethiol monolayers formed on Au(*111*) adopt a commensurate $(\sqrt{3}\times\sqrt{3})R30°$ structure, with a nearest neighbor spacing of 0.497nm, which corresponds to a surface area of $0.2nm^2$ per chain [Hautman and Klein, 1989], figure 13.

The self-assembly mechanism of alkanethiol C_n, $[-SH(CH_2)_{n-1}CH_3]$, molecules on Au(*111*) has also been studied using several techniques including scanning force microscopy techniques [Delamarche et al., 1994; Poirier, 1997; Takano et al., 1999; Guo et al., 2002].AFM sliding friction experiments on C_n-alkanethiol SAMs showed a friction behavior dependent by chain-length [Lio et al., 1997]. The coefficient of friction decreased as the chain length was increased and the low friction of methylic group terminated SAMs is due largely to low surface energy coming from their hydrophobic tail groups and high load bearing capacity due to their compact structure minimizing van der Waals interaction energy.

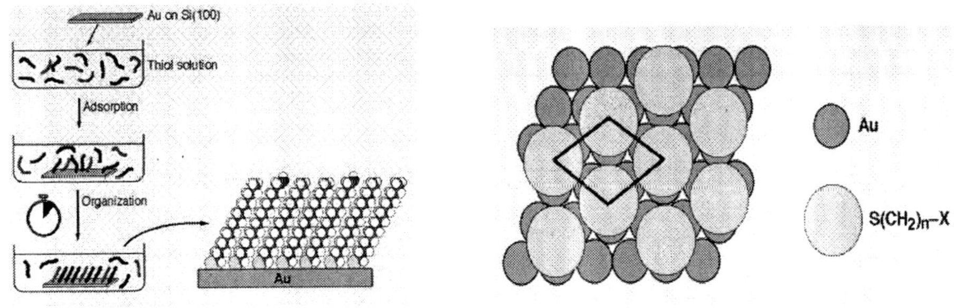

(a) (b)

Figure 13. (a) Schematical sketch for the preparation of SAMs. The substrate, Au on Si, is immersed into an ethanol solution of the desired thiol(s). Initial adsorption is fast (seconds); then an organization phase follows which should be allowed to continue for >15 h for best results. A schematic of a fully assembled SAM is shown on the right. (b) A schematic model of the $(\sqrt{3}\times\sqrt{3})R30°$ structure formed by alkanethiolate SAMs on Au(*111*).

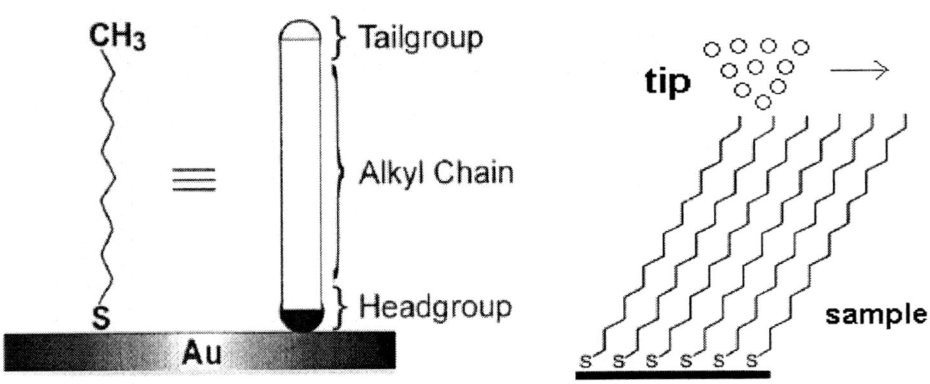

(a) (b)

Figure 14. (a) Single C_{15}-alkanethiol chain, and the three main components composing a chain: the headgroup, the alkyl chain or carbon backbone and the tailgroup, that in our case is a methyl group. (b) Schematic sketch of the AFM probe tip-SAM system. The contact with the probe tip involves several possible deformations for the alkanethiol chains. The close-packed arrangement of molecules should hamper the large deformations because these require space and are energetically expensive.

Onset wear tests can be simulated on a SAM surface taking as an example an experimental AFM test consisting in many scanning cycles on the same area and acquiring data of such progressive damaged area scan after scan. In the simulated approach, the percentage of the worn area can be studied as a function of several parameters, such as applied loads, lateral friction force and chain lengths, and curves are plotted indicating evolution of progressive damages. Consequently, the progressive damaging of the surface due to the continuous passage of the AFM tip can be monitored scan by scan. Because SAM lattice structure, this implies the possibility of a 2D quantification for the wear damage rather than classical macroscopic volume test quantification. In a wear test making use of an AFM, the regular lattice of SAM surface represents an excellent system that can display the time evolution of damages induced by the continuous passage of the probe tip. For this reason, we can quantify the progressive wear formation on the SAM surface observing how the primitive regular lattice area is progressively changed and reduced during the wear test. When the applied load is not enough low, deformations of the SAM structure or defects or formation of free radicals due to applied stress lead to changes in the 2D regular structure. The connection between dissipation mechanisms and wear is essentially given by the fact that initial surface defects correspond to the points where the lateral force and exchange of energy between tip and sample are higher.

It is well known that alkanethiol monolayers on Au(*111*) exhibit a distribution of pit-like defects [Endinger et al., 1993]. These pit-like defects should be a consequence of already existing defects of underlying Au surface. Other defects can be formed during alkanethiol monolayer assembly. They diminish locally the physical and chemical properties of alkanethiols chains, other than to change the height profiles. During the AFM induced compression, the molecular spring assembly can experience orientation and compression under normal load [Salmeron, 2001], so, the SAM chains experiment further deformations that can be considered in the class of defects. Generally, three types of AFM induced viscoelastic defects can be considered. The first type involves the creation of gauge distorsions at the free end groups. This results in a 120° rotation of the methyl group around the C-C backbone. The height shortening of the chain is expected by about 0.1nm, and the energy activation for such rotation is around 0.1×10^{-19}J. The second type of AFM induced defect is internal gauge deformations, which require displacement of the neighboring molecules. The third type is given by the rigid chain tilts. These two last types of defects are more difficult to produce, because their energy activation is considerable higher. Nevertheless, in the surrounding of structural initial defects such as pit-like defects, the alteration of the molecular packing density can lower their energy activation. Barrena *et al.* observed that the heights of self-assembled alkysilanes decrease in discrete amounts with normal load [Barrena et al., 1999]. This step-like behavior is due to the discrete molecular tilts, which are dictated by the geometrical requirements of the close packing of molecules. Only certain angles are allowed due to the zigzag arrangement of the carbon atoms.

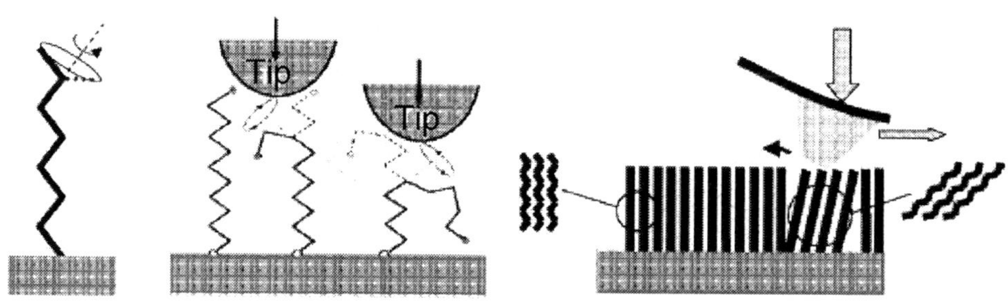

(a) (b) (c)

Figure 15. Schematic representation of three types of vicoelastic deformations of alkane molecules proposed by Salmeron, [Salmeron, 2001]. (a) Terminal gauche distorsions, (b) internal gauche distorsions, which require displacement of the neighboring molecules, and (c) rigid molecular tilts. The requirement of displacing neighboring molecules is less stringent in this case due to the concerted motion. (Reprinted from [Salmeron, 2001], with permission).

It is reasonable suppose that the energy exchanged between the probe tip and a SAM sample involves the initial defects points for nucleation activation for larger defects when the scanning probe tip slides more and more times. The idea is that at the edge of a defect the energy cohesion due to van der Waals interaction and responsible of chain packaging should be lowered by the extra-space available due to the presence of defects. The van der Waals energy cohesion ε_{vdW} is nearly 0.6-1.3×10^{-20}J, and the van der Waals radius of the methyl groups is approximately 0.2nm in a SAM packed structure [Salmeron, 2001]. The bound energy of C-C is nearly 1÷2×10^{-19}J, and S-C and S-Au binding energy fall in the range 2÷4×10^{-19}J [Lavrich et al., 1998].

Under opportune applied loads, the shearing stress induced by the passages of the probe tip should provoke the enlargement of the defects area provoking modifications of period lattice with subsequent changes of orientation of the chains and, over threshold loads, the breaking of covalent bonds with the formation of holes. These holes are due to the stripping off of radicals, such as CH_3-, macroradicals such as $(CH_2)_x$-CH_3-, or simply H-, here subscript x denotes the fraction stripped off from the original chain backbone. But, it is reasonable to suppose that the AFM induced de-bonding regards essentially C-C connections. Any carbon end group can be considered in a well potential and the passage of probe tip should make possible to overcome this potential barrier with the probability of transition enhanced by the stress applied, figure 12. Generally, the formation of active debris should involve the adhesive attachment of such SAM debris to the sliding probe tip changing the shape of the probe tip (adhesion tip wear) and as a consequence the real area of contact. Adhesion tip wear due to the attachment of progressive formed debris to the probe tip could be describe satisfactory by a double well potential, where any well potential corresponds to one surface in dynamic contact with other one, as described in the preview section on activated process [Bassani and D'Acunto, 2000; D'Acunto, 2003; D'Acunto, 2004]. Although the adhesion tip wear is a common occurring wear mechanism, in this review, we will not include its description, limiting ourselves to study general condition for progressive damaging of SAM surface without considering the dynamics of SAM debris when detached by the carbon backbone.

It is reasonable to suppose that in an AFM abrasive test consisting of a certain number of scans the density of defects should increase before gross wear will take place. This process is essentially a process of nucleation of defects forming holes increasing in time, where the time is the product of the residence time of the probe tip for the number of scanning cycles. Nucleation is an activated process [Dickinson et al. 1996; Dickinson et al. 1997; Dickinson et al., 1999; Scudiero et al., 1999], as such its rate has an exponential dependence on the free-energy barrier to nucleation defects. If at the *i-th* site of the SAM surface the free energy barrier is ΔU_i, and the frequency of attempts at unfolding is ν_i(~10^{13}Hz), then the rate of nucleation at the *i-th* site is

$$\Gamma_i \approx \nu_i \exp\{-[\varepsilon_{0,i} - \delta\varepsilon_i(F_N, F_l)]/k_B T\} \tag{13}$$

where $\Delta U_i = \varepsilon_{0,i} - \delta\varepsilon_i(F_N, F_l)$ is expressed by the mean value ε_0 of the C-C bond and the stress-induced deviation represented by the term $\cdot\delta\varepsilon(F_N, F_l)$ that is a function of the applied load F_N and lateral force F_l, k_B is the Boltzmann's constant, T is the temperature. The stress-induced deviation is a combined effect of compression and shear stresses and it lowers the energy barrier making possible the carbon stripping off from the chain backbone. The lateral stress is generally higher in correspondence of defects respect to regular lattices, as a consequence, the rate of nucleation must be calculated locally and cannot be considered homogeneous. The number of defects produced in the absence of stresses is rather small, under the tip stress the maximum possible rate of debris production, $\Delta N_i/\Delta t$ at the i-*th* point contact, should be given by the relation for thermal activated detachment [Weeks et al., 2002]:

$$\frac{\Delta N_i}{\Delta t} \cong \frac{R^2}{a^2}\Gamma_i \tag{14}$$

where $\Delta t=(A_r/A_{\mathrm{image}})t_{\mathrm{image}}$ is the residence time of the probe tip on the sample, A_r is the real area of contact, A_{image} is the area to be scanned and t_{image} is the image required for an image [Kopta and Salmeron, 2000], R is the radius of the real area of contact between the tip and the SAM, a is the van der Waals average radius of single methyl groups, and Γ_i is the rate of activation for the creation of AFM induced defects. Here, we assume that the continuum mechanics could give a good approximate value for the real area of contact and the depth of penetration even if the contact is discrete [Leng and Jiang, 2000]. For a paraboloid shape of the probe tip in non-adhesive contact with a flat surface one finds that the real area of contact and penetration depth are respectively [Johnson, 1985]:

$$A_r = \pi\left(\frac{3F_N\rho}{4E^*}\right)^{2/3} \qquad z_d = \frac{1}{2}\left(\frac{3F_N}{4\rho^{1/2}E^*}\right)^{2/3} \tag{15}$$

where $E^*=\left[\left(1-v_1^2\right)/E_1+\left(1-v_2^2\right)/E_2\right]^{-1}$ is the reduced modulus, $E_{1,2}$ are the Young moduli and $v_{1,2}$ the Poisson's ratios, F_N is the applied load and ρ (=1nm) is the tip radius. For typical moduli of Young and an applied load of 2nN, we obtain a value for the reduced modulus of nearly 29.5Gpa [Leng and Jiang, 2000], as a consequence the real area is approximately 0.4nm^2, a single C_n-alkanethiol head group formed by -CH_3 has an area of 0.2nm^2, this means that approximately 2 final head groups are simultaneously in contact with the probe tip and support the applied load. It should be noted that differently to friction behavior, SAM films show chain-length independence of elastic modulus [Leng and Jiang, 2000].

We suppose that the decrease of activation energy $\delta\varepsilon(F_N,F_l)$ in eq. (13) has two major contributions depending by the normal stress or shearing stress. The first contribution, due essentially to the normal stress, $\delta\varepsilon_1$ (divided by the C-C bond energy, $\varepsilon_{C\text{-}C}$) can be estimated by multiplying the force acting on the methyl group carbon atoms involved simultaneously in contact with the tip by the resulting bond stretch

$$\frac{\delta\varepsilon_1}{\varepsilon_{C-C}} \approx \frac{a^2}{R^2}\frac{z_d}{\left(A_r/\pi\right)^{1/2}} \tag{16}$$

where z_d is the penetration depth, and A_r is the real area of contact, R is the radius of the real area of contact between the tip and the SAM, and a is the van der Waals average radius of single methyl groups. The relation (16) can be justified by the assumption that the applied load induced bond stretch should be on the same order as the Hertzian penetration depth z_d, divided by the characteristic length of the contact area, which is on the order of the contact radius. The ratio between the areas of contact and the van der Waals region of single methyl group takes in account the number of methyl groups supporting the tip-SAM contact. The relation (16) shows that the contribution of applied load stress to decrease in activation energy changes as $F_N^{1/3}$. In the second term of the decrease of activation energy, the shearing stress plays an important role and should contribute for a lowering factor as:

$$\frac{\delta\varepsilon_2}{\varepsilon_{C-C}} \approx \frac{d_{C-C}}{(A_l/\pi)^{1/2}} \frac{z_d \times stickingforce}{\varepsilon_{vdW}} \tag{17}$$

where $d_{C\text{-}C}$=0.153nm is the carbon-carbon bond length along chain axis, A_l is the lateral real area calculated at the edge of penetration depth, ε_{vdW} is the van der Waals contribution to the energy of cohesion between chains, and sticking force is the maximum value of the lateral force in sticking phase during the typical stick-slip motion of the probe tip sliding on the SAM surface. The existence of a penetration depth implies that the carbon atoms set at the edge of the penetration depth should support entirely the shearing stress provoked by the coming forward probe tip along the sliding direction [Nie et al., 2005]. The cohesion energy ε_{vdW} between chains involves a resistance to the shearing inducing debris production on the edge of the penetration depth as the tip slides. An advantage of a probe scanning based wear test is that the state of a progressively worn surface can be monitored scan after scan. The

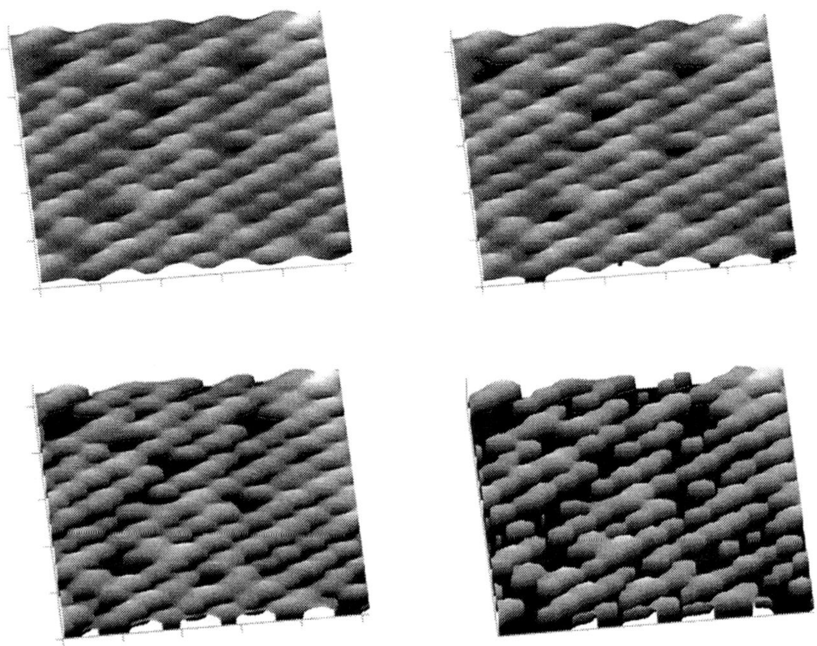

Figure 16. C_{18}-chain progressive damage of simulate methyl group end surface. The initial image (left upper) is based on an AFM experimental acquired data on a C_{18} sample, the typical commensurate c(4×2) uperlattice of (√3×√3)R30° arrangement is slightly altered by the applied load. The final image (right down) shows that nearly 20% of methyl groups have been stripped off by the AFM tip after 100 scans with an applied load of 5nN. The black portions represent the wear domains.

time evolution of such worn areas progressively damaged can be visualized as in a real AFM experiment. For example, figure 16 proves the progressive AFM induced damaging for a C_{18} surface for an applied load of 5nN.

4. CONCLUSION

The chapter has been devoted to give a review about experimental data and models on the onset wear occurring on nanoscale. Wear is the progressive loss of mass from the operating surface of a body as a result of relative motion at the surface. The occurrence of different wear mechanisms involves the fundamental mechanics of molecular and supra-molecular scale. As a consequence, a fundamental understanding of surface properties on nanoscale level should be generated to have a satisfactory knowledge of responses of materials and machine components also at macrometer scale. Investigation of the fundamental characteristics of wear at the micro-scale is complicated by some factors and forces that act on a nanoscale level have not yet been addressed in the tribology of macro-systems. Since these forces are sensitive to the environment and surface condition of the specimens, it is quite difficult to determine them accurately. Further, quantification of wear is not straightforward since the amount of wear is often too small to be detected by surface-sensitive instruments. Nano-wear measurements have been object of rapid development of precise measuring tools following to the introduction of atomic force microscopy (AFM). The advent of AFM opened new opportunity to study wear mechanisms on nanoscale, making possible to simulate one surface sliding on other one and controlling many physical parameters during the contact dynamics. Further, AFM made possible to a have a direct connection between experimental data and theoretical models.

Muscovite mica, ionic crystals, self-assembled monolayers represent ideal samples where to study the fundamental features for wear on nanoscale.

REFERENCES

Archard, J.F.; 1953, *J. Appl. Phys.*, **24**, 981-8.

Bader, R.F.; 2000, *Physical Review* **B**, **61**, 7795-7802.

Bagnold, R.; *The Physics of Blown Sand and Desert Dunes*, Methuen, London, 1941.

Barrena, E.; Kopta, S.; Ogletree, D.F.; Charych, D.H.; Salmeron, M.; 1999, *Phys. Rev. Lett.*, **82**, 2880-3.

Barthel, E.; Roux, S.; 2000, *Langmuir*, **16**, 8134-38.

Bassani R.; and D'Acunto, M.; 2000, *Tribology International*, **33**, 434-443.

Bhushan, B.; Israelachvili, J.N.; Landman, U.; 1995, *Nature*, **374**, 607-16.

Bhushan, B.; and Koinkar V.N.; 1994, *J. Appl. Phys.* **75**, 5741-8.

Bhushan, B.; and Liu, H.; 2001, *Physical Review B*, **63**, 245412(1-11).

Bhushan, B.; *Springer Handbook of Nanotechnology*, Springer-Verlag, Berlin, 2003.

Bliznyuk, V.N.; Everson, M.P.; and Tsukruk, V.V.; 1998, *J. of Tribology*, **120**, 489-96.

Carpick, R.W., Agrait, N.; Ogletree, D.F.; Salmeron, M.; 1996a, *J. V. Sci. Technol.* **B 14**, 1289-95.

Carpick, R.W., Agrait, N.; Ogletree, D.F.; Salmeron, M.; 1996b, *Langmuir*, **12**, 3334-40.

Carpick, R.W.; and Salmeron, M.; 1997, *Chemical Reviews*, **97**, 1163-94.
Chung K.H.; Kim D.E.; 2003, *Tribology Letters*, **15**(2), 135-144.
D'Acunto, M.; 2004, *Nanotechnology*, **15**, 793-804.
Delamarche, E.; Michel, B.; Gerber, C.; Anselmetti, D.; Guentherodt, H.J.; Wolf, H.; Ringsdorf, G.E.; 1994, *Langmuir*, **10**, 2869-71.
Dickinson, J.T.; Park, N.S.; Kim M.W.; Langford, S.C.; 1997, *Tribology Letters* **3**, 69-80.
Dickinson, J.T.; Hariadi, R.F.; Scudiero, L.; Langford, S.C.; 1999, *Tribology Letters* **7**, 113-119.
Edinger, K.; Golzhauser, A.; Demota, K.; Woll C.; Grunze, M.; 1993, *Langmuir,* 9, 4-8.
Erlandsson, R.; Hadziioannou, G.; Mate, C.M.; McClelland, G.M.; Chiang, S.; 1988, *J. Chem. Phys.* **89**, 5190-93.
Gåhlin, R.; and Jacobson S.; 1998, *Wear,* **222**, 93-102.
Gnecco, E.; Bennewitz, R.; Meyer, E.; 2002, *Phys. Rev. Lett.*, **88**, 215501(1-4).
Guo, Q.; Sun, X.; Palmer, R.E.; 2002, *Phys. Rev.* **B, 71**, 035406.
Haussling, L.; Ringsdorf, H.; Schmitt, F.J.; Knoll, W.; 1991, *Langmuir***,** **7**, 1837-40.
Hautman J.; and Klein, M.L.; 1989, *J. Chem. Phys.,* **91**, 4994-5001.
Helt, J.M.; and Batteas, J.D.; 2005, *Langmuir,* **21**, 633-639.
Hu, J.; Xiao X.D.; Ogletree, D.F.; Salmeron, M.; 1995, *Surface Science*, **327**, 358-70.
Hutchings, I.M.; *Tribology: Friction and Wear of Engineering Materials*, CRC Press, 1992.
Israelashvili, J.N.; 1992, *Intermolecular and Surfaces Forces*, 2nd ed., Academic Press, NY.
Jiang, S.; 2002, *Molecular Physics*, **100**, 2261-75.
Johnson, K.L.; *Contact Mechanics*, Cambridge University Press, 1985.
Jun, S.; Zhang, H.; Bechhoefer, J.; 2005, *Phys. Rev.* **E, 71**, 011908(1-6).
Kim, Y.; Huang, W.N.; and Lieber, C.M.; 1991, *Applied Phys. Lett.,* **59**, 3404-6.
Kopta, S.; and Salmeron, M.; 2000, *J. Chemical Physics*, **113**, 8249-52.
Kramers H.A., 1940, *Physica* **7,** 284-304.
Lavrich, D.J.; Wetterer, S.M.; Bernasek, S.L.; Scoles, G.; 1998, *J. Phys. Chem.* **B, 102**, 3456-65.
Leng, Y.S.; and Jiang, S.; 2000, *J. Chem. Phys.*, **113**, 8800-6.
Lio, A.; Charych, D.H.; Salmeron, M.; 1997, *J. Phys. Chem.* **B, 101**, 3800-5.
Ludema K.; *Friction, Wear and Lubrication, a Textbook in Tribology,* CRC Press, 1996.
Ludema K., in Fundamentals of Tribology and Bridging the Gap between the Macro- and Micro/Nanoscale, Bhushan ed., *Kluwer Academic Publishers*, 2001.
Miyake, S.; 1994, *Appl. Phys. Lett.* **65**, 980-2.
Nie, X.; Zhang, P.; Wiener A.M.; Cheng, Y.T.; 2005, *Nano Letters,* **5**, 1992-6.
Nuzzo, R.G.; and Allara, D.L.; 1983, *J. Am. Chem. Soc.,* **105**, 4481-3.
Park, N.-S.; Kim, M.-W.; Langford, S.C.; Dickinson, J.T.; 1996a, *Langmuir,* **12**, 4599-604.
Park, N.-S.; Kim, M.-W.; Langford, S.C.; Dickinson, J.T.; 1996b, *J. Appl. Phys.,* **80**, 2680-6.
Poirier, G.E.; 1997, Langmuir, **13**, 2019-26.
Risken, H.; *The Fokker-Planck Equation. Methods of Solution and Applications.* Springer-Verlag, Berlin Heidelberg, 1984.
Salmeron, M.; 2001, *Tribology Letters,* **10**, 69-79.
San, A.; and Wortis, M.; 2004, *Phys. Rev.* **E, 70**, 031102(1-11).
Sang, Y.; Dubè, M.; Grant, M.; 2001, *Phys. Rev. Lett.,* **87**, 174301(1-4).
Schessler, H.M.; Karpovich, D.S.; Blanchard, G.J.; 1996, *J. Am. Chem. Soc.*, **118**, 9645-51.
Schöfer, J.; and Santner, E.; 1998, *Wear,* **222**, 74-83.

Scudiero, L.; Langford, S.C.; Dickinson, J.T.; 1999, *Tribology Letters* **6**, 41-55.
Sear, R.P.; 2004, *Phys. Rev.* **E, 70**, 021605(1-6).
Shevade A.V.; Zhou, J.; Zin M.T.; Jiang, S.; 2001, *Langmuir* **17**, 7566-7562.
Sokoliuc, A.; Bennewitz, R.; Gnecco, E.; Meyer, E.; 2004, *Phys. Rev. Lett.,* **92**, 134301(1-4).
Sokolov, I.Y.; 2002, *Tribology Letters,* **12**, 131-134.
Stein, D.L.; Doering, C.R.; Palmers, R.G.; van Hemmen, J.L.; McLaughlin, R.M.; 1990, *J. Phys. A, Math. Gen.* **23**, L203-L208.
Takano, H.; Kenseth, J.R.; Wong, S.; O'Brien, J.C.; Porter, M.D.; 1999, *Chem. Rev.* **99**, 2845-90.
Tidwell, H.C.D.; Ertel, S.I.; Ratner, B.D.; Tarasevinch, B.J.; Atre, S.; Allara, D.L.; 1997, *Langmuir*, **13**, 3404-13.
Ulman, A.; *Introduction to Ultrathin Organic Films: From Langmuir-Blodgett to Self-Assembly*, Academic Press, San Diego, CA, 1991.
Ulman, A.; 1996, *Chemical Review*, **96**, 1533-54.
Unertl, W.N.; 1999, *J. Vac.Sci. Technol.* **A, 17**, 1779-86.
van Kampen, N.G.; *Stochastic Processes in Physics and Chemistry.* North-Holland Publishing Company, 1983.
Wang, W.; Lee T.; Reed, M.A.; 2003, *Phys. Rev. B*, **68**, 035416(1-7).
Weeks, B.L.; Noy, A.; Miller, A.E.; De Yoreo, J.J.; 2002, *Phys. Rev. Lett.,* **88**, 255505(1-4).
Xu, L.; Salmeron, M.; 1998, *Langmuir*, **14**, 2187-90.

In: Frontal Nanotechnology Research
Editor: Martin V. Berg, pp. 189-219

ISBN: 1-60021-212-3

Chapter 8

IMPROVING DIELECTRIC PROPERTIES OF POLYMER NANOCOMPOSITES THROUGH NANOPARTICLE SURFACE MODIFICATION

Dongling Ma,[a,1] Linda S. Schadler,[a] Richard W. Siegel,[a] Nianqiang Wu,[b] Diana F. Rogers,[c] and Brian C. Benicewicz[c]

[a] Department of Materials Science and Engineering and Rensselaer Nanotechnology Center, Rensselaer Polytechnic Institute, Troy, NY 12180, USA

[b] Keck Interdisciplinary Surface Science Center, NU*ANCE*, Northwestern University, IL 60208, USA

[c] Department of Chemistry and NYS Center for Polymer Synthesis, Rensselaer Polytechnic Institute,Troy, NY 12180, USA

ABSTRACT

In this study, the potential of using nanoparticle surface chemistry to tune the dielectric properties of polymer nanocomposites is explored. Composites with 5 wt% of titania (TiO_2) nanoparticles and low-density polyethylene (LDPE) were fabricated by melt mixing. Both polar (N-(2-aminoethyl)3-aminopropyl-trimethoxysilane (AEAPS)) and nonpolar (decyltrichlorosilane (DCS)) silane coupling agents were used to modify the nanoparticle surface to yield either hydrophilic or hydrophobic surfaces. Two coating methods were employed and the resulting surface structures were characterized in detail. It was found that the surface structure was controlled primarily by the experimental conditions of the coating process. In addition, a previously un-described surface reaction that resulted in the formation of Si-CH_3 bonds was proposed, which may be due to the reaction of AEAPS with oxygen vacancies on the TiO_2 nanoparticle surface. A large observed change in the chemical shift (over 40 ppm) in the ^{29}Si NMR and the −2.3 ppm chemical shift in ^{13}C NMR supported this conclusion. The method leading to a "dry" compact monolayer surface structure was chosen for preparing TiO_2 nanoparticles with either polar or non-polar surfaces for composite sample preparation and subsequent electrical studies. In both cases, surface modification had a positive effect on dielectric breakdown strength although the possible underlying mechanisms were different. Compared to the uncoated TiO_2 filled samples, AEAPS-TiO_2 and DCS-TiO_2

[1] E-mail address: ma@emt.inrs.ca. Phone: 1-450-929-8120. Fax: 1-450-929-8102. Corresponding author.

filled samples showed a 35% and 43% increase in the short-term Weibull breakdown strength, respectively. The space charge distribution patterns were also clearly different among the samples. The microstructure characteristics of the composite samples, associated with different TiO_2 surfaces, were studied and correlated to the corresponding dielectric properties.

INTRODUCTION

Polymers play an important role in electrical insulating technology. Polyethylene (PE) is one of the most attractive cable insulating materials due to its unique characteristics such as high dielectric strength, ease of fabrication, low cost and simple maintenance and, thus, has long been of research interest [1-5]. There are problems, however, related to the application of PE as an insulating material that still have not been resolved successfully. First is the inherent dependence of PE resistivity on temperature and electrical field [1]; the other is the reliability and long-term life of PE. Degradation [6,7], which can be thermal, oxidative, photo-oxidative or electrical, results in material failure.

Additives/fillers are mixed with polymer matrices to modify their high field stability, mechanical and thermal–mechanical properties, and resistance to degradation [8-10]. But in most cases, the introduction of additives/fillers has a significant negative effect on dielectric breakdown strength due to increased field distortion. Therefore, one important goal is to improve other properties without negatively affecting or even improving the electrical properties. To achieve this goal, it is important to carry out detailed investigations of the electrical properties and structure of the composite materials in order to gain sufficient understanding about the system. Most of the studies so far have focused on conventional micron-size additives/fillers, while in this work, the focus is on nanoparticles.

Nanomaterials are attracting more and more attention because their small size of their constituent domains leads to unique properties. The constituent domains of nanomaterials have at least one dimension less than 100 nm in size. They are intermediate in size between molecular and microscopic (micron-size) structures [11]. In this regime, material properties are dominated by quantum and other confinement behavior that makes their characteristics fundamentally different from their micron-scale counterparts [12-17]. In addition, nanoscale structures have high surface area and high surface-volume ratio. As the particle size decreases, the surface area to volume ratio increases dramatically when the length scale falls below 100 nm. Therefore, at the same particle loading, nanocomposites possess much more interfacial area than microcomposites. This large surface area, in combination with enhanced surface reactivity of nanoparticles, suggests a great potential of tailoring the macroscopic properties of nanocomposites by modifying the nanoparticle surface at the nanoscale. To be specific, for example, it is known that particle filler dispersion, interfacial compatibility and matrix morphology are all possible factors in affecting the dielectric breakdown strength. For a specific polymer, the interfacial compatibility is solely determined by the surface chemistry of the nanoparticles and the other two factors are also closely related to the surface chemistry of the nanoparticles. Therefore, the surface chemistry must have an important effect on the dielectric breakdown behavior. Despite this large potential, only recently have studies emerged on the effects of surface modification of the nanoparticles on the dielectric properties of the nanocomposites [18-20]. In our previous work [18,21], we have demonstrated that the surface states of TiO_2 nanoparticles had non-negligible effects on the dielectric breakdown strength and space charge distribution in polyethylene nanocomposites.

In the present work, we further study the effect of different surface modifications of the TiO_2 nanoparticles, with either polar or non-polar silane coupling agents, on the permittivity, dielectric breakdown strength and space charge distribution of LDPE. TiO_2 was chosen because it is becoming increasingly important in composite applications.

EXPERIMENTAL

Materials

Low-density (0.922 g/cc) polyethylene (LDPE) DOW 6811 from Dow Chemical was used as the matrix, with the number-average and weight-average molecular weight being 14000 and 70000, respectively. TiO_2 nanoparticles with a surface area of 42 m^2/g were provided by Nanophase Technologies Corporation. The average particle diameter was 23 nm as determined by transmission electron microscopy (TEM) as shown in Figure 1.

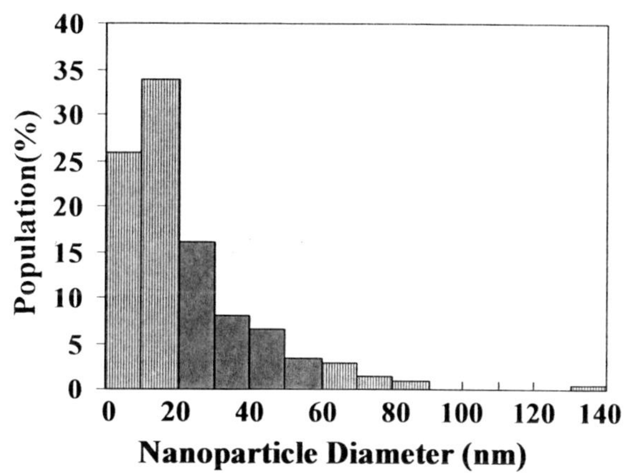

Figure 1. TiO_2 nanoparticle size distribution determined from TEM.

The particle diameters ranged from less than 10 nm to over 100 nm, with about 70% of the nanoparticles smaller than 30 nm in diameter. X-ray diffraction confirmed that the particles were principally anatase. N-(2-aminoethyl)3-aminopropyl-trimethoxysilane (AEAPS) ($(CH_3O)_3Si(CH_2)_3NH(CH_2)_2NH_2$) and decyltrichlorosilane (DCS) ($CH_3(CH_2)_9SiCl_3$) were purchased from Gelest, Inc. and used as received. Toluene (water content <0.03%), calcium hydride (CaH_2, purity 93%) and potassium bromide (purity 99+%) were purchased from Mallinckrodt Baker Inc., Acros Organics and Aldrich Chem. Co., respectively.

Coating Procedures

The coating procedures were derived from those described by Tsubokawa *et al* [22]. In coating method I, 10 g of nanoparticles were dispersed in toluene via sonication (Sonics Vibracell, 750 W) at 70% power for 5 minutes. The solid full wave probe was 13 mm in diameter and 254 mm in length. To avoid a sharp temperature rise of the suspension during sonication, a beaker containing the suspension was surrounded by ice. 0.015 mol of silane coupling agents was added slowly to the nanoparticle-toluene mixture and stirred for one hour at room temperature. Then the mixture was refluxed for 48 hrs under dry air atmosphere. The resultant slurry was centrifuged at 6,000 rpm for 10 minutes and the deposited particles were washed with fresh toluene. The centrifuge-toluene wash procedure was repeated two more times and the silane-coated particles were dried in *vacuo* for 24 hrs at 30 °C.

Coating method II was similar to method I with two modifications. The toluene was dried over CaH_2 powder in a nitrogen atmosphere, and then purified by distillation. The TiO_2 nanoparticles were dried in *vacuo* at 195 °C for 24 hrs prior to their immediate use in the coating procedures.

Sample Preparation

5 wt% of as-received or surface modified TiO_2 nanoparticles were melt-mixed into LDPE at 130 °C using a Haake System 90 Buchler Mixer for 10 minutes. Pellets for permittivity and thin films (10 to 40 μm in thickness) for dielectric breakdown measurement were formed by hot compression at 160 °C in a Carver Press (Model 3912). Polytetrafluoroethylene films (McMaster-Carr Supply Company) were used as backing. The diameter of these prepared samples was 58 mm. Processing details can be found in ref [23]. The pressure was changed slightly among the samples to accommodate the viscosity of the mixtures. Each sample was kept in a desiccator for at least one day for relaxation before any measurements were carried out. The neat LDPE went through exactly the same process to exclude the factor of different processing on the properties of pure polymer and filled polymer samples.

Plaque samples with a thickness of 2 mm for space charge measurements were pre-press molded, then pressed against two 1 mm-thick semicon electrodes [18]. The semicon electrodes were composed of polyethylene copolymer and carbon black particles, and were also prepared by hot compression. All the press moldings were performed at 130 °C and polyethylene terephthalate films were used as backing during these processes.

Structural characterizations were made with thin film samples prepared for dielectric breakdown testing except for those used for atomic force microscopy (AFM), which required an extra step. AFM samples were prepared by placing a pre-pressed film on a hot, clean Si wafer in the mold at 160 °C without pressure. The fully melted film was slowly cooled to below 50 °C in the mold. In this case, Kapton (DuPont) was used as backing films. The free surface of thus pressed films was examined by tapping mode AFM.

Characterization Techniques

Solid-state ^{29}Si and ^{13}C cross-polarization magic-angle-spinning (CPMAS) nuclear magnetic resonance (NMR) measurements were made on a Chemagnetics CMX 360 instrument operating at a frequency of 71.25 and 90.53 MHz, respectively. The ^{29}Si spectra were obtained using a 4.5 μs 90° pulse width, a contact time of 6 ms, a 5 s recycle delay, and a spinning speed of 3 kHz; tetrakis (trimethylsilyl) silane was used as an external reference. The ^{13}C spectra were obtained using a 4.5 μs 90° pulse width, a contact time of 1 ms, a 5 s recycle delay, and a spinning speed of 3 kHz; hexamethyl benzene was used as an external reference.

Fourier-transform infrared spectra (FTIR) were obtained on a Magna-IR 560 spectrometer in transmission mode. Potassium bromide (KBr) pressed pellets were prepared by first drying the KBr in an oven at 200 °C for 3 hrs followed by mixing with about 3 wt% of coated TiO_2 nanoparticles. The spectrum of pure AEAPS silane was obtained by placing a couple of drops of the coupling agent between two pressed pure KBr pellets. The spectra of composite samples were obtained by using as-molded thin films directly. The signal was averaged over 512 scans with a resolution of 1 cm^{-1}.

Thermogravimetric analysis (TGA) was performed on a Mettler TGA/SDTA 851e at a heating rate of 20 °C/min in air for coated particles, and in nitrogen for water content determination of uncoated particles.

The particle dispersion in the composite samples was observed with a JEOL JSM-6330F field emission scanning electron microscope (FE-SEM). The samples were broken in liquid nitrogen and the cross section was examined after being sputter-coated with either gold or platinum.

Time-of-flight secondary ion mass spectroscopy (SIMS) analysis was performed with a Physical Electronics PHI TRIFT III, which is equipped with a pulsed Ga^+ liquid ion gun operated at 15 kV. The ion source was operated with a current of 600 pA. A pulsed low-energy electron flood gun neutralized the charging effect. The secondary ions were accelerated to ±3 kV by applying a bias on the sample.

Differential scanning calorimetry (DSC) experiments were carried out on a Mettler DSC 822e under N_2 atmosphere with a ramp rate at 10 °C/min. The weight of samples ranged from 3 to 6 mg. The original DSC curves were normalized to the net weight of LDPE and then used for analysis. The degree of crystallinity was estimated using $W_c^{DSC} = \Delta H_f / \Delta H^o$, where ΔH_f is the heat of fusion obtained after integrating the area under the normalized melting curve and ΔH^o = 293 J/g is the reference heat of fusion for 100 % crystalline polyethylene.

X-ray diffraction patterns (XRD) were recorded at room temperature with a Scintag XDS 2000 diffractometer using a wavelength of 0.154 nm. The signal was obtained in step mode with the step size of 0.02° and the collection time of 1 s. The peak analysis was done by using a mixed Gaussian and Lorentzian function. For comparison, the original diffraction patterns were normalized to their own (110) reflection intensities.

A NanoScope III (Digital Instruments Veeco Metrology Group) AFM was operated in tapping mode at room temperature to visualize composite morphology. Both topographic and phase images were simultaneously obtained using Si cantilevers with a spring constant of 40 N/m.

Permittivity measurements were carried out at room temperature with a HP 4192A LF Impedance Analyzer. The fixture had two parallel disk electrodes with the top electrode having a guard ring. During the testing, the top electrode was grounded and the bottom electrode was connected to a high potential. Sample plates were inserted between the two electrodes without an air gap. Capacitance was read at 1 kHz for the calculation of permittivity.

To measure the dielectric breakdown strength, a sphere-plane geometry was employed. A stainless steel sphere with a diameter of ¼ " was connected to high potential while the plane electrode was grounded. The test was performed with a Glassman High Voltage FC60P02 at room temperature using a stepwise increased voltage with steps of 100 V and application periods for each step of approximately 1 second. Between steps the high voltage electrode was grounded and then the next higher voltage was applied. The process proceeded until a sharp current increase was observed indicating electrical failure of the sample. Dielectric breakdown strength was calculated as the electrical voltage causing the sample failure divided by the sample thickness.

A pulsed electro-acoustic (PEA) technique was used to measure the space charge distribution. The tests were carried out at 50 °C under the voltage of +40 kV (20 kV/mm). The space charge distribution was measured at 0, 15, 30, 60, 180, and 360 minutes after the voltage application and a 2.2 kV pulse voltage was used. A calibration run was done at low electrical field, where no space charge build-up was assumed to be present, and used as a reference signal. The relative space charge density is presented here for a comparative evaluation. Details of the basic PEA measurements can be found elsewhere [24].

PARTICLE SURFACE MODIFICATION

Two similar coating methods I and II were employed to coat TiO_2 nanoparticles using the AEAPS silane. The resultant coating structures were characterized in detail using TGA, FTIR, ^{29}Si NMR and ^{13}C NMR. Based on the coating structure analysis, and accordingly the expectation for the influence of coating structure on the insulating properties, coating method II was chosen for chemical modification of the nanoparticles by DCS.

TiO_2 surfaces contain water and hydroxyl groups, both of which are involved in the coating process. Surface water can hydrolyze the silane, while hydroxyl groups contribute directly to the formation of coating structures by reacting with other hydrolyzed silane [25]. Similarly, the hydrolyzed product of the silane coupling agent may hydrogen bond and/or covalently bond to the surface [25]. Hydrogen bonds are formed between hydrolyzed silane and the TiO_2 surface via surface hydroxyl groups or adsorbed molecular water. Further condensation reactions may produce Ti-O-Si bonds, in which the silane is categorized as covalently bonding to the surface. Also, $-OCH_3$ groups of the silane can exchange with surface hydroxyl groups to form a covalently bound coating structure. The bonding status and organization of silane on the surface affect the performance of the composite materials [25]. In this study, a covalent bond was preferred. Detailed structure characterization is very important towards achieving controlled synthesis.

Characterization of Nanoparticle Surfaces Coated with AEAPS

The water content and hydroxyl group density of the particles before the reaction were approximated from the TGA thermograms reported previously in ref [21]. For as-received TiO_2 nanoparticles, approximately 13 μmol/m^2 of water and 4.5 μmol/m^2 of surface hydroxyl groups were present on the surface. The drying process had efficiently removed most of the surface water. However, the density of the hydroxyl groups (approximately 12 μmol/m^2) on the dried particles was more than double that of as-received TiO_2 nanoparticles, which may be due to dissociation of trapped water before it evaporates. A similar surface –OH content of 13.7 μmol/ m^2 on dry TiO_2 has also been reported [26]. The calculation was based on the surface area value (42 m^2/g) provided by the producer of the nanoparticles.

The TGA thermograms for coated particles (Figure 2) showed a significant mass loss, indicating the presence of coupling agent. Approximately, 4.9 and 2.8 wt%, or 12.1 and 6.8 μmol/m^2 surface coverage, of AEAPS silane were present on the particles after coating with methods I and II, respectively. Drying had a significant effect on the amount of coating on the nanoparticle surfaces: more silane was present for coating method I compared to method II. The amount of silane present on the surface was more than the hydroxyl group density, indicating possible silane multi-layers resulting from method I. It was also noted that particles prepared by coating method I showed a sudden and more evident decrease in weight at about 210 °C than those prepared by coating method II, which may have been due to the loss of trapped water.

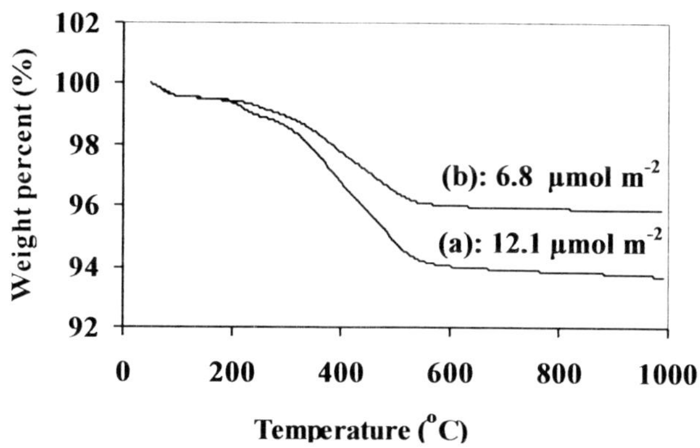

Figure 2. Thermogravimetry of AEAPS silane modified TiO_2 by (a) coating method I, and (b) coating method II.

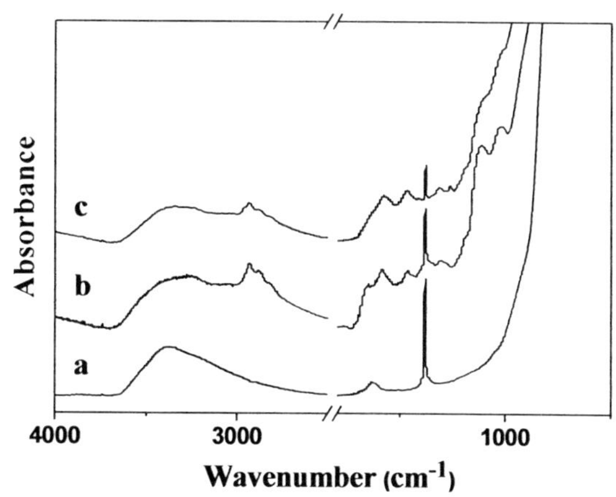

Figure 3. IR spectra of (a) as-received TiO_2, AEAPS silane modified TiO_2 by (b) coating method I, and (c) coating method II.

The FTIR spectra for untreated and silane coated TiO_2 particles are shown in Figure 3. The broad O-H stretching band from 2800-3700 cm^{-1} arises from surface hydroxyl groups and the weak 1640 cm^{-1} peak from molecular water on pure TiO_2 [27,28]. Upon surface modification, the molecular water band shifted to 1660 cm^{-1} for method I, and was evidently diminished for method II. This peak should not be confused with two nearby peaks at 1480 and 1590 cm^{-1}, which were present for both coating methods I and II and were due to the reaction product of $-NH_2$ of AEAPS silane with CO_2 in the air [29].

A broad and strong peak of the coated particles ranging from 2500 to 3700 cm^{-1} was again due to the formation of the amine bicarbonate [30]. The symmetric and asymmetric C-H stretches for both coating methods are observed at 2940 and 2910 cm^{-1}. The 1390 cm^{1} peak in as-received TiO_2 was believed to be due to organic contaminants of the particles during packaging and most of them can be removed by the drying, as evidenced by SIMS. Three sharp bands for Si-O-CH_3 [30] at 2840, 1200, and 1090 cm^{-1} present for pure AEAPS coupling agent were not present for either coating method, indicating that the methoxy groups had been hydrolyzed in coating method I or exchanged in method II. Two broad peaks centered at 1040 and 1130 cm^{-1}, respectively, indicate the formation of a long or branched Si-O-Si structure [30] and were much stronger in coating method I than II. An interesting peak at 1270 cm^{-1} was present for modified TiO_2 prepared by coating method II. The band was attributed to the Si-C stretching mode of a silane methyl group [31,32].

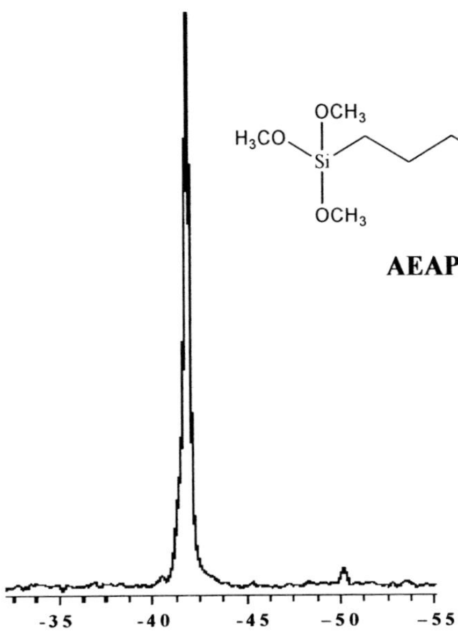

Figure 4. ^{29}Si NMR of AEAPS liquid coupling agent.

^{29}Si NMR was employed to determine the chemical environment around the silicon atoms. The ^{29}Si NMR spectrum of as-received AEAPS silane coupling agent, with a major peak at −41 ppm and a minor peak at −50 ppm, is shown in Figure 4. The latter peak was due to the hydrolysis of the AEAPS silane during storage and the amount of hydrolysis was estimated to be less than 2%. Figure 5 shows the spectra of AEAPS silane coated TiO_2 particles. A broader resonance had been observed for coated particles than that of pure AEAPS silane, indicating reduced mobility and surface attachment. The lower spectrum (coating method I) displayed a single peak at −67 ppm, which was attributed to the anchoring of all three oxygen atoms in the AEAPS silane with the TiO_2 surface or with nearby silane molecules [33-35]. The upper spectrum (coating method II) clearly contained two distinct resonances. As described earlier, the peak at −67 ppm is ascribed to the coupling of all three oxygen atoms of the silane molecule with the TiO_2 surface or other silane molecules. In addition, a shoulder at −49 ppm was present, which was assigned to the mono-coupling of the silane with the TiO_2 surface or other silane molecules. The resonance at −19 ppm, however, was unexpected.

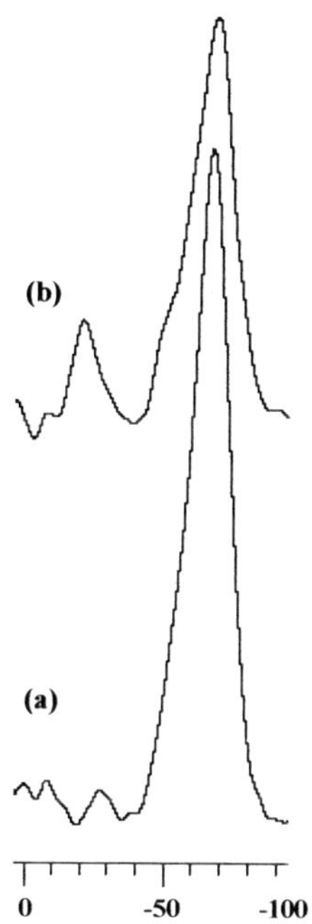

Figure 5. ^{29}Si NMR of AEAPS silane treated TiO_2 (a) prepared by method I, and (b) prepared by method II.

This peak could not be attributed to the pure coupling agent, which had a chemical shift of −41 ppm, nor could it be attributed to hydrolyzed forms of the molecule according to our ^{29}Si NMR observation of hydrolyzed AEAPS silane. Hydrolyzed AEAPS silane (not shown here) showed chemical shifts at −41, −49, −58 and −67 ppm; they are assigned to $RSi(OR')_3$, $RSi(OSi{\equiv})(OR')_2$, $RSi(OSi{\equiv})_2OR'$, and $RSi(OSi{\equiv})_3$, respectively, where R represents the organic group in the AEAPS silane and R′ is H or $-CH_3$. In fact, the ^{29}Si chemical shift at −19 ppm was not consistent with silicon having three oxygen atoms as its nearest neighbors. The chemical shift was, however, consistent with silicon bound by two oxygen atoms and two carbon atoms [36].

The ^{13}C NMR spectra are shown in Figure 6. The chemical shifts were labeled for each carbon atom of the AEAPS silane according to [34] and the additional peak at 168 ppm is attributed to the amine bicarbonate salt formed from reaction of the primary amine with carbon dioxide [37]. The sharp resonance at −2.3 ppm in the ^{13}C NMR spectrum for the silane coated particles prepared by method II is attributed to $Si\text{-}CH_3$ where Si was bound by two oxygen atoms and two carbon atoms directly [36].

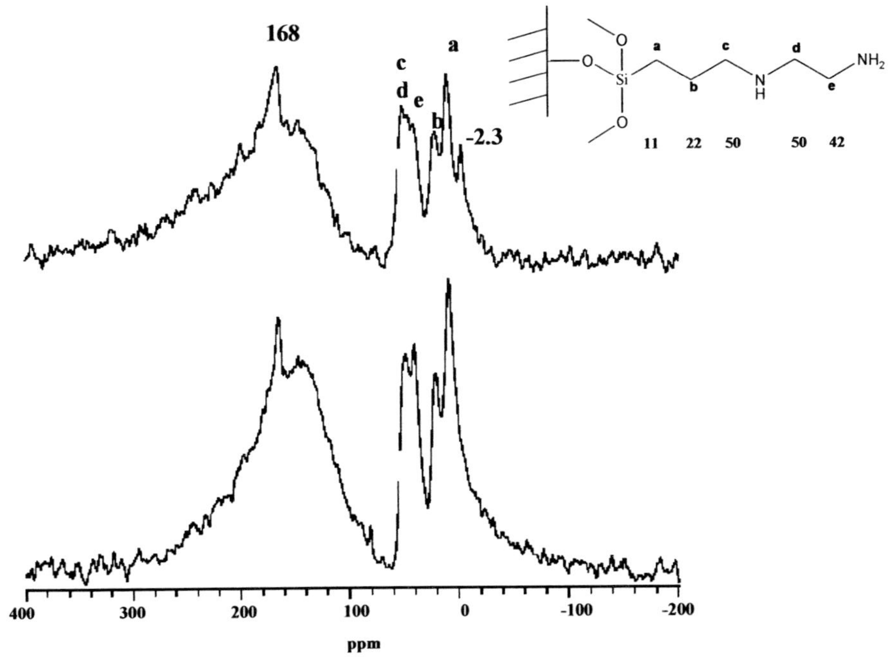

Figure 6. ^{13}C NMR of AEAPS silane treated TiO_2 (a) prepared by method I, and (b) prepared by method II.

Determination and Comparison of AEAPS Coating Structures

It is crucial to determine whether the silane was covalently bound (i.e., forming Si-O-Ti) or hydrogen-bound (i.e., forming Si-OH---OH-Ti) to the particle surface because the former provides a stronger interface and is preferred. In the case of coating method I, both water on the nanoparticle surface and in the as-received toluene can hydrolyze the silane. Therefore, hydrolysis of the Si-OCH_3 bond proceeds prior to condensation with either surface hydroxyls or other silanols. The resulting Si-OH can form a hydrogen bond with the Ti-OH or surface water on the TiO_2 nanoparticles or further condensation may occur to form Si-O-Ti covalent bonds at the interface.

As for coating method II, the reaction mechanism could differ from that for method I because of the absence of water in the system. Most of the surface moisture on the TiO_2 surface was removed during the drying treatment and also most of water in toluene was removed. Thus, theoretically, most of the AEAPS silane was initially available to form Si-O-Ti chemical bonds via direct alcoxolation. It is still possible that small amounts of Si–OH, either hydrolyzed during storage (Figure 4) or by small amounts of water left in the anhydrous toluene, form hydrogen bonds with the Ti-OH on the nanoparticle surface.

Mass spectrometry (MS) is commonly used to study interfacial bonding. According to a previous report [38], silicon-containing fragments could be detected by MS during the heating process if the silane was hydrogen-bound to the surface. These authors also reported that silicon-containing fragments would not be detected if the silane was covalently bonded to the surface. However, this method is only suitable for silanes with relatively low boiling points. During the heating process, two reactions at the interface between particles and a silane could occur and are competitive: one is evaporation of the hydrolyzed (totally or partially) silane and the other is the condensation reaction between Si-OH and surface hydroxyls. Thus, if the silane has a much higher boiling point than the corresponding condensation temperature, a condensation reaction may occur first and covalently bond the silane to the particles. In this case, the absence of silicon-containing fragments is not conclusive proof that the original interfacial bonding is covalent.

Here we employed ^{29}Si NMR, which is a very useful technique to identify interfacial structure and to investigate the nature of bonding at the interface. According to Maciel *et al.* [33], the chemical shift due to hydrolysis of Si–OCH_3 to Si-OH could not be resolved by ^{29}Si solid-state NMR. However, reaction of Si-OCH_3 or Si-OH to form a Si-O-Si bond resulted in a chemical shift of approximately 10-ppm upfield. The upfield shift of 10 ppm for the formation of Si-O-Si bonds was consistent with other literature reports [34,35] as well as our data of the hydrolyzed AEAPS. As noted earlier, the chemical shifts at –41, -49, -58, and –67 ppm corresponded to $RSi(OR')_3$, $RSi(OSi\equiv)(OR')_2$, $RSi(OSi\equiv)_2OR'$, and $RSi(OSi\equiv)_3$, respectively. For coated particles prepared by method I, the absence of -41, -49 and –58 ppm shifts (Figure 5(a)) indicated that Si-OH and Si-OCH_3 were not present in the coating structure, which was also supported by the FTIR results. Therefore, these results confirmed the covalent bonding of AEAPS coupling agent to the particle surface, as well as condensation reactions resulting in Si-O-Si bonds. For coated particles prepared by method II, the presence of the small shoulder at –49 ppm in the ^{29}Si NMR spectrum (Figure 5(b)) indicated a small amount of Si-OCH_3 or Si-OH was present in particles prepared by coating method II, which was not detected by FTIR. However, the major peak at –67 ppm in the ^{29}Si NMR spectrum indicated that most of the AEAPS coupling agent bonded either to the particle surface covalently forming Si-O-Ti bonds, or other silane molecules, as evidenced by the weak Si-O-Si FTIR absorbances, which appeared as two shoulders in the region of 1000-1170 cm^{-1}. The difference between Si-O-Si and Si-O-Ti could not be resolved by ^{29}Si NMR.

Coating coverage was calculated from the TGA results (Figure 2). It can be seen that nearly twice as much silane was coated onto the particle surfaces using coating method I compared to coating method II. By comparing the surface hydroxyl concentration (4.5 $\mu mol/m^2$) of the as-received TiO_2 nanoparticles with the silane coverage (12.1 $\mu mol/m^2$) coated with method I, it was concluded that silane multilayers were formed in coating method I. This would result from an excess of Si-OH groups formed by the reaction of the silane coupling agent with readily available water in both toluene and on the TiO_2 surface. In addition, the silane molecules in the layer nearest to the TiO_2 surface may only form a single Ti-O-Si bond. Subsequent condensation between neighboring coupling agents that are only singly bonded to the TiO_2 would result in the formation of a Si-O-Si network, which has been suggested by two broad peaks centered at 1040 and 1130 cm^{-1} (Figure 3(b)). Again due to the high concentration of Si-OH groups, the final coating coverage would be restricted primarily by spatial hindrance due to the large organic group, and secondarily by the availability of the hydrolyzed groups. As for the particles modified with method II, because the concentration of surface hydroxyls (12 $\mu mol/m^2$) was nearly twice that of the coating coverage (6.8 $\mu mol/m^2$), it could be concluded that a compact monolayer was formed and, on average, each silicon formed two covalent bonds with the nanoparticle surface. It may be argued that a "patchy" multilayer surface structure could be formed in the above situation, instead of a relatively uniform monolayer structure. However, the FTIR results did not support this argument. The two Si-O-Si shoulders in the FTIR spectra were quite weak and inconsistent with a patchy multilayer surface structure. Thus, the anhydrous conditions of method II resulted in lower coating amounts as compared to method I, and a predominantly monolayer structure.

The sudden weight loss at approximately 210°C was likely due to water trapped by the silane-layered structure. MS at that temperature showed no difference for coated particles prepared by both methods. Since MS only can detect ions with molecular weights higher than 20 and our expected silane fragments have a molecular weight higher than 20, the loss could not be due to an organic fragment. FTIR results supported this conclusion where particles prepared by coating method I showed a broad peak at 1660 cm^{-1}, which was assigned to residual water and a weak contribution from the amine bicarbonate salt [29]. This result further supported the conclusion that coating method I yielded a multi-layered structure with some water trapped between silane layers and the particle surface. FTIR and TGA studies did not detect any trapped water in the case of coating method II.

A previously unreported [29] Si NMR resonance appearing at −19 ppm was observed for particles prepared by coating method II (Figure 5(b)), which suggested the occurrence of a previously unobserved reaction during the coating process. The resonance was consistent with silicon bonded only to two oxygen atoms. Based on this analysis, the following reaction mechanism was proposed. After the particle drying treatment, the TiO_2 surface could become oxygen deficient. In addition, it is also known that oxygen vacancies are present even in as-received TiO_2 nanoparticles [39], which are preferred sites of impurity or gas adsorption. Drying then exposes these "old" vacancies to the silane by removing small molecule impurities. Thus, when the surface was treated with the silane, a silane methoxy group may fill the surface oxygen vacancy thereby breaking a silicon-oxygen bond. Therefore, the number of first nearest neighbors of Si changed from three to two oxygen atoms. The silicon atom may then react with a methyl group from either a neighboring silane compound or the surface methoxy group producing a Si-CH_3 bond in place of the silane methoxy group, consistent with the −19 ppm shift in ^{29}Si NMR. Additionally, the presence of the −2.3 ppm resonance in ^{13}C NMR and 1270 cm^{-1} in FTIR supported this hypothesis.

In summary, the surface coating structures were largely controlled by the experimental conditions of the coating process. The presence of water at the interface in particular was found to exert a large effect on the nature of the surface structure. Coating method I led to a multilayer silane structure on the TiO_2 nanoparticle surface with residual water trapped under the silane layers. The multilayer structure formed covalent bonding (Ti-O-Si) to the titania surface, as well as Si-O-Si bonding within the multilayer. For coating method II, a primarily compact monolayer surface structure was formed with each silane molecule forming, on average, two primary bonds to the surface. Two different structures were formed based on the reactivity with adsorbed water on the nanoparticle surfaces. In addition, a new reaction was detected on the titania nanoparticle surface in the coating method II. The formation of Si-CH_3 bonds possibly resulted from the previously unreported reaction of the silane with oxygen vacancies on the dried nanoparticle surface.

Both coating methods yielded covalent-bonded coating structures. Compared to hydrogen-boned coating structures, these structures are more stable and more resistant to water attack. But for the purpose of modifying the nanoparticle/polymer interface, generally, monolayer coating, yielded through coating method II, is sufficient. In addition, trapped water was not detected for the particles coated by method II, while it was detected for those coated by method I. As known, the presence of water at the interface can electrically degrade the insulation. Therefore, coating method II was more suitable for surface modifying the nanoparticles for insulating application. Accordingly, the TiO_2 particles modified with the AEAPS silane (AEAPS-TiO_2) via the coating method II were chosen for composite sample preparation and characterization.

Characterization and Determination of DCS Coating Structure

As stated above, coating method II was preferred for coating nanoparticles. Therefore, only coating method II was applied to coating the nanoparticles with DCS ($CH_3(CH_2)_9SiCl_3$). TGA of the DCS coated particles (DCS-TiO_2) is shown in Figure 7. The amount of DCS coating was 1.8 wt%, or 3.1 $\mu mol/m^2$. Similar to the AEAPS, the DCS, which was coated onto the particle surface by coating method II, presented a monolayer coating structure. The lower coverage of DCS compared to the AEAPS was believed to result from the large spatial hindrance caused by its long hydrocarbon chain.

^{29}Si NMR results of neat DCS silane and DCS coated particles are shown in Figure 8. The DCS silane exhibited a single peak at −13 ppm indicating that no observable hydrolyzation or condensation occurred during storage. The spectrum of the DCS coated particles contained two nearby peaks; and as expected, the resolved two peaks were centered at the same positions as the AEAPS coated particles, i.e., −58 and −67 ppm, respectively. This can be easily understood because the chemical component difference in the long side chains in both silanes began from the fourth atom away from the central Si atom; thus, the first different atom was too far away to influence the ^{29}Si NMR spectra of the coated particles. Based on what was described earlier about the correlation of the ^{29}Si NMR peaks with chemical bonding in the section on AEAPS coated nanoparticles, it was determined that the DCS was primarily covalently bonded to the TiO_2 surface, or with itself. No trapped water was detected from TGA or FTIR.

In summary, similar to the nanoparticles coated with the AEAPS silane via method II, the DCS silane formed a "dry", mainly covalently bound monolayer structure on the TiO_2 nanoparticles.

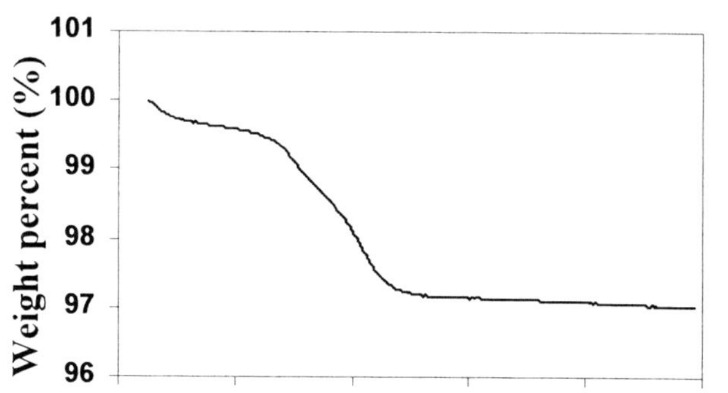

Figure 7. Thermogravimetry of DCS silane modified TiO_2 by the coating method II.

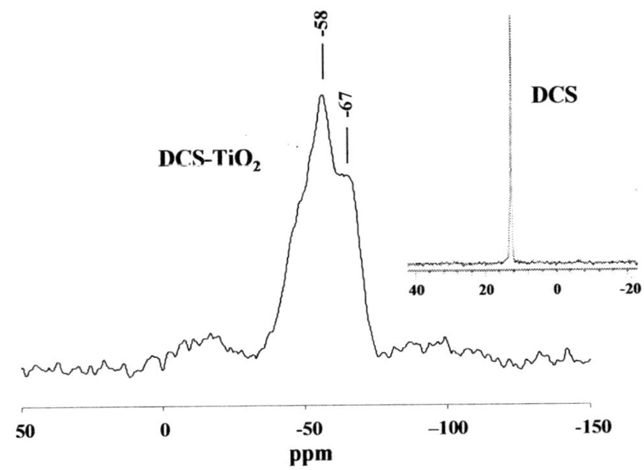

Figure 8. ^{29}Si NMR of the DCS silane and DCS modified TiO_2 nanoparticles.

MICROSTRUCTURE CHARACTERIZATION OF POLYMER NANOCOMPOSITES

Dispersion of nanoparticles in a polymer and the morphology of resulting composites have non-negligible effects on dielectric properties. For example, it is known that particles act as local field enhancing defect centers in filled polymers [1,40,41], which causes a decrease in the dielectric breakdown strength as compared to the neat polymer. The extent of the local field enhancement, and hence the dielectric breakdown strength, is directly related to the size of the particles (or agglomerates) and their dispersion in the matrix. In addition, the motion of charges could be impeded at the interface between the particles and the matrix, and hence the space charge distribution could also be affected by the particle dispersion. On the other hand, polymer morphology can affect the dielectric properties. LDPE is a semi-crystalline polymer, so it consists of an ordered crystalline phase and a disordered non-crystalline phase comprising a partially ordered layer adjacent to the crystalline phase [42] and a completely disordered amorphous region. The degree of crystallinity, the crystallites' size and their distribution, and the organization of the crystallites with respect to the non-crystalline phase govern various physical properties [42], and have long been of research interest. In order to gain a fundamental understanding of the dielectric properties of interesting systems, it is crucial to carry out a thorough microstructural investigation of the prepared nanocomposites. In our previous work [23,43], we have combined a variety of methodologies to characterize the crystalline structure of polyolefin nanocomposites, including a deep exploration of the use of small-angle light scattering (SALS) in the nanocomposites. Herein, we will include some of our previous work and also additional work with a different nanocomposite system.

Dispersion

In order to fully characterize the dispersion of the nanoparticles, FE-SEM and SIMS were employed together. Figure 9 (a) and (b) show FE-SEM micrographs of the bulk dispersion of the as-received TiO_2 (AR-TiO_2) and AEAPS-TiO_2 nanoparticles in the LDPE, respectively. The dispersion of the DCS-TiO_2 was similar to that of the AEAPS-TiO_2. Qualitatively, the as-received nanoscale TiO_2 exhibited a much better dispersion in the matrix than did the coated nanoparticles. Except for individual particles, agglomerates were observed in all samples, but the coated TiO_2 nanoparticles formed more and larger agglomerates in the matrix. Some agglomerates can even be seen without the aid of a microscope. They appeared as "chunks" under polarized reflection optical microscopy. To complement the dispersion characterization, the time-of-flight SIMS technique was used (Figure 10). One advantage of SIMS over FE-SEM to observe the dispersion is that it will not distort the dispersion, which possibly occurs during preparation of the SEM samples. It can be seen that although the dispersion of the nanoparticles in the matrix was not very homogeneous at the nanoscale and agglomerates were also present, "globally", the particles were uniformly distributed in all three composites. Converging all the above observations, it was clear that the nanoparticles were dispersed in the composites at different length scales, from nano to micro, in the form of individual nanoparticles, small aggregates and large agglomerates, respectively. The coated particle filled samples, no matter whether the particle surface was more hydrophilic or hydrophobic as compared to the AR-TiO_2, had more and larger agglomerates than the uncoated particle filled composites. Generally, the dispersion was closely related to the compatibility of particles with the polymer matrix. The increasing order of surface hydrophobicity of all the particles studied was AEAPS-TiO_2 < AR-TiO_2 < DCS-TiO_2 expected from the silane chemical structure. Because the LDPE molecules were hydrophobic, the compatibility between the particles and the matrix would order in the same way, and thus the dispersion would be expected to improve in the same order. The inconsistency arose from the coating process, which formed some agglomerates that could not be broken during shear mixing. Agglomerates may also form during melt mixing because of enhanced incompatibility in the AEAPS-TiO_2 / LDPE.

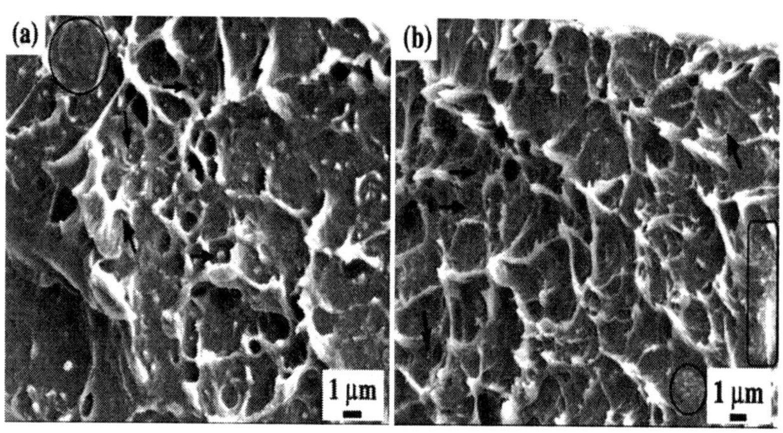

Figure 9. FE-SEM images showing the dispersion of (a) as-received TiO_2 nanoparticles and (b) AEAPS coated TiO_2 nanoparticles in LDPE. Circles and a rounded rectangle indicate agglomerates and arrows point to particles. Reprinted with permission from [18]. Copyright (2005) IOP Publishing Ltd.

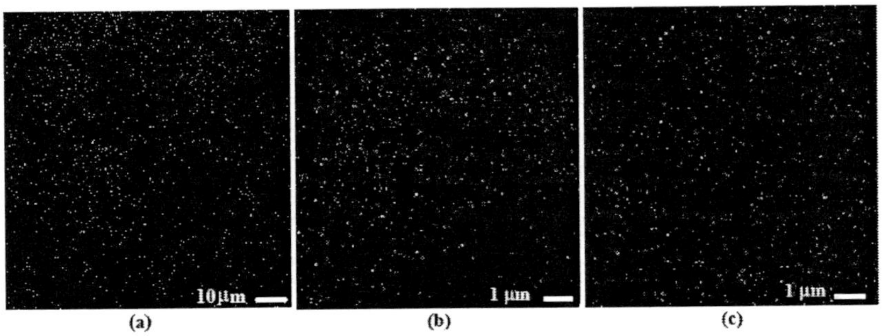

Figure 10. SIMS images showing Ti^+ map in the LDPE composite samples filled with (a) as-received, b) AEAPS coated, and (c) DCS coated nanoparticles.

Morphology

DSC was employed to measure the degree of crystallinity. The values calculated from normalized DSC thermograms (Figure 11) varied from 45 to 48% among all the samples, indicating that the incorporation of the nanoparticles and their different surface states did not evidently affect the degree of crystallinity of the polymer matrix. The same conclusion was made from XRD although the absolute values obtained from this technique were systemically lower than those from DSC, mainly due to the sensitivity difference of each technique to complex crystalline morphology [44].

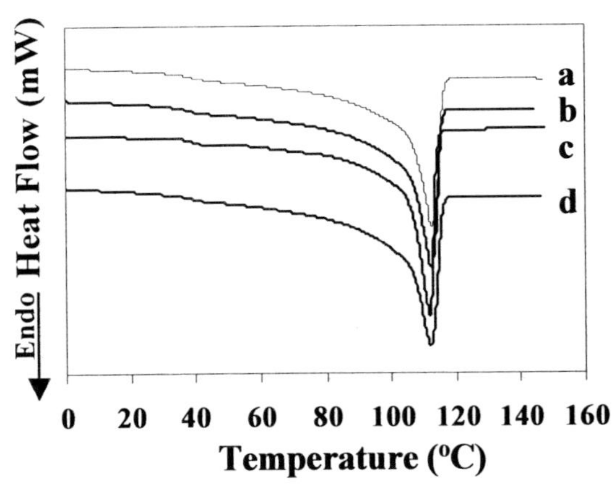

Figure 11. DSC melting curves for (a) neat LDPE, (b) AR-TiO_2 / LDPE, (c) AEAPS-TiO_2 / LDPE, and DCS-TiO_2 / LDPE.

The unit cell parameters were determined from XRD. All normalized diffraction patterns for the samples studied are shown in Figure 12. The amorphous scattering appeared as a broad peak under the strong and sharp diffraction pattern of the crystallites. The diffraction peaks centered near 22° and 24° are characteristic of the (110) and (200) planes in the crystalline region of the LDPE. Among all the samples, the shifts of crystalline peak positions were within experimental errors, which meant that the unit cell dimension was not altered by the nanoparticles and their surface modification. However, the widths of the (110) and (200) diffraction peaks narrowed in the nanocomposites (except for the (200) of the DCS-TiO_2 / LDPE) as compared to those of the neat LDPE. Therefore, the lateral dimension of the PE lamellae might increase with the incorporation of nanoparticles. It should be pointed out, however, that the above discussion was based on an assumption that the peak width is only related to crystallite size. The internal strain and degree of perfection of the crystallites could also alter the peak width. Although it is still not clear which was the most direct reason for the changes of the peak widths, it is likely that both nanoparticles and their different surface features contributed to this variation. The differences were not significant, however.

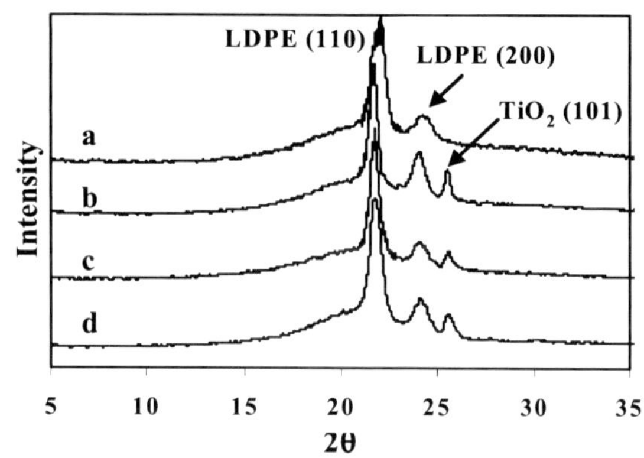

Figure 12. XRD scans of (a) neat LDPE, (b) AR-TiO_2 / LDPE, (c) AEAPS-TiO_2 / LDPE and (d) DCS-TiO_2 / LDPE.

The average lamellar thickness (i.e., the third dimension of the lamellae) and thickness distribution were studied by AFM. Representative phase AFM images of the neat LDPE and the AR-TiO_2 / LDPE are presented in Figure 13. The width of the linear patterns, which is a projection of the lamellae width, can represent the lamellar thickness for relative comparison. All lamellar thickness as observed by AFM varied in the range of 20-40 nm, with the most probable thickness around 25 nm. Although the AFM observations were not able to provide absolute values of lamellar thickness, it still can be reasonably concluded that the average lamellar thickness did not change significantly with the addition of nanoparticles. In addition, DSC analysis also yielded useful information about the lamellar thickness. All DSC thermograms revealed similar melting shape, peak melting temperatures and final melting temperatures, suggesting similar lamellar size distribution. Furthermore, the same final melting temperatures indicated the same thickness of the largest crystals (lamellae) formed in all samples according to the Thompson-Gibbs equation [45,46].

The macroscopic lamellae arrangement (i.e., the suprastructure) was studied by cross-polarized light microscopy. All samples showed similar fine crystallite structure (not shown here) [18], which appeared as bright spots with similar sizes. Small-angle light scattering (SALS) confirmed that they were spherulites with different degrees of perfection [23]. AFM was employed to directly demonstrate the lamellar arrangement. Figure 14 shows the phase images of the neat LDPE film, the AR-TiO_2 / LDPE, the AEAPS-TiO_2 / LDPE, and the DCS-TiO_2 / LDPE, respectively.

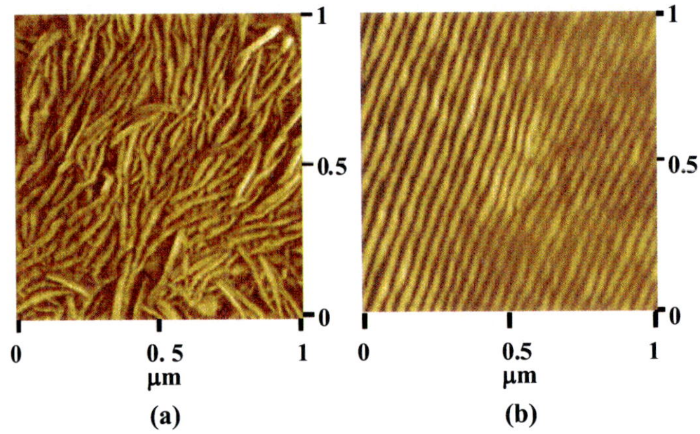

Figure 13. AFM phase images showing lamellae in (a) neat LDPE and (b) AR-TiO_2 / LDPE. Reprinted with permission from [23]. Copyright (2005) Wiley Periodicals, Inc.

It was clear that the neat LDPE exhibited a well-defined impinging spherulite texture (Figure 14(a)). The interspherulitic boundaries could be resolved easily and the diameter of the spherulites ranged from several microns to over 10 microns, consistent with that observed from optical microscopy. Curved individual lamellae or lamellar-bundles were observed. All the nanocomposites exhibited coarser spherulitic texture, increasing the complexity of resolving individual spherulites, lamellar bundles or individual lamellae. Despite this difficulty, it was still clear that all the nanocomposites exhibited textures characteristic of poorly developed spherulites, with the AEAPS-TiO_2 / LDPE displaying the most severely disrupted spherulites with the highest disorder of internal organization of lamellae or their bundles. In addition, micrometer-scale nanoparticle aggregates were easily observed surrounding lamellar bundles. Not like AFM, which only yields surface texture information, SALS studies the bulk crystalline structure and has been carried out on the neat LDPE, the AR-TiO_2 / LDPE, and the AEAPS-TiO_2 / LDPE [23]. From the SALS, relatively quantitative information about the degree of spherulite perfection was achieved. It was found that among these three samples, the AEAPS-TiO_2 / LDPE possessed the worst developed spherulites. This finding was in a good agreement with what was observed at the surface in AFM, confirming that the surface and the bulk had similar crystalline structure.

Based on the above investigations about morphology, the TiO_2 nanoparticles, with various surface conditions, did not evidently change the degree of crystallinity, the unit cell dimension, the average lamellar thickness, and the average spherulite size. However, they definitely affected the degree of perfection of the supramolecular structure by influencing the relative arrangement and orientation of the lamellae within the spherulites and the particle concentrations at the interspherulite boundaries. The AEAPS-TiO_2 / LDPE exhibited the most disrupted and poorly developed spherulite structure. This is associated with surface treatments. The surface of the AEAPS-TiO_2 was more hydrophilic as compared to those of the AR-TiO_2 and the DCS-TiO_2, and thus less compatible with the polymer matrix. This incompatibility led to repelling of particle aggregates to boundary regions and also caused most severe interference to spherulite organization.

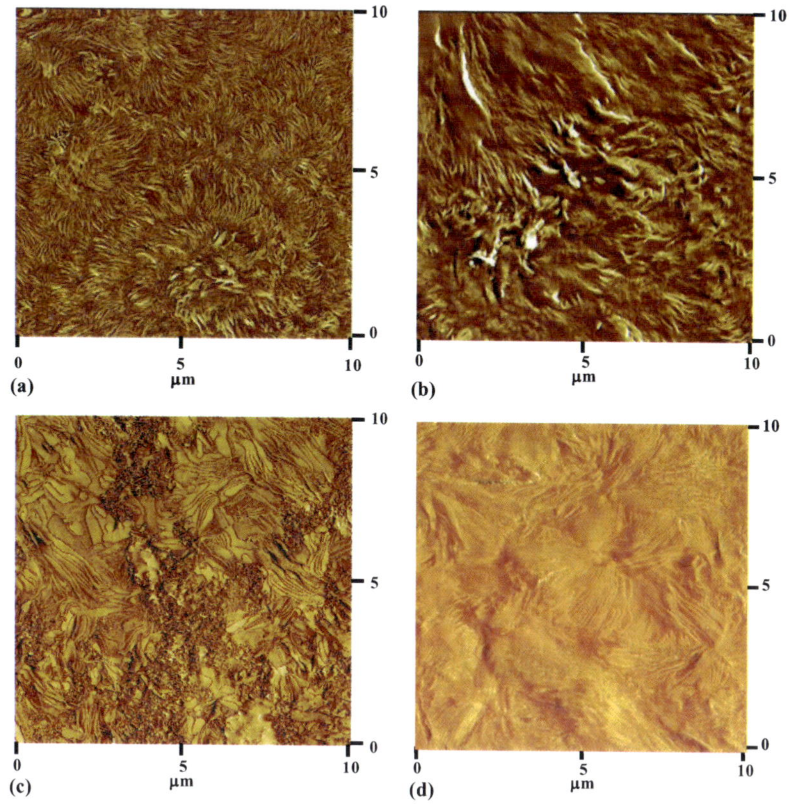

Figure 14. AFM phase images of (a) neat LDPE, (b) AR-TiO_2 / LDPE, (c) AEAPS-TiO_2 / LDPE and (d) DCS-TiO_2 / LDPE. (a), (b) and (c) are reprinted with permission from [23]. Copyright (2005) Wiley Periodicals, Inc.

DIELETRIC BEHAVIOR

In this section, the dielectric properties of the nanocomposites filled with the TiO_2 nanoparticles with different surfaces are presented. To be specific, permittivity, dielectric breakdown strength, and space charge distribution were measured and analyzed. It was found that tailoring the nanoparticle surface by silane modification had a significant effect on the dielectric breakdown strength and space charge distribution, while having little effect on permittivity.

Permittivity

The permittivity of the neat LDPE and LDPE nanocomposite samples is shown in Figure 15. The permittivity of the neat LDPE is 2.26 and all the other data fall into the same region between 2.0 to 2.5. The experimental error was 10 %. The difference in permittivity in those samples was not significant. It was because the nanoparticle loading, 5 wt% (about 1.1 vol %), was too low for the nanoparticles to show their influence on the permittivity of the composite samples. When samples filled with the as-received TiO_2 nanoparticles at different loadings were examined, however, it was found that the overall permittivity increased nearly linearly with increasing TiO_2 volume fraction (Figure 16). This increase could be easily understood because TiO_2 nanoparticles have a much higher permittivity than neat LDPE. This linear relationship indicated that the composite permittivity was mainly affected by the bulk property, instead of by the surface property, of the nanoparticles. Otherwise, the permittivity would have shown a linear relationship with d^2, instead of d^3 (proportional to the volume fraction), where d is the average diameter of the nanoparticles. For comparison, the permittivity of the microcomposite samples filled with as-received TiO_2 microparticles was also measured, which yielded nearly the same permittivity value as the nanocomposites. The above results indicate that there is not much opportunity to change permittivity by doing surface modification in our studied system.

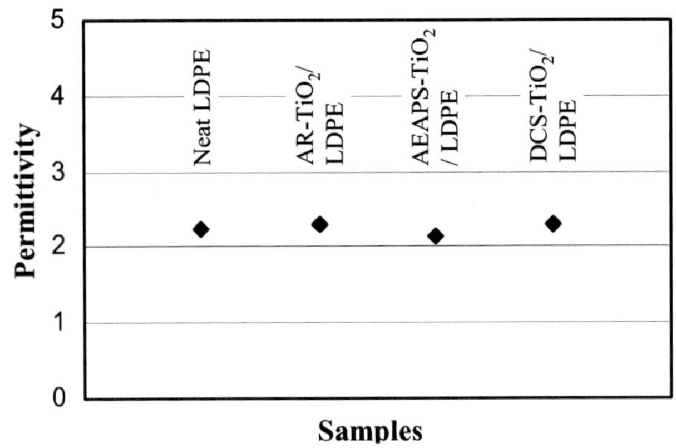

Figure 15. Permittivity of neat LDPE and LDPE samples filled with as received (AR), AEAPS coated, and DCS coated TiO_2 nanoparticles.

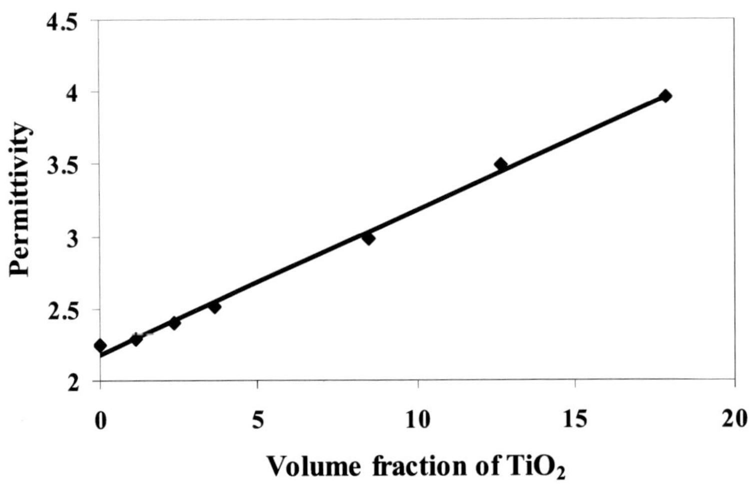

Figure 16. Permittivity linearly increased with increasing volume fraction of as-received TiO_2 nanoparticles. A trendline is shown.

Dielectric Breakdown Strength

Dielectric breakdown strength is defined as the electric field intensity that causes an "insulator-to-conductor" transition in a material and is accompanied by a sharp increase in current. It is known that even if variables such as morphology, additives and test conditions are under exact control, scatter in the dielectric breakdown is inevitable, which increases the difficulty of data analysis. Therefore various statistical methods are used to treat data. A Weibull distribution [1,47] is used in this work to present breakdown strength data. The cumulative probability P of the electrical failure takes the form of [1,47]:

$$P = 1 - \exp[-(\frac{E}{E_0})^{\beta}] \tag{1}$$

where β is a shape parameter that represents the data scatter and E_0 is a scale parameter that represents the breakdown strength at the cumulative failure probability of 63.2%. E_0 is also called Weibull breakdown strength and frequently used to evaluate the dielectric strength of different materials. P is calculated from [47]:

$$P_i = \frac{i - 0.5}{n + 0.25} \tag{2}$$

where i is the i-th result when the values of E are sorted in ascending order and n is the number of measurements. Both β and E_0 were obtained from linear regression of the linearized Equation (1) by taking the logarithm to it. Figure 17(a) shows the Weibull plots of the dielectric breakdown strength of neat LDPE, the AR-TiO_2 / LDPE and the AEAPS-TiO_2 / LDPE samples. The scattered data of the DCS-TiO_2 / LDPE fell in between those of the AEAPS-TiO_2 / LDPE and the neat LDPE. In order to avoid too crowded data and to highlight the difference in the Weibull strength E_0 of different samples more clearly, the Weibull plot of the DCS-TiO_2 / LDPE was not included in Figure 17(a), but shown in Figure 17(b) separately. For comparison, the Weibull plot of the neat LDPE is also shown in Figure 17(b).

The data points were not aligned very linearly. The rapid decrease at relatively low breakdown strength values are thought to be due to the presence of agglomerates [18,21]. Despite this, the Weibull plots were sufficient for comparing differences between our samples and identifying general trends in the results. Neat LDPE had an E_0 of 3.8±0.4 MV/cm. With the incorporation of 5 wt% of AR-TiO_2, E_0 decreased by about 40% to 2.3±0.3 MV/cm. The large negative impact of small amounts of additive particles to the dielectric breakdown strength has been reported previously. For example, Khalil *et al.* [48] showed that the addition of 1 wt% of TiO_2 caused a 9% decrease in step dc breakdown strength of LDPE. Ueki *et al.* [47] reported that the incorporation of 2.5 wt% of carbon black N220 into high-density polyethylene led to a reduction of about 50% in dc breakdown strength. After particle surface modification, E_0 of the nanocomposite samples was still lower than that for the neat LDPE. However, both coated samples showed evidently higher dielectric breakdown strength than that for the AR-TiO_2 / LDPE. The dielectric breakdown strength of the AEAPS-TiO_2 / LDPE (3.1±0.3 MV/cm) was nearly 40% higher than that for the AR-TiO_2 / LDPE. The DCS-

TiO_2 / LDPE showed an even higher Weibull breakdown strength of 3.3±0.4 MV/cm. Evidently, the nanoparticle surfaces affected the dielectric breakdown strength significantly. Furthermore, it appeared that both the hydrophilic surface (coated with polar AEAPS) and the more hydrophobic surface (coated with nonpolar DCS) showed an enhanced Weibull breakdown strength compared to the nanocomposites filled with the uncoated TiO_2 nanoparticles. It is important to understand this interesting phenomenon by investigating the microstructure and correlating it with dielectric breakdown fundamentals.

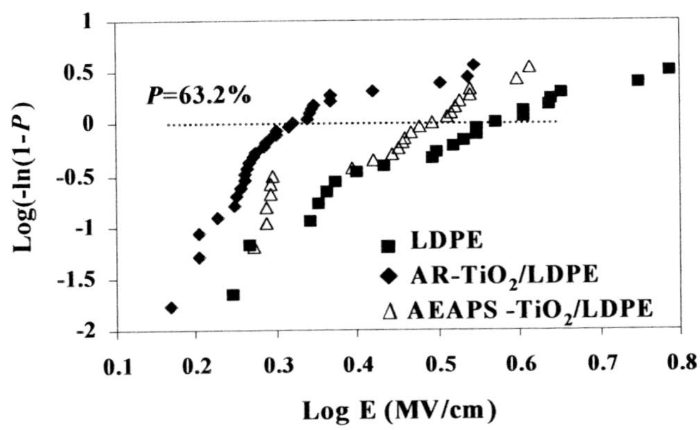

a)

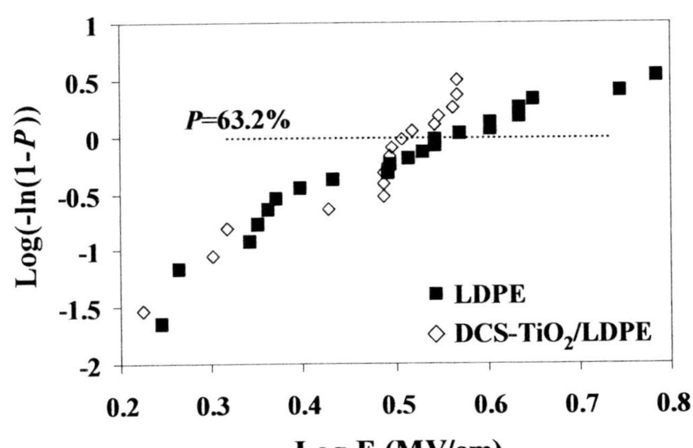

b)

Figure 17. Weibull plots of dielectric breakdown strength of (a) neat LDPE, AR-TiO_2 / LDPE and AEAPS-TiO_2 / LDPE, and (b) neat LDPE and DCS-TiO_2 / LDPE. (a) is adapted with permission from [18]. Copyright (2005) IOP Publishing Ltd.

It is known that dielectric breakdown in polymers is a complicated process, and could be dominated by more than one mechanism [49-52]. The process is influenced by many factors such as temperature [53], molecular weight [53], degree of crystallinity [53], and spherulite size and structure [54]. With the introduction of the nanoparticles, the complexity is further increased. Nanoparticles could not only influence the polymer morphology, but also could act as defect centers electrically and introduce many unique interfacial areas as well. Therefore, in addition to those factors in the neat polymer, the nature of the particles, their dispersion and surface features all could directly affect the dielectric breakdown process. In our case, for example, the reported conductivity of nanocrystalline TiO_2 (35 nm) is on the order of 10^{-6} to 10^{-7} $(\Omega\text{-cm})^{-1}$ at 500 °C and under atmospheric pressure [55], while that for neat polyethylene at room temperature is in the range of 10^{-15}-10^{-17} $(\Omega\text{-cm})^{-1}$ [56]. The difference between these two conductivities exceeds 9 orders of magnitude. Therefore, electric field distortion around the TiO_2 nanoparticles is inevitable. If two particles happen to be quite close to each other so that their distorted field superimposes, a higher field enhancement will occur and the polymer between the two particles will suffer a much higher local electrical stress compared to the bulk polymer. Therefore, the polymer between the two particles will fail at a lower external electric field first and then this local failure will initiate macroscopic breakdown. Thus, the TiO_2 nanoparticles can act as electrical defect centers in the composites and their poor dispersion would significantly reduce the dielectric breakdown strength further.

In our previous work [18], we analyzed the effects of the AEAPS modification on the dielectric breakdown strength in detail. Herein, in order to compare how different surface modification influenced the dielectric breakdown strength with different mechanisms, it is necessary to give a brief summary. The AEAPS-TiO_2 / LDPE exhibited a worse particle dispersion, had a more incompatible interface with the LDPE matrix and revealed spherulites with less organized internal structure, as compared to those of the AR-TiO_2 / LDPE. All these facts would theoretically lead to lowering of the dielectric breakdown strength. On the contrary, the AEAPS-TiO_2 / LDPE had Weibull breakdown strength nearly 40% higher than that of the AR-TiO_2 / LDPE. The improved dielectric breakdown strength was associated with modification of electrical features at the interface between the coated nanoparticles and the matrix: the polar groups could act as charge scattering centers, traps or excitation sites. Therefore, the incorporation of polar groups can decrease the acceleration of electrons, thus decreasing the chance of them initiating dielectric breakdown.

This mechanism cannot explain the behavior for the DCS-TiO_2 / LDPE as no polar groups are presented. The DCS silane has a 10-C hydrocarbon chain, and thus the presence of it on the nanoparticle surface could change the nanoparticle surface from hydrophilic to hydrophobic. Therefore, the interfacial compatibility between the nanoparticles and the hydrophobic LDPE matrix should be improved as compared to that of the AR-TiO_2 / LDPE and the AEAPS-TiO_2 / LDPE. The improved compatibility in the DCS-TiO_2 / LDPE was evidenced by using solution mixing and sonication, which were able to break the agglomerates formed during the coating process. These agglomerates were not broken up by melt-mixing. In this case, the dispersion of the nanoparticles would theoretically be a direct reflection of the interfacial compatibility. It was found that the DCS-TiO_2 / LDPE revealed a nice dispersion of the nanoparticles, while the AEAPS-TiO_2 / LDPE showed inhomogeneous dispersion of the nanoparticles after solution mixing.

The DCS-TiO_2 / LDPE prepared by melt mixing also had worse dispersion than the AR-TiO_2 / LDPE. In characterizing the morphology, the DCS-TiO_2 / LDPE and the AR-TiO_2 / LDPE had a similar degree of crystallinity, unit cell parameters, and average lamellar thickness and degree of perfection of spherulite structure. Therefore, the over 40% increase in the dielectric breakdown strength of the DCS-TiO_2 / LDPE over the AR-TiO_2 / LDPE was mainly due to improved interfacial compatibility despite the worse dispersion in the former sample.

Based on the above discussion, it could be predicted that there was still a possibility to further enhance the Weibull breakdown strength by improving the particle dispersion in the DCS-TiO_2 / LDPE. In order to test this hypothesis, breakdown measurements were carried out on solution mixed DCS-TiO_2 / LDPE and compared with that of the melt-mixed DCS-TiO_2 / LDPE. As expected, the Weibull strength of the solution-mixed DCS-TiO_2 / LDPE was increased by 11% and, thus, quite close to that of the neat LDPE. Investigation of the microstructure of the solution processed samples did not show any significant differences in crystallinity and spherulite structure compared to the melt mixed DCS-TiO_2/LDPE composite, indicating that the main difference was the nanoparticle dispersion between the DCS-TiO_2 / LDPE samples prepared by melt mixing and solution mixing. Therefore, by carrying out particle surface modification and adjusting the processing conditions, the Weibull breakdown strength of the nanocomposites approached that of the neat LDPE.

On the other hand, β is an important parameter and need to be taken into account. β values were 3.5, 5.1, 3.3 and 5.1 for the neat LDPE, AR-TiO_2 / LDPE, AEAPS-TiO_2 / LDPE and DCS-TiO_2 / LDPE, respectively. β has been used to evaluate the data scatter and, in polymer composites, the data scatter is closely related to the size distribution of the defects or filler particles. Higher β values indicate a narrower data distribution. Worse particle dispersion in the AEAPS-TiO_2 / LDPE caused relatively greater data spread in this sample than in the AR-TiO_2 / LDPE. However, it is not clear at this moment why the DCS-TiO_2 / LDPE showed the much decreased data scatter than the AEAPS-TiO_2 / LDPE and the similar data scatter as the AR-TiO_2 / LDPE although the AR-TiO_2 / LDPE exhibited better particle dispersion. Better interface in the DCS-TiO_2 / LDPE could be a reason.

Space Charge Distribution

Space charge accumulation is a common phenomenon in polymer materials and has been extensively studied [24,57-60]. The build-up of space charge can ruin both short and long term performance of polymers in electrical applications [1,61,62]. It is imperative to include a study of space charge distribution, if one would like to evaluate a polymer insulating material comprehensively. It should be pointed out, however, because space charge distribution is much affected by the electrode materials [24,57], the results and discussion of the space charge distribution will be only restricted to the specific measuring system investigated. The underlying meaning of this statement is that both material properties of a sample and interaction between the sample and electrodes account for the space charge distribution.

The space charge distribution of the neat LDPE, the AR-TiO_2 / LDPE, the AEAPS-TiO_2 / LDPE and the DCS-TiO_2 / LDPE is presented in Figure 18(a), (b), (c), and (d), respectively. Figure 18(a) illustrates development of the space charge distribution versus time in the neat LDPE sample. The building-up of a small amount of hetero space charges, i.e. the charges of the opposite polarity to the nearby electrodes, was completed within the first 30 minutes. Then, the space charges were nearly constant with time. The distribution of the space charge was symmetric with positive and negative space charges being of the same density near the two electrodes, indicating little or no charge injection from either electrode [62]. These hetero space charges possibly originated from impurity dissociation and migration under the field.

Figure 18(b) shows the space charge distribution in the AR-TiO_2 / LDPE. Its space charge distribution was completely different from that of the neat LDPE. The nature of the charges changed. The space charges were of the same polarity as the nearby electrodes, i.e. primarily homo. This indicates enhanced charge injection from the electrodes with the introduction of the nanoparticles. It was also noted that the charge distribution in the sample was not symmetric. The accumulation of positive charges near the anode was more evident than that of the negative charges near the cathode, which spread out evenly, indicating that the electrons/negative charge carriers injected from the cathode had a higher mobility than that of holes/positive charge carriers injected from the anode. The space charge became stabilized after 30 minutes. The above observations were different from those previously reported [63]. In that work, LDPE revealed homo space charge distribution, but the TiO_2 / LDPE sample revealed negative space charges all through the sample. Therefore, the space charge distribution is affected significantly by the nature of both the matrix and the filler. Specifically, it could be influenced by the surface state and crystal structure of the filler.

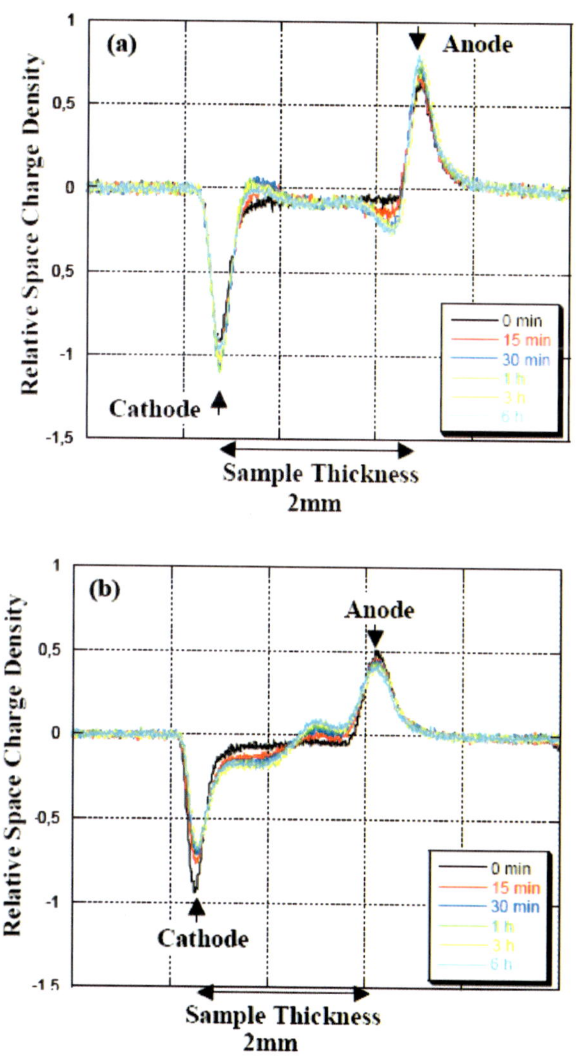

Figure 18. continued on next page.

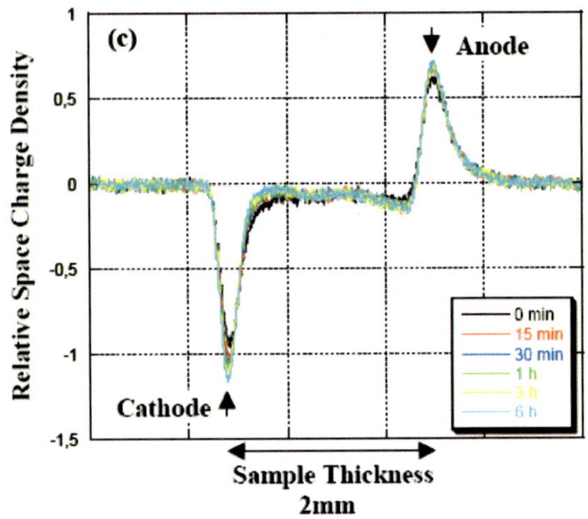

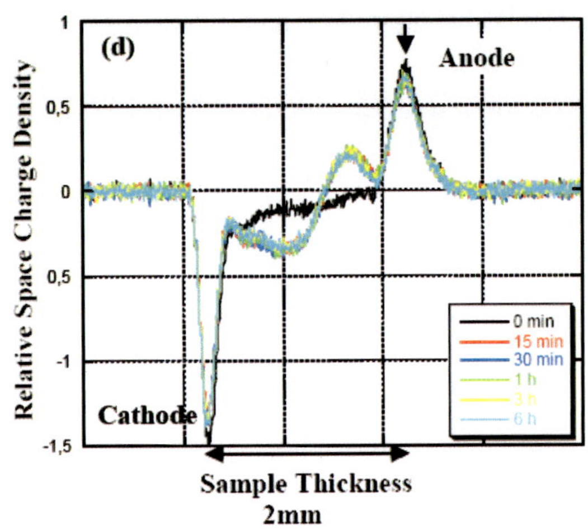

Figure 18. Space charge distribution in (a) neat LDPE, (b) AR-TiO_2 / LDPE, (c) AEAPS-TiO_2 / LDPE, and (d) DCS-TiO_2 / LDPE at different times after application of an electric field of 20 kV/mm. (a), (b) and (c) is reprinted with permission from [18]. Copyright (2005) IOP Publishing Ltd.

Figure 18(c) demonstrates the clearly suppressed space charge distribution in the AEAPS-TiO_2 / LDPE. Only a small number of homo space charges were observed in the entire sample and during the observation period of 6 hr, no appreciable increase in the space charge was found. A small number of space charge together with a constant density (flat curve) in the entire composite are beneficial to both short and long term breakdown strength. The evident improvement of the space charge distribution in this sample was the result of introduction of polar groups at the interface between the nanoparticles and the polymer matrix, as explained in our previous work [18].

The space charge distribution of the DCS-TiO_2 / LDPE is revealed in Figure 18(d). The homo space charge increased remarkably during the initial 15 min, then remained fairly stable. Both the positive and negative charges formed the peaks near the two electrodes. No "flat" region was observed in this sample. The DCS-TiO_2 / LDPE had the largest net intensity of the space charge among all the samples studied here.

The variation of the space charge distribution among four samples was easily observed. As shown in Figure 18, the nature of the space charge distribution changed from hetero to homo with the incorporation of 5 wt% of the AR-TiO_2 nanoparticles into the LDPE. Charge suppression and charge amplification were observed for the AEAPS-TiO_2 / LDPE and the DCS-TiO_2 / LDPE, respectively. Moreover, the nature of charges was different: hetero in the AEAPS-TiO_2 / LDPE, homo in the DCS-TiO_2 / LDPE. Clearly, the introduction of the nanoparticles and their surface modification both had significant effects on the space charge distribution.

Space charge distribution is determined by comprehensive trapping/detrapping and transport behavior of charges in insulating polymers. Additive particles could affect the space charge distribution directly by annihilating the trap centers or by introducing new traps [62,64,65], or indirectly by modifying the polymer morphology [59,64,66], and thus varying trap density, trap depth distribution or charge carrier mobility. However, the microstructure of the polymer had perhaps second-order effects on the space charge distribution, while electrical features of the filler particles themselves played a more important role [24,62]. In view of the observation that the morphology difference in our samples was even less than what was previously reported [24,62], here we discuss the difference in the space charge distribution from the aspect of electrical features of the particles and particle/electrode interaction, disregarding the morphology variations in the composite samples due to different particle surface conditions.

The reason that the AR-TiO_2 / LDPE showed a completely opposite space charge distribution pattern to that of the neat LDPE is probably because the TiO_2 nanoparticles changed the surface chemistry of the electrodes, allowing a massive charge injection to form a homo space charge distribution [62]. It also could not be excluded that ionic conduction of Ti ions contributed to the overall space charge distribution. The DCS-TiO_2 / LDPE showed a similar homo space charge pattern. However, the space charge density was much higher indicating more charge injection from electrodes related to the surface modified nanoparticles. Moreover, the more symmetric positive and negative charge distribution suggested similar mobility of both charges in the DCS-TiO_2 / LDPE, while in the AR-TiO_2 / LDPE the mobility was different for positive and negative charges. Therefore, DCS surface modification apparently not only enhanced the electrode charge injection, but also altered the charge mobility.

The AEAPS-TiO_2 differed from the AR-TiO_2 and DCS-TiO_2 because of the presence of polar molecules on the surface. These polar molecules or dipoles at the particle/polymer interface may affect charge carrier transport and trapping. The observed tiny number of hetero space charges was from deep traps because the charge build-up seen in Figure 18(c) was still present for over 6 hr after discharge. The suppression of the space charge accumulation therefore indicates a reduced number of charges trapped in deep traps in this sample. This could be due to a reduced number of deep traps or an increased number of shallow traps formed when introducing polar groups [18], and hence the decreased chance of charges being trapped by deep traps. Except for positive effects on the space charge accumulation, it should also be noted that the nature of the space charge changed from homo to hetero through modifying the TiO_2 surface with the AEAPS. This finding suggests reduced or no charge injection from the electrodes as compared to the samples filled with the AR-TiO_2 and DCS-TiO_2 nanoparticles. This was quite likely due to the change of interaction between the sample and the electrode materials after the nanoparticle surface modification with the polar AEAPS.

Another possibility was that the AEAPS surface modification of the nanoparticles encouraged local charge transfer at the polymer/electrode interface [67]. The energy of the donor states could be raised due to the existence of a large polarization energy related to the polar group [67]. Therefore, polar groups on the nanoparticle surface may play a multiple role in the mechanism of space charge generation.

In summary, the surface modification of the nanoparticles had different effects on the different dielectric properties of the nanocomposites. Permittivity seems not to be influenced by the particle surface modification appreciably at the low loading. Both polar and non-polar particle surface modification improved the dielectric breakdown strength significantly as compared to the composite filled with the uncoated nanoparticles. On the other hand, the surface modification influenced the space charge build-up in a complicated way; not only could the amount of charges, but also the nature of charges could be affected.

CONCLUSION

Surface modifications of the nanoparticles were carried out by using two silanes, which had quite different features. They were N-(2-aminoethyl)3-aminopropyl-trimethoxy-silane(AEAPS) having polar groups, and decyltrichlorosilane(DCS) having a relatively long hydrocarbon chain, which changed the nanoparticle surface from hydrophilic to hydrophobic. Surface coating structures were very sensitive to coating conditions. By using a dry nanoparticle / dry toluene reflux coating process, a desired coating structure was obtained. It was found that particle surface modification played a very important role in determining the dielectric breakdown strength. The AEAPS-TiO_2 / LDPE had Weibull breakdown strength 35% higher than that of the AR-TiO_2 / LDPE, associated with modification of electrical features at the interface between the coated nanoparticles and the matrix: the polar groups could act as charge scattering centers, traps or excitation sites in complicated electrical processes. The dielectric breakdown strength of the DCS-TiO_2 / LDPE was further increased, as compared to the AEAPS-TiO_2 / LDPE, mainly attributed to the improved interfacial compatibility between the particle fillers and the LDPE matrix. Although both polar and nonpolar surface modification had beneficial effects on the dielectric breakdown strength, their effects on the space charge distribution were quite different. The AEAPS-TiO_2 / LDPE exhibited hetero space charge distribution, with evidently suppressed charges, while the DCS-TiO_2 / LDPE revealed similar homo space charge distribution to that of the AR-TiO_2 / LDPE, with a largely enhanced charge accumulation. The evident improvement in the space charge formation in the AEAPS-TiO_2 / LDPE was again due to modification of electrical features at the interface between the coated nanoparticles and the matrix, and hence modification of the trap distribution in this material. The increased homo charges in the DCS-TiO_2 / LDPE were due to enhanced charge injection from the electrode because of particle/electrode interaction. This work has clearly demonstrated that the dielectric properties of polymer nanocomposites can be tuned by doing particle surface modification, a finding that can be used in insulating material optimization for industrial applications.

ACKNOWLEDGEMENTS

The work was supported by the ABB Group Service Center and the Nanoscale Science and Engineering Initiative of the National Science Foundation under NSF Award Number DMR-0117792. We would also like to acknowledge receipt of the Mettler-Toledo Thermal Analysis Education Grant 2001, and the TGA and DSC equipment. In addition, we would like to gratefully acknowledge Anna Christerson and Dmitri V. Zagorevski for their work on space charge distribution and mass spectroscopy, respectively. Finally, we would like to give our thanks to Eva Mårtensson, Carina Önneby, Benjamin J. Ash, and Tom Apple for their very useful discussions of this work.

REFERENCES

[1] Khalil, M. S. *IEEE T. Dielect. El. In.* 2000, 7, 261.

[2] Zhu, Y.; Yoon, H. G.; Suh, K. S. *IEEE T. Dielect. El. In.* 1999, 6, 164.

[3] Tanaka, Y.; Ohnuma, N.; Katsunami, K.; Ohki, Y. *IEEE T. Electr. Insul.* 1991,26, 258.

[4] Morshuis, P. H. F.; Kreuger, F. H.; Leufkens, P. P. *IEEE T. Electr. Insul.* 1988, 23, 1051.

[5] Wintle, H. J. *IEEE T. Electr. Insul.* 1990, 25, 27.

[6] Ramu, T. S. Conference Record of the 1990 *IEEE international Symposium on Electrical Insulation,* Toronto, Canada, 1990, 21.

[7] Peacock, A. J. Handbook of Polyethylene: Structures, Properties and Applications, *Marcel Dekker*: New York, NY, 2000.

[8] Ashcraft, A. C.; Eichhorn, R. M.; Shaw, R. G. Conference Record of the 1976 IEEE *International Symposium on Electrical Insulation,* Montreal, Canada, 1976, 213.

[9] Gächter, R.; Müller, H. (Editor) *Plastics Additives Handbook,* Hanser Publishers: Munich, Germany, 1990.

[10] Ryfšavý, D.; Tkadlečková, H. Polym. *Degrad. Stabil.* 1992, 37, 19.

[11] Siegel, R. W.; Hu, E.; Roco, M. C. (Editor) WTEC Panel Report on Nanostructure Science and Technology: Rand D Status and Trends in Nanoparticles, Nanostructured Materials, and Nanodevices, *International Technology Research Institute*: Dordrecht, Baltimore; Kluwer Academics Publishers: Boston, 1999.

[12] Buuren, T. V.; Dinh, L. N.; Chase, L. L.; Siekhaus, W. J.; Terminello, L. *J. Phys.Rev. Lett.* 1998, 80, 3803.

[13] Ren, S. Y.; Dow, J. D. *Phys. Rev.* 1992, B 45, 6492.

[14] Frenkel, J.; Dorfman, *J. Nature* 1930, 126, 274.

[15] Mahoney, W.; Schaefer, D. M.; Patil, A.; Andres, R. P.; Reifenberger, *R. Surf. Sci.* 1994, 316, 383.

[16] Rastogi, A. C.; Sharma, S. N.; Kohil, S. *Semicond. Sci. Technol.* 2000, 15, 1011.

[17] Rao, C. N. R.; Kulkarni, G. U.; Thomas, P. J.; Edwards, P. P. *Chem Eur. J.* 2002, 8, 28.

[18] Ma, D.; Hugener, T. A.; Siegel, R. W.; Christerson, A.; Mårtensson, E.; Önneby, C.; Schadler, L. S. *Nanotechnology* 2005, 16, 724.

[19] Roy, M.; Nelson, J. K.; Reed, C. W.; MacCrone, R. K.; Keefe, R. J.; Zenger, W.; Schadler, L. S. *IEEE T. DIELECT. EL. IN.* 2005, in press.

[20] Roy, M.; Reed, C. W.; MacCrone, R. K.; Schadler, L. S.; Nelson, J. K.; Keefe, R.; Zenter, W. 2005 *International Symposium on Electrical Insulating Materials*, Tokyo, Japan, 2005.
[21] Ma, D.; Siegel, R. W.; Hong, J-I.; Schadler, L. S.; Mårtensson, E.; Önneby, C. *J. Mater. Res.* 2004, 19, 857.
[22] Tsubokawa, N.; Kogure, A.; Maruyama, K.; Sone, Y.; Shimomura, M.; Polym. *J. 1990*, 22, 827.
[23] Ma, D.; Akpalu, Y. A.; Li, Y.; Siegel, R. W.; Schadler, L. S. J. Polym. Sci. B: Polym. Phys. 2005, 43, 488.
[24] Tanaka, Y.; Chen, G.; Zhao, Y.; Davies A. E.; Vaughan, A. S.; Takada, T. 2003 *IEEE Trans. Dielectr. Electr. Insul.* 2003, 10, 148.
[25] Plueddemann, E. P. Silane Coupling Agents, Plenum: New York, 1982.
[26] Munuera, G.; Rives-Arneau, V.; Saucedo, A. *J. Chem. Soc. Faraday Trans.* 1979, 75, 736.
[27] Finklea, H. O.; Vithanage, R. *J. Phys. Chem.* 1982, 86, 3621.
[28] Maira, A. J.; Coronado, J. M.; Augugliaro, V.; Yeung, K. L.; Conesa, J. C.; Soria, J.; *J. Catal.* 2001, 202, 413.
[29] Culler, S. R.; Ishida, H.; Koenig, *J. L. Appl. Spectrosc.* 1984, 38, 1.
[30] Launer, P. J. Infrared Analysis of Organosilicon Compounds, Spectra-structure Correlations, in: *Silicone Compounds Register and Review,* Petrarch Systems Catalogue, or www.gelest.com.
[31] Flett, M. St. C. Characteristic Frequencies of Chemical Groups in the Infra-Red, *Elsevier Publishing Company*: Amsterdam, London, New York, 1963.
[32] Kim, Y.; Kim, H. J. *J. Korean Phys.* Soc. 2002, 40, 94.
[33] Yang, J. J.; El-Nahhal, I. M.; Maciel, G. E. *J. Noncryst. Solids* 1996, 204, 105.
[34] Yang, J. J.; El-Nahhal, I. M.; Chuang, I-S.; Maciel, G. E. *J. Noncryst. Solids* 1997, 209, 19.
[35] da Fonseca, M. G.; Airoldi, C. *J. Mater. Chem.* 1999, 9, 1375.
[36] Sindorf, D. W.; Maciel G. E. *J. Am. Chem. Soc.* 1983, 105, 3767.
[37] Zaper, A. M.; Koenig, *J. L. Polym. Compos.* 1985, 6, 156.
[38] Abboud, M.; Turner, M.; Duguet, E.; Fontanille, M. *J. Mater. Chem.* 1997, 7, 1527.
[39] Parker, J. C.; Siegel, R. W. *J. Mater. Res.* 1990, 5, 1246.
[40] Chen, G.; Davies A. E. *IEEE Trans. Dielectr. Electr. Insul.* 2000, 7, 401.
[41] Fujita, S.; Ruike M.; Baba, M. 1996 *Annual Report of the Conference on Electrical Insulation and Dielectric Phenomena, Millbrae, CA*, 1996, 2, 738.
[42] Peacock, A. J. Handbook of Polyethylene: Structures, Properties and Applications, *Marcel Dekker:* New York, 2000.
[43] Akpalu, Y. A.; Ma, D.; Siegel, R. W.; Schadler L. S. PMSE Preprints, *American Chemical Society, Division of Polymeric Materials: Science and Engineering,* 2004, 91, 918.
[44] Mark, J. E.; Eisenberg, A.; Graessley, W. W.; Mandelkern, L.; Samulski, E. T.; Koenig, J. L.; Wignall, G. D. *Physical Properties of Polymers, American Chemical Society:* Washington DC, 1993.
[45] Sanchez, I. C.; Eby, R. K. *Macromolecules* 1975, 8, 638.
[46] Crist, B.; Howard, P. R. *Macromolecules* 1999, 32, 3057.
[47] Ueki, M. M.; Zanin, M. *IEEE T. Dielec. El.* In. 1999, 6, 876.

[48] Khalil, M. S.; Henk, P. O.; Henriksen, M. Conference Record of the 1990 *IEEE International Symposium on Electrical Insulation*, Toronto, Canada, 1990, 268.
[49] Blythe, A. R. *Electrical Properties of Polymers*, Cambridge University Press: Cambridge, London, New York, Melbourne, 1979.
[50] Wagner, K. W. *AIEE Trans.* 1922, 41, 288.
[51] Stark, K. H.; Garton, C. G. *Nature* (London) 1955, 176, 1225.
[52] Artbauer, J.; Kolloid Z. und Z. *Polymere* 1965, 15, 202.
[53] Ieda, M. *IEEE T. Electr. Insul.* 1980, 15, 206.
[54] Kolesov, S. N. Vysokomol. *Soed.* 1968, 10, 582.
[55] Knauth, P.; Tuller, H. L. *J. Appl. Phys.* 1999, 85, 897.
[56] Callister, W. D. Materials Science and Engineering: an Introduction, John Wiley and Sons, Inc.: New York, Chichester, Brisbane, Toronto, Singapore, 1990.
[57] Chen, G.; Tay, T. Y. G.; Davies, A. E.; Tanaka, Y.; Takada, T. *IEEE Trans. Dielectr. Electr. Insul.* 2001, 8, 867.
[58] Matsui, K.; Tanaka, Y.; Takada, T.; Fukunaga, K.; Maeno, T. *Proceedings of the 7th International Conference on Properties and Applications of Dielectric Materials*, Nagoya, Japan, 2003, 867.
[59] Fleming, R. *J. Brazilian J. Phys.* 1999, 29, 280.
[60] Zhang, C.; Mizutani, T.; Kaneko, K.; Ishioka, M. J. Phys. D: *Appl. Phys.* 2002, 35 1875.
[61] Laurent, C.; Massines, F.; Mayoux, C. *IEEE Trans. Dielectr. Electr. Insul.* 1997, 4, 585.
[62] Cartwright, G. A.; Davies, A. E.; Swingler, S. G.; Vaughan, A. S. *IEE Proceedings Science, Measurement and Technology* 1996, 143, 26.
[63] Kahali, M. S.; Zaky, A. A.; Hansen, B. S. 1985 *Annual Report of the Conference on Electrical Insulation and Dielectric Phenomena*, Buffalo, NY, 1985, 143.
[64] Khalil, M. S.; Montanari, G. C.; Fabianl, D.; Palmieri, F. *XIIIth International Symposium on High Voltage Engineering,* Delft, Netherlands, 2003, 1.
[65] Perlman, M. M.; Haridoss, S. *IEEE Trans. Electr. Insul.* 1987, 22, 9.
[66] Li, X. ; Du, Q.; Kang, J.; Tu, D. *J. Polym. Sci. B: Polym. Phys.* 2002, 40, 365.
[67] Lewis, T. *J. IEEE Trans. Electr. Insul.* 1984, 19, 210.

INDEX

C

D

E

I

J

K

L

M

N

O

P

T

U

V

W